The Cretaceous-Tertiary Boundary in the San Juan and Raton Basins, New Mexico and Colorado

Edited by

James E. Fassett
U.S. Geological Survey
903 National Center
Reston, Virginia 22092

J. Keith Rigby, Jr.
Department of Earth Sciences
University of Notre Dame
Notre Dame, Indiana 46556-1020

THE GEOLOGICAL SOCIETY OF AMERICA · 1888 ·

SPECIAL PAPER

209

Published by The Geological Society of America, Inc.
3300 Penrose Place, P.O. Box 9140, Boulder, Colorado 80301

GSA Books Science Editor Campbell Craddock

Printed in U.S.A.

Library of Congress Cataloging-in-Publication Data

The Cretaceous-Tertiary boundary in the San Juan and
Raton Basins, New Mexico and Colorado.

(Special paper ; 209)
Bibliography: p.
1. Geology, Stratigraphic—Cretaceous. 2. Geology,
Stratigraphy—Tertiary. 3. Geology—San Juan Basin
(N.M. and Colo.) 4. Geology—Raton Basin (Colo, and
N.M.) I. Fassett, James E. II. Rigby, J. Keith, Jr.
III. Series: Special paper (Geological Society of
America) ; 209.
QE685.C89 1987 551.7'7'097896 86-31850
ISBN 0-8137-2209-8

On the cover: A scene during latest Late Cretaceous (or earliest Paleocene?) time somewhere in the Western Interior of North America. A *Tyrannosaurus* is hot on the heels of a juvenile *Struthiomimus,* and an early multituberculate waits in the foreground. The plants are cycadeoids on the right and the angiosperm *Sapindosis* on the left. This scene is from the Tower of Time mural located in the Dinosaur Hall of the Museum of Natural History, Smithsonian Institution, Washington, D.C. The painting is by John A. Gurche of the Smithsonian. Photography of the mural was by Chip Clark of the Smithsonian and the cover layout was by Scotty Livingston of the U.S. Geological Survey. Permission to reproduce this painting was granted by the Smithsonian Institution.

Contents

Raton Basin

Relationship of San Juan and Raton basins to other areas

Geological Society of America
Special Paper 209
1987

Introduction

James E. Fassett
U.S. Geological Survey
903 National Center
Reston, Virginia 22092

J. Keith Rigby, Jr.
Department of Earth Sciences
University of Notre Dame
Notre Dame, Indiana 46556-1020

LOOKING BACK

It has been less than a decade since the landmark paper by Alvarez, Alvarez, Asaro, and Michel was published in *Science* (v. 208, p. 1095–1108) in 1980. That paper (somewhat outrageously, according to some geologists) put forth the idea that the extinction of many forms of life at the end of the Cretaceous Period was caused by the impact of an asteroid, 10 km in diameter, upon the earth. Supposedly this impact created an earth-enveloping dust cloud that stayed in the atmosphere for several years, turned out the light on the surface of the earth, lowered temperatures, and caused large numbers of plants and animals on the land and in the sea to die. Even though the evidence is growing for an extraordinary event having occurred at the Cretaceous-Tertiary boundary, the biological effects of such an impact are still being hotly debated.

The asteroid-impact idea of the Alvarez group was based on their discovery at the Cretaceous-Tertiary boundary—first in Gubbio, Italy, and subsequently in Denmark and New Zealand—of a thin layer of rock enriched in iridium from 20 to 160 times the background level. (The Alvarez and others paper indicated that the background level for iridium in the marine carbonate rocks above and below the enriched zone was a bit less than 0.3 parts per billion.) Knowing that some meteorites are enriched in iridium, the Alvarez group did some simple arithmetic and calculated backward from the total amount of iridium they figured was originally present at the K-T boundary around the world to the size that a "type 1 carbonaceous chondrite" would have to have been to provide that amount of iridium. The answer: a 10-km-diameter asteroid!

One can question the rigor of this procedure with a number of "what-ifs." For example: What if the asteroid was not a type 1 carbonaceous chondrite and contained more, or less, or even no iridium; what if all, or even some, of the iridium was liberated from within the crust by the impact; what if there was no asteroid, and the iridium originated on earth and was dispersed by a large number of volcanic eruptions; what if the impactor was a comet; and on and on. No one yet knows the answers to these questions, yet today the 10-km-diameter asteroid seems to have developed a life of its own. Some other, possibly more directly answerable questions have been asked about the mass-extinction-by-asteroid idea, such as: How abrupt or "instantaneous" was the end-of-the-Cretaceous extinction; why were some forms of life—principally marine—apparently killed "instantly," whereas other life forms—both marine and continental—survived the Cretaceous-Tertiary boundary event; and where was the crater?

SNOWBIRD CONFERENCE

In an attempt to answer some of these kinds of questions, a meeting of scientists was convened at Snowbird, Utah, in October 1981 to evaluate the possibility that periodic mass extinctions of life could have been caused by extraterrestrial objects impacting on the earth. Many of the papers presented at that meeting were theoretical; relatively few presented hard data resulting from studies specifically designed to test for mass extinctions at carefully selected sites. A volume of papers presented at the Snowbird meeting was published in 1982 by the Geological Society of America (GSA) as Special Paper 190.

The Snowbird meeting clearly demonstrated that the mass-extinction-by-extraterrestrial impact idea had burst upon the geoscience scene like a brilliant new comet; we were seeing another of the great debates of geology unfolding before our eyes. Not since the beginning of the plate tectonics debate had the geoscience community been so challenged to test a new concept. (The concept itself—of mass extinction by extraterrestrial impact—was not new, having roots in both mythology and science fiction; the proposition had even been suggested seriously by scientists. What was new, and exciting, was the iridium layer—the first tangible evidence that something extraordinary had taken place at the end of the Cretaceous Period.)

Unfortunately for the early proponents of the 10-km-asteroid idea, the Snowbird meeting validated the seriousness of many of the early questions about the impact-extinction concept, and most thoughtful scientists came away from Snowbird with more doubts than answers. It was clear that more intensive, multifaceted studies of the rocks adjacent to mass extinction boundaries had to be conducted wherever possible.

SALT LAKE CITY SYMPOSIUM

An ideal area in which to concentrate such an effort was northern New Mexico and southern Colorado, where Upper Cretaceous and Paleocene rocks are preserved in the San Juan and Raton basins. In May 1983, a symposium was held on the Cretaceous-Tertiary boundary problem of the San Juan Basin and northern New Mexico at a joint meeting of the Rocky Mountain and Cordilleran sections of the GSA in Salt Lake City. At that meeting, papers on the stratigraphy, sedimentology, vertebrate paleontology, palynology, radiometric age, magnetostratigraphy, and geochemistry of the rocks near the boundary in the San Juan and Raton basins were presented. Following that meeting, the convenors (the editors of this volume) contacted all of the authors who had presented papers at the symposium and asked them to prepare written versions of their reports for this Special Paper. The papers making up this volume represent the findings of those workers who agreed to participate in this effort.

SAN JUAN BASIN

Knowledge

Because of the San Juan Basin's rich endowment of energy resources in the form of coal, oil and gas, and uranium, the area has been intensively studied by industry and government geologists for many years. The outcropping rocks rimming the basin have been mapped and remapped. The basin's subsurface geology has been studied in exquisite detail using geophysical logs, well cuttings, cores, and other subsurface data provided by the more than 16,000 oil and gas wells drilled in the basin.

Since the late 1800s, paleontologists had studied the basin's large numbers of terrestrial fossil vertebrates, both mammalian and reptilian. Because the San Juan Basin is one of the few places on earth where such fossils occur in abundance in Upper Cretaceous and lower Tertiary continental rocks, the basin became a mecca for paleontologists to come and study the transition of life across this period and era boundary.

Beginning in the 1970s, and continuing into this decade, a new wave of geologists has come into the basin to study the rocks adjacent to the Cretaceous-Tertiary boundary. These recent workers' studies of the vertebrate paleontology, paleomagnetism, palynology, and geochemistry of these rocks have greatly increased our knowledge of the basin.

Problems

In spite of the superb exposures of the rocks and the well-documented and detailed basinwide information about their geometry or physical stratigraphy, there have been serious disagreements about the rocks adjacent to the Cretaceous-Tertiary boundary in the San Juan Basin. These disagreements can be focused into two questions: (1) is there an unconformity at the Cretaceous-Tertiary boundary, and (2) what are the ages and what are the stratigraphic definitions of the rocks adjacent to the Cretaceous-Tertiary boundary?

The San Juan Basin papers in this volume provide a multifaceted data set that can be used to try to resolve these questions. When applied to the first question of whether there is an unconformity at the Cretaceous-Tertiary boundary, these papers provide a convincing response: Yes, there is a substantial time gap at or near the boundary, and it encompasses all or most of the Maastrichtian (uppermost Cretaceous) and probably some part of the early Paleocene. There is still disagreement, however, as to exactly where the hiatus is located, and it has been suggested that there could be more than one hiatus at or near the Cretaceous-Tertiary boundary. The amount of geologic time represented by the unconformity (or unconformities) appears to be on the order of more than 8 million years.

The question of the age and definition of the rocks adjacent to the boundary is not resolved by these papers. This question presents difficulties because the critical basal Ojo Alamo Sandstone contact is at, or very close to, the Cretaceous-Tertiary boundary. Thus, if the rock-stratigraphic contact is placed at one level, the Ojo Alamo Sandstone is Tertiary; if placed at a lower level, it is Cretaceous and Tertiary. To further complicate an already complex situation, the Cretaceous-Tertiary boundaries defined on the basis of palynology and vertebrate paleontology do not seem to quite coincide in the San Juan Basin. One of the papers in this book suggests that in spite of its contained dinosaurs, the Ojo Alamo Sandstone is Tertiary in age based on pollen analysis—in other words, the dinosaurs are Tertiary in age. Because of these problems, all authors must carefully define their terms when discussing the rock units adjacent to the boundary in the San Juan Basin.

RATON BASIN

Knowledge

The Raton Basin has also been studied by geologists since the 1800s, mainly because of its coal beds in both uppermost Cretaceous and lower Tertiary rocks. Unlike the San Juan Basin, deposition across the Cretaceous-Tertiary boundary appears to have been continuous in the Raton Basin; however, the terrestrial rocks near the boundary in the Raton Basin have been a veritable wasteland for vertebrate paleontologists. Another important difference between the two basins is that the Raton has produced almost no oil or gas. As a consequence of the paucity of well control, the physical stratigraphy of the rocks in the subsurface of the Raton Basin is not as well known as that in the San Juan Basin.

The Cretaceous-Tertiary boundary in the Raton Basin was originally placed at the base of the Raton Formation, on the basis of plant fossils. The boundary was later placed stratigraphically higher, within the lower part of the Raton Formation, again on the basis of fossil plants. The exact placement of this boundary remained somewhat controversial until detailed palynological studies in the late 1960s and 1970s made possible a precise location of the boundary within the lower part of the Raton Formation.

In the early part of this decade, the exciting discovery of an iridium-enriched zone was made at precisely the level of the palynological Cretaceous-Tertiary boundary in the Raton Basin. This discovery, reported by Orth and others in 1981 in *Science* (v. 214, p. 1341–1343), was of great importance to North American stratigraphers. If this iridium-enriched zone was isochronous with the iridium-enriched zone of Gubbio, Italy, it was possible, for the first time, to correlate the marine Cretaceous-Tertiary microfossil extinction level with the continental palynological Cretaceous-Tertiary boundary.

Problems

At the Gubbio, Italy, site the iridium-enriched layer is within a section of rock that has been determined to have been deposited during a time of reversed magnetic polarity. This time of reversed polarity has been labeled polarity chron 29 r. (For a discussion of polarity chrons, see *A Geologic Time Scale* by Harland and others, Cambridge University Press, New York, 1982.) The first published report on the polarity of the rock section containing the Cretaceous-Tertiary boundary in the Raton Basin indicated that the iridium-enriched zone at the boundary was in a *normal* magnetic polarity interval, suggesting that the boundary and iridium-enriched zone there were *not* isochronous with the boundary and iridium-enriched zone at Gubbio. A paper in this volume on the magnetic polarity of the boundary rocks in the Raton basin concludes that these rocks are indeed of reversed polarity and suggests that the earlier experiment, showing a normal polarity at the boundary, was not performed rigorously enough to remove modern normal-polarity overprinting.

CONCLUSIONS

The papers in this volume provide new information about the rocks adjacent to the Cretaceous-Tertiary boundary in the San Juan and Raton basins. They show quite conclusively that in the San Juan Basin, deposition in latest Cretaceous and early Tertiary time was not continuous. Instead, there is a significant time gap (or gaps) in the rock record at or near the Cretaceous-Tertiary boundary totaling more than 8 million years. In addition, dinosaur bone has been found above the palynological Cretaceous-Tertiary boundary. These data would seem to argue against a widespread terrestrial mass extinction at the boundary in this area. Furthermore, the apparent rapid faunal turnover at the boundary is the result of missing rock record and is not the result of a mass extinction.

In the Raton Basin, where deposition across the boundary was continuous, the magnetic polarity of the rocks containing the Cretaceous-Tertiary boundary has been shown to be reversed, as is the case at other sites around the world where the iridium-enriched zone has been found at the boundary. There is therefore no reason to doubt that the iridium-enriched zone in the Raton Basin is isochronous with other such sites. As to a mass extinction at the boundary in the Raton Basin, to date no vertebrate fossils have been found there; thus, the fate of vertebrate animals in the Raton Basin area at that time is unknown. Palynological evidence seems to indicate a severe shock to the plant life in the Raton Basin at the Cretaceous-Tertiary boundary. Most of the Cretaceous plants, however, reappear not far above the boundary. There was no mass extinction of plants at the end of the Cretaceous Period in the Raton Basin.

In summary, the studies represented by the papers in this volume do not support an impact-associated mass extinction in either the San Juan or Raton basin at the end of the Cretaceous Period. Clearly, much more work is needed, not only in these two basins but in other basins where the boundary rocks are present, to resolve this question. But even if there was no worldwide catastrophic mass extinction of life at the end of the Cretaceous Period associated with the iridium-zone event, the discovery of this worldwide isochronous marker bed has provided geologists with a powerful new correlation tool with which to fine-tune the stratigraphy of the rocks adjacent to the Cretaceous-Tertiary boundary.

Clearly, the possibility of catastrophic mass extinctions has caught the attention of the public (scientists and nonscientists, alike) because if it happened before, it could happen again—to us! Even though this possibility is an important question, especially as it may have affected the course of evolution of life on earth, we suggest that the identification of a worldwide isochron, the first ever discovered on earth, will turn out to be the really great contribution of the Alvarez group.

Manuscript Accepted by the Society March 24, 1986

Printed in U.S.A.

Geological Society of America
Special Paper 209
1987

The ages of the continental, Upper Cretaceous, Fruitland Formation and Kirtland Shale based on a projection of ammonite zones from the Lewis Shale, San Juan Basin, New Mexico and Colorado

James E. Fassett
U.S. Geological Survey
903 National Center
Reston, Virginia 22092

ABSTRACT

The Kirtland Shale or Fruitland Formation directly underlies the Cretaceous-Tertiary boundary throughout most of the San Juan Basin of northwest New Mexico and southwest Colorado. These formations have been known to be Late Cretaceous in age since the early 1900s. Now, with the greatly renewed interest in rocks adjacent to mass extinction boundaries, it is important to place more precise ages on such rock units as the Fruitland and Kirtland. Deposition of the Fruitland and Kirtland was closely related to deposition of the underlying marine-regressive Pictured Cliffs Sandstone. Because the Pictured Cliffs was deposited as a strandline sandstone in a subsiding seaway, its stratigraphic expression, when related to a time horizon (the Huerfanito Bentonite Bed), is a series of rising-to-the-northeast, time transgressive, stair steps. Thus, time lines (or horizons) drawn parallel to the Huerfanito cut through the marine Lewis Shale, the strandline Pictured Cliffs Sandstone, and the continental Fruitland Formation and Kirtland Shale.

Ammonites have been collected and identified from various stratigraphic levels within the Lewis Shale around the northwest, north, and east sides of the San Juan Basin. These fossils can be tied in to the established ammonite zonation of the Western Interior seaway. Because some of these ammonite zones have been radiometrically dated outside the San Juan Basin, it is possible to project these dated faunal zones from the Lewis Shale along time lines into the Fruitland Formation and Kirtland Shale and thereby estimate the age of those rocks. Based on these projections the part of the Fruitland and Kirtland laterally time-equivalent to the Lewis Shale is estimated to range from 73.2 ± 0.7 Ma to 73.9 ± 0.8 Ma. The average age for this interval based on these dates is 73.5 ± 0.5 Ma; the maximum range of the interval at the 95 percent confidence level is 71.8 to 75.5 Ma. This age range puts these rocks in the Campanian Stage of the upper Cretaceous in the San Juan Basin.

INTRODUCTION

The continental Fruitland Formation or Kirtland Shale directly underlies the Cretaceous-Tertiary boundary in the San Juan Basin of northwest New Mexico and southwest Colorado. These formations crop out around the rim of the San Juan Basin and, with the exception of two small areas on the east side, are present throughout the entire basin area (Fig. 1). The age of the Fruitland and Kirtland interval has been known to be Late Cretaceous since the early 1900s (Schrader, 1905). Recently, Alvarez and others (1980) have suggested that the end of the Cretaceous was punctuated by a globally catastrophic event caused by the impact of a 10-km (6-mi) diameter asteroid. One result of this somewhat outrageous suggestion has been a more careful evaluation of the rocks adjacent to the Cretaceous-Tertiary boundary, worldwide.

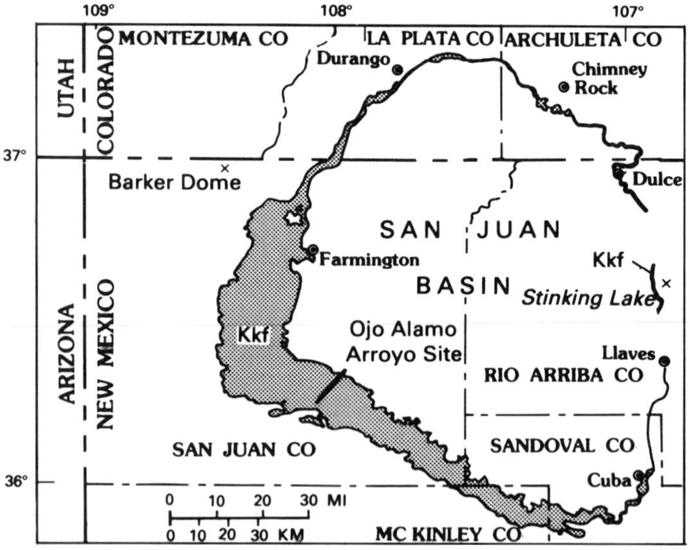

Figure 1. Index map showing the location of the San Juan Basin as outlined by the outcrop of the Fruitland Formation and Kirtland Shale (patterned area). These units, labeled Kkf, are shown undivided.

In the San Juan Basin, a controversy has developed over the past few years as to the nature of the Cretaceous-Tertiary boundary: Was deposition continuous across the boundary as suggested by Lindsay and others (1978) or does a major unconformity separate Cretaceous and Tertiary rocks (Fassett, 1973, 1979, 1982, 1985; Fassett and Hinds, 1971; Fassett and others, this volume)? The focus of this paper is the age of rocks that directly underlie the boundary, based on a projection of radiometrically dated ammonite zones from laterally time-equivalent marine rocks into continental rocks. Clearly, a more precise determination of the age of Fruitland and Kirtland rocks should help resolve the unconformity dispute at the Cretaceous-Tertiary boundary in the San Juan Basin.

GEOMETRY OF THE ROCKS

The rocks discussed in detail in this report are the Lewis Shale (including the Huerfanito Bentonite Bed), the Pictured Cliffs Sandstone, the Fruitland Formation, and the Kirtland Shale. The stratigraphic relations of these rocks are shown on Figure 2. These rocks, throughout the subsurface and surface of the San Juan Basin, were discussed and illustrated in great detail in an earlier paper (Fassett and Hinds, 1971), from which are derived most of the data and figures I use to show the geometry of these rocks throughout the basin.

The subsurface evaluation of the Fruitland and Kirtland formations and the underlying Pictured Cliffs Sandstone was greatly facilitated by the discovery of the continuous and distinctive (on geophysical logs) Huerfanito Bentonite Bed. This marker bed is present in the marine Lewis Shale, which underlies the Pictured Cliffs Sandstone throughout the San Juan Basin. Figure 2 shows three stratigraphic cross sections across the basin, all trending northeast. All three cross sections are constructed using the Huerfanito Bentonite Bed as a datum. These sections show that the Pictured Cliffs Sandstone rises northeastward across the basin. On all of the sections the base of the Pictured Cliffs is close to the level of the Huerfanito Bentonite Bed in the southwest part of the basin, whereas it is far above the Huerfanito in the northeast part of the basin.

Figure 3 is an isopach map of the interval from the Huerfanito Bentonite Bed to the top of the Pictured Cliffs Sandstone. This illustration shows in map view what Figure 2 shows in cross section—that is, that the top of the Pictured Cliffs rises some 335 m (1100 ft) northeastward across the San Juan Basin (from the 150-ft isopach to the 1250-ft isopach). Because the isopachs on this map parallel a time horizon that was laid down on a flat, gently sloping surface, they must also depict a close approximation of shoreline trends as the Pictured Cliffs sea retreated northeastward out of the San Juan Basin area. As the map shows, the Pictured Cliffs shoreline trended generally northwest.

WALTHER'S LAW

The geometric relations of the Lewis Shale, Pictured Cliffs Sandstone, Fruitland Formation, and Kirtland Shale represent a textbook example of Walther's Law of Correlation of Facies: "The various deposits of the same facies areas and similarly the sum of the rocks of different facies areas are formed beside each other in space, though in cross section we see them laying on top of each other" (Walther, 1894, p. 979, as quoted in Middleton, 1973). Figure 4 is a diagrammatic representation in cross section of the sequence of events that resulted in deposition of the above-listed rocks. At time A, a volcanic ash fall (the Huerfanito Bentonite Bed) has just been deposited on the sea floor. At time B the Pictured Cliffs strandline has regressed northeastward, part way across the San Juan Basin area. At time C the sea has retreated to the northeast edge of the basin area. Panel B shows deposition at the same time of continental rocks on the land surface, beach and shoreface sandstone adjacent to the shoreline, and marine shale on the sea floor. A profile or slice of the rocks nearly anywhere on panel C will show that rocks that actually formed as lateral time equivalents are indeed superimposed in a vertical sequence as Walther's Law states.

The dashed, dotted, and solid lines in Figure 4 represent time-lines that can be seen to parallel both the Huerfanito Bed and the plane of deposition (the land surface and the sea floor). It is thus clear that at time C, when the beach and shoreface sands of the Pictured Cliffs were being laid down in the northeast part of the basin, continental rocks of the same age were being deposited in the southwest part of the basin at a level many hundreds of meters above the buried Pictured Cliffs sands. If we could somehow put a time scale along the right edge of panel C, we would not only know the ages of the marine shales present there, but by projection along time lines parallel to the Huerfanito Bed, we could also determine the age of the laterally equivalent beach and shoreface sands and continental deposits in other parts of the basin.

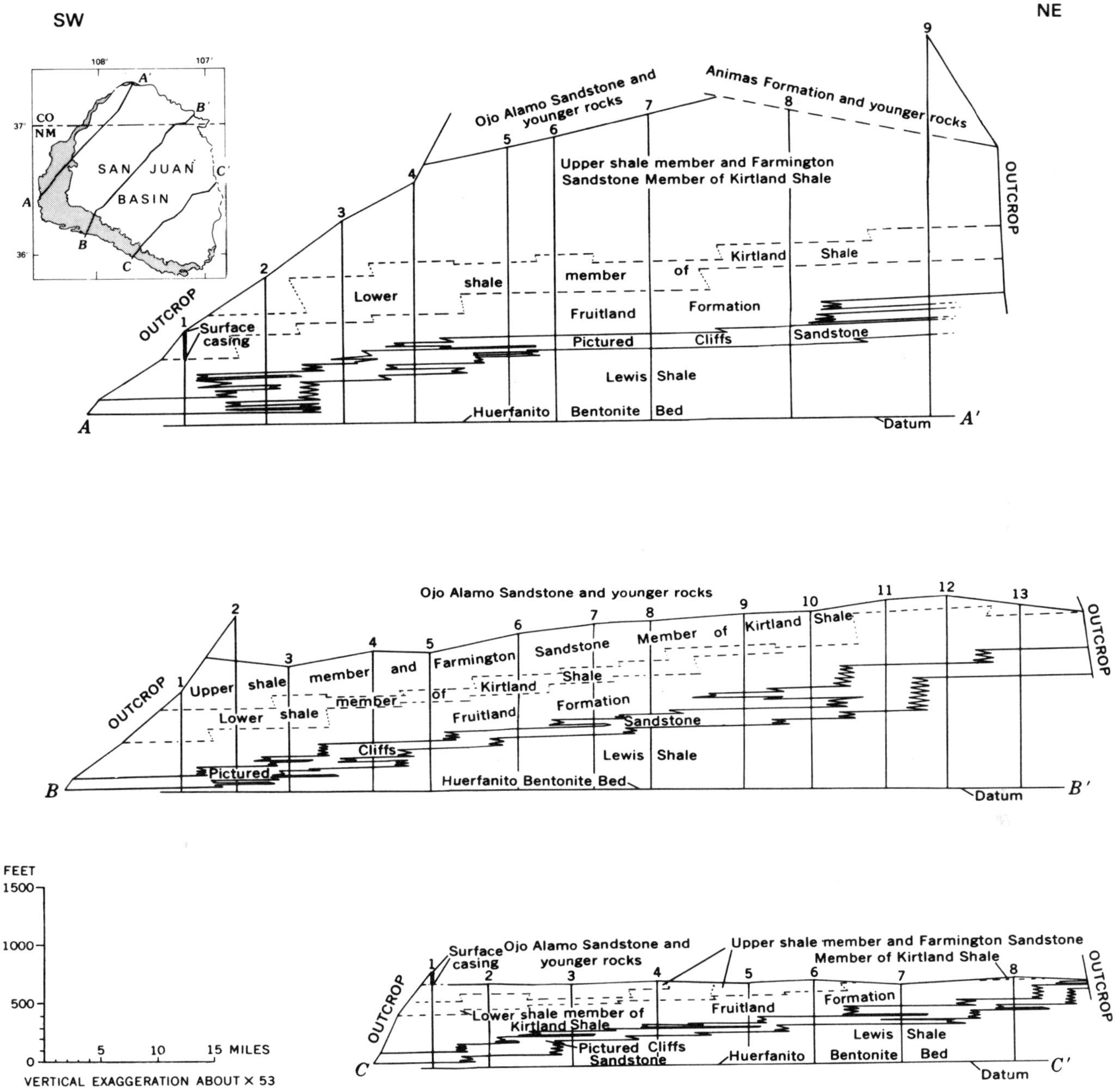

Figure 2. Three northeast-trending stratigraphic cross sections showing the relations of the upper Lewis Shale, Pictured Cliffs Sandstone, Fruitland Formation, Kirtland Shale, and overlying rocks. Datum is Huerfanito Bentonite Bed of the Lewis Shale. Modified from Fassett and Hinds (1971, plate 2).

Figure 4 is a depiction of the model for regressive deposits originated by Sears and others (1941, Fig. 20) based on their study of older Cretaceous rocks in the southern part of the San Juan Basin. The point to remember from this diagrammatic illustration is that time lines or time planes are parallel to the Huerfanito Bentonite Bed. This conclusion is, of course, not valid if differential tectonic movement took place between the time of deposition of the Huerfanito Bentonite Bed and deposition of the overlying rocks. There is no evidence, however, for such movement in these rocks, except possibly in the northwest part of the basin, where the 700-ft depression-contour is shown on Figure 3, and possibly along the extreme eastern edge of the basin.

AMMONITES FROM THE LEWIS SHALE

In an earlier report Fassett and Hinds (1971, p. 16) suggested, on the basis of two ammonites from the northern and

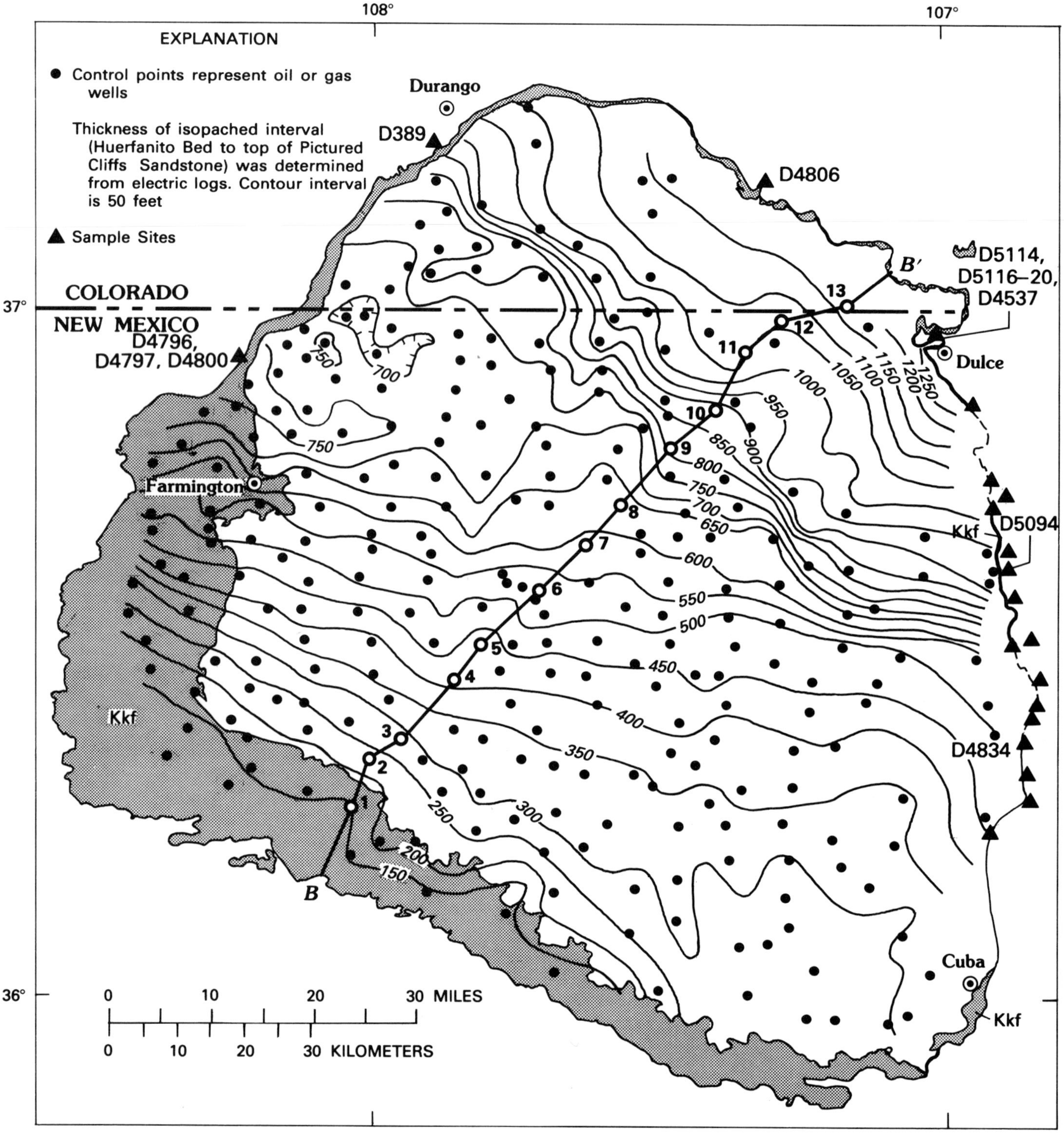

Figure 3. Isopach map of the interval between the Huerfanito Bentonite Bed of the Lewis Shale and the top of the Pictured Cliffs Sandstone. Black triangles are fossil-collection localities in the Lewis Shale; fossil localities are from Cobban (1973) and Cobban and others (1974). Isopachs are from Fassett and Hinds (1971). Fossils from numbered localities are stratigraphically tied to the Pictured Cliffs Sandstone and are listed in Tables 1 and 2. The line of cross section B-B′ of Figures 2 and 5 is shown.

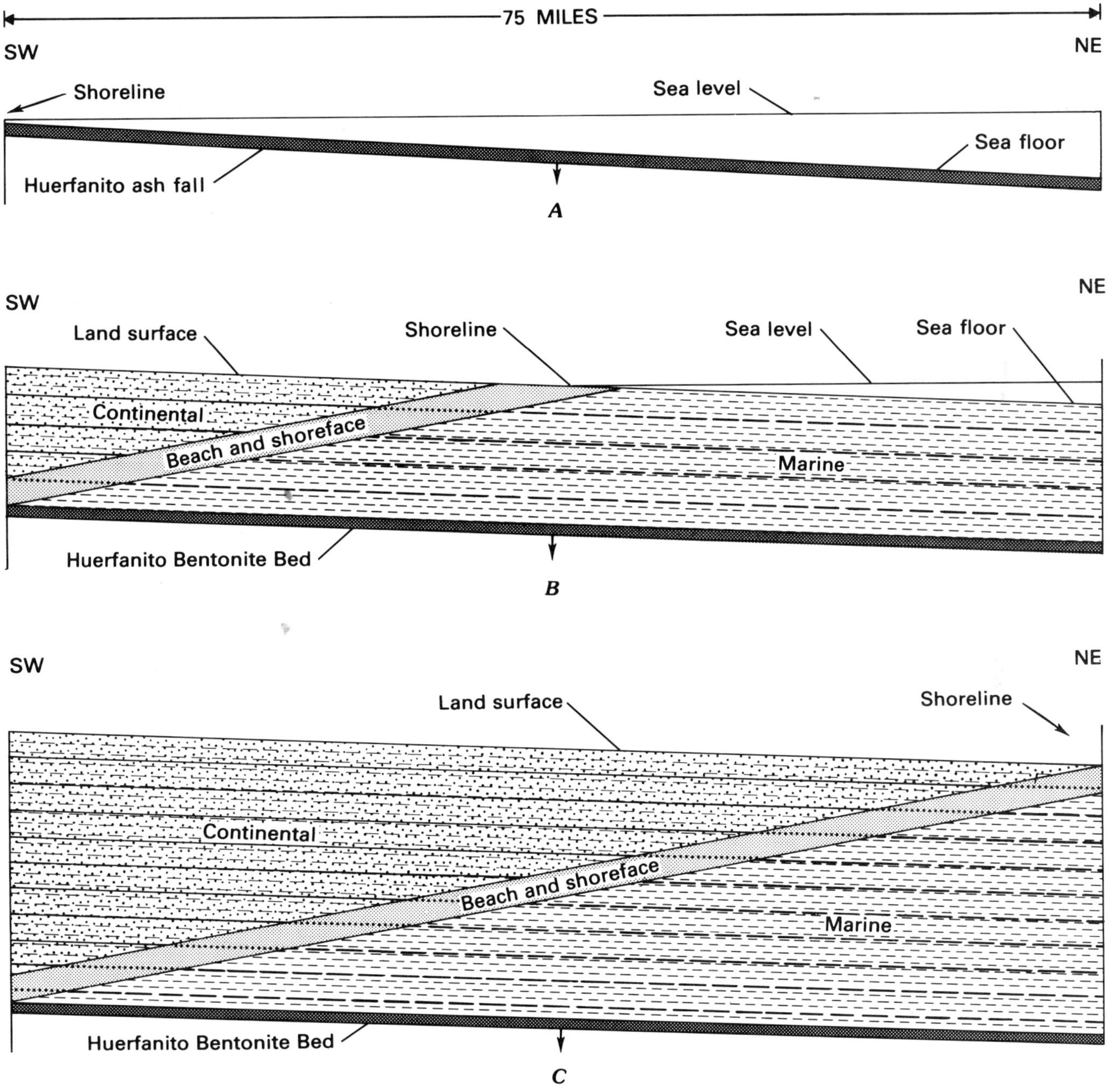

Figure 4. Diagrammatic cross sections trending northeast (normal to the shoreline) across the San Juan Basin area showing deposition at times A, B, and C (oldest to youngest) of marine, beach and shoreface, and continental rocks. Dashed, dotted, and solid lines represent time lines through the rocks deposited in the three environments of deposition. The arrows beneath the Huerfanito Bed indicate continued subsidence of this area as these rocks were being deposited.

eastern parts of the San Juan Basin, that because the ages of these ammonite zones had been determined or inferred in Wyoming by Gill and Cobban (1966), these dated ammonite zones could be used to date the regression of the Pictured Cliffs Sandstone by extrapolation. It was concluded in that report that on the basis of such extrapolation, the Pictured Cliffs Sandstone regression across the San Juan Basin area probably took around three million years. In 1975, Obradovich and Cobban published a revised time chart for the Western Interior ammonite zones that decreased the time elapsed between the two San Juan Basin ammonites to one-third of what it had been. On the basis of these new data, the time for the Pictured Cliffs regression suddenly changed from three to one million years.

Subsequently, additional ammonites were collected from the Lewis Shale at many more localities around the east, north, and northwest rims of the San Juan Basin. These collections were described by Cobban (1973) and Cobban and others (1974). Figure 3 shows all of the fossil localities from these two reports; Figure 5 is a map showing the localities that are stratigraphically tied in to the Pictured Cliffs sandstone with a high degree of certainty.

Table 1 is a listing of the stratigraphically tied-in ammonite collection localities. The localities are keyed to Figures 3 and 5 of this report by locality numbers. Each numbered locality has been stratigraphically tied to the Huerfanito Bentonite Bed by subtracting the distance of the fossil locality below the top of the Pictured Cliffs Sandstone from the thickness of the interval between the top of the Pictured Cliffs and the Huerfanito Bentonite Bed at each locality. The Pictured Cliffs–Huerfanito Bed interval thicknesses were taken from a projection of the isopach lines of Figure 3 to the numbered ammonite localities. The ammonite collection levels from Cobban and others (1974) were related to the top of the Lewis Shale (base of the Pictured Cliffs Sandstone). I added 30 m (100 ft) to the Cobban and others (1974) numbers at the Dulce localities and 8 m (25 ft) at the Stinking Lake locality (the approximate thicknesses of the Pictured Cliffs Sandstone at those places).

Table 2 is a listing of the Western Interior ammonite zones from Table 1 showing the stratigraphic level for each collection site within each zone. On the basis of this table, it is possible to estimate the stratigraphic level of the boundary between ammonite zones in terms of distance above the Huerfanito Bentonite Bed. The ammonite zone boundary levels are bracketed within varying footages, as Table 2 shows.

The boundary between *Baculites compressus* and *Didymoceras cheyennense* is between 349 and 335 m (1145 and 1100 ft) above the Huerfanito bed; thus I show the boundary at 335 m (1100 ft). There is an apparent overlap of the lower part of the *Didymoceras cheyennense* zone and the underlying *Exiteloceras jenneyi* zone as shown on Table 2 at localities D4800 and D5094. Because the fossils at locality D4800 were fragmentary and their identification not absolutely certain (Cobban, 1973, p. 150), the 229-m (750-ft) level indicated for *D. cheyennense* at that site is questionable as a constraint for placement of the zone boundary. The fossils collected at locality D5094, however, seem to represent a positive identification of the zone fossil even though its stratigraphic level appears to be anomalously low.

Cobban and others (1974) suggested a southward thinning of the Lewis on the east side of the San Juan Basin. I suggest that this southward thinning occurs only locally along the extreme eastern edge of the basin and represents tectonic uplift east of the present basin area at the time the Lewis Shale was being deposited, as discussed by Fassett and Hinds (1971, p. 12). The apparently low stratigraphic position of *Didymoceras cheyennense* at fossil locality D5094 is thus a localized phenomenon occurring only along the east-central edge of the San Juan Basin. Throughout most of the basin, the ammonite zones in the Lewis Shale are tabular with parallel upper and lower boundaries. In this report, the boundary between *D. cheyennense* and *E. jenneyi* is placed at the 290-m (950-ft) level between the highest occurrence of *E. jenneyi* at 278 m (911 ft) and the lowest, more representative, basinwide occurrence of *D. cheyennense* at 290 m (952 ft) at locality D5119 (Table 2).

The boundary for the top of the *Didymoceras stevensoni* zone was placed at a level of 213 m (700 ft) above the Huerfanito Bed, more or less midway between the lowest level of *E. jenneyi* and the highest level of *D. stevensoni.* The level for the top of the *Didymoceras nebrascense* zone was placed at 152 m (500 ft), and that for the base of the zone at 61 m (200 ft) in order to make this zone as thin as possible and still allow a slight cushion above and below the highest and lowest occurrence of this zone fossil. (Cobban and others, 1974, p. 282, state that this ammonite zone is only 30 m—100 ft—thick near Pueblo, Colorado.) The lower limit of the *Baculites scotti* zone must be at least 15 m (50 ft) beneath the Huerfanito Bed, but there is no fossil control for the base of this zone. Cobban and others (1974, p. 282) indicated that the *B. scotti* zone is consistently thick in eastern Colorado—152 to 213 m (500 to 700 ft)—and could be 183 m (600 ft) thick in the San Juan Basin.

Figure 5 is a map, here named a bio-isochrone map, that shows Western Interior ammonite zones as northwest-trending bands. The contacts of each band represent projections to a horizontal plane of lines where the ammonite zone boundaries intersect the top of the time-transgressive Pictured Cliffs Sandstone (top of the stratigraphically highest marine rocks) in the subsurface of the San Juan Basin. The bio-isochrones are: (1) lines of equal biological value (biozone boundaries), (2) lines of equal thickness of rock (each line is an isopach of the interval from the Huerfanito Bentonite Bed to the top of the Pictured Cliffs Sandstone), and (3) lines connecting points of equal time value.

RADIOMETRIC AGE DATES

The ages of some of the Western Interior ammonite zones, three of which are present in the Lewis Shale over the stratigraphic interval spanned by the time-transgressive Pictured Cliffs Sandstone in the San Juan Basin, were presented in a paper by Obradovich and Cobban (1975). These ages were determined by

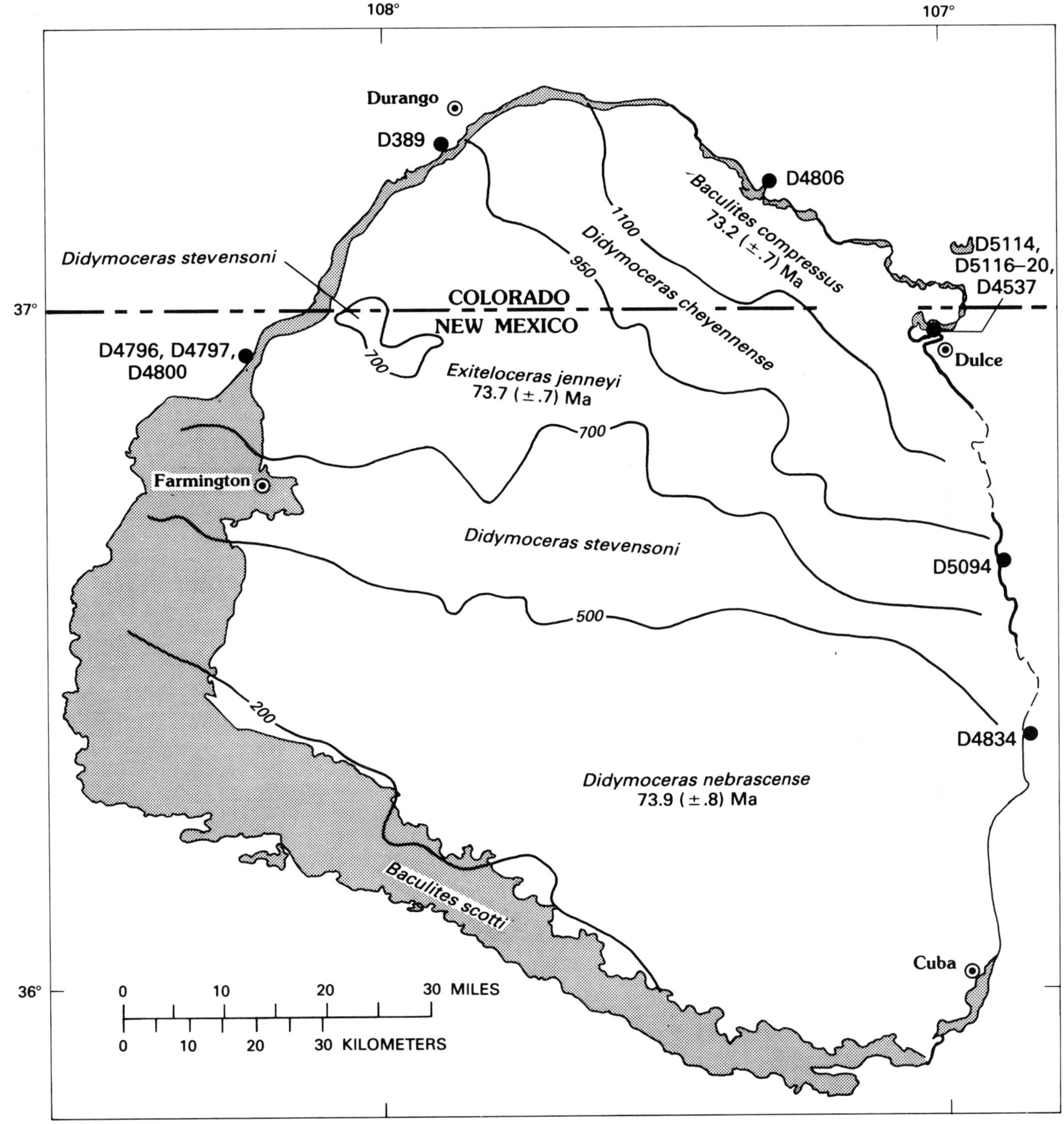

Figure 5. Bio-isochrone map of the San Juan Basin showing the projections to a horizontal plane of the intersections of the boundary horizons of the Western Interior ammonite zones with the top of the time-transgressive Pictured Cliffs Sandstone (the highest marine unit in the basin). The bio-isochrones are: (1) lines of equal biological value (biozone boundaries), (2) lines of equal thickness of rock (each line is an isopach of the interval from the Huerfanito Bentonite Bed to the top of the Pictured Cliffs Sandstone), and (3) lines connecting points of equal time value. The corrected radiometric ages for three of these zones (from Obradovich and Cobban, 1975) are shown. This map provides a useful tool for estimating the time of deposition of the Pictured Cliffs Sandstone, or laterally equivalent rocks, at any point within the Pictured Cliffs outcrop in the San Juan Basin. It should also have value for predicting which of the various Western Interior ammonites (or equivalents) will be present in the Lewis Shale near the base of and at specific stratigraphic levels below the Pictured Cliffs around its outcrop in the San Juan Basin.

TABLE 1. WESTERN INTERIOR AMMONITE-ZONE FOSSILS (OR EQUIVALENTS) COLLECTED FROM THE LEWIS SHALE, AT SITES IN THE NORTHERN AND EASTERN SAN JUAN BASIN*

Collection locality name	U.S. Geological Survey fossil locality number	Western interior ammonite-zone fossil collected (Zone fossil equivalent collected)	Stratigraphic level (Feet below top of Pictured Cliffs)	Thickness of Pictured Cliffs-Huerfanito interval (in feet)	Feet above or below Huerfanito Bed	Source of Data
Barker Dome	D4800	Didymoceras cheyennense? (Didymoceras fragments)	75	775	700	Cobban (1973)
Barker Dome	D4797	Didymoceras nebrascense	515	775	260	Cobban (1973)
Barker Dome	D4796	Baculites scotti (Didymoceras n. sp.)	825	775	-50	Cobban (1973)
Durango	D389	Didymoceras stevensoni	350	900	550	Cobban (1973)
Chimney Rock	D4806	Didymoceras cheyennense	100	1200 (est.)	1100	Cobban (1973)
Dulce NE	D4537	Baculites compressus (Inoceramus sagensis, Placenticeras sp.)	100	1300 (est.)	1200	Cobban and others (1974)
Dulce N	D5120	Baculites compressus (Placenticeras meeki)	130	1275 (est.)	1145	Cobban and others (1974)
Dulce N	D5116	Didymoceras cheyennense (Baculites aff. B. rugosus)	210	1275 (est.)	1065	Cobban and others (1974)
Dulce N	D5114	Didymoceras cheyennense	280 ("float")	1275 (est.)	995	Cobban and others (1974)
Dulce N	D5119	Didymoceras cheyennense (Baculites n. sp.)	323	1275 (est.)	952	Cobban and others (1974)
Dulce N	D5118	Exiteloceras jenneyi (Baculites rugosus)	364	1275 (est.)	911	Cobban and others (1974)
Dulce N	D5117	Exiteloceras jenneyi	450	1275 (est.)	825	Cobban and others (1974)
Stinking Lake	D5094	Didymoceras cheyennense (Inoceramus aff. I pertenuis)	175	925	750	Coban and others (1974)
Llaves	D4834	Didymoceras nebrascense	100	500	400	Cobban and others (1974)

*Identifications and field geographic and stratigraphic locations are from Cobban (1973) and Cobban and others (1974); USGS fossil locality numbers are shown on Figures 3 and 5.

radiometric analysis of minerals (biotite and sanidine) from bentonite beds found within the faunal zones at sites in Montana, Colorado, and Wyoming (for a detailed discussion of those age determinations the reader is directed to that report). Figures 5 and 6 and Table 2 show the ages of these ammonite zones as presented in Obradovich and Cobban (1975) and corrected mathematically to the new decay constants (John Sutter, written communication, 1984).

On the basis of these radiometrically dated ammonite zones, it is possible to determine the ages of the laterally equivalent Fruitland-Kirtland rocks. Figure 6 is a northeast-trending cross section that has been greatly exaggerated vertically to show the relationships of the ammonite zones to the physical geometry of the rocks. (This cross section was drawn using the same subsurface data as cross section B-B′ of Figure 2, but at different horizontal and vertical scales.) On the right side of the section, the Western Interior ammonite zones are shown; also shown are the ages of the three dated zones. The boundaries of the ammonite zones, time lines, are shown as solid lines through the marine rocks and as dashed lines where projected through continental rocks. As in Figure 4, these time lines pass through three different facies: the offshore marine Lewis Shale, the beach and shoreface regressive-marine Pictured Cliffs Sandstone, and the continental fluvial, paludal, and lacustrine Fruitland Formation and Kirtland Shale. The age of the rocks in each facies is the same along these time lines.

It is thus possible to state that the age of the Fruitland Formation in the southwest part of the San Juan Basin, where it directly overlies the Pictured Cliffs Sandstone, is equivalent to the age of the *Baculites scotti* ammonite zone and is in excess of 73.9 ± 0.8 Ma. Rocks at the stratigraphic level of dated volcanic ash beds in the Kirtland and Fruitland (at about the 152-m—500-ft—level) in the southwest part of the basin are equivalent in age to the boundary between the *Didymoceras stevensoni* and the *D. nebrascense* ammonite zones and are consequently between 73.7 ± 0.7 and 73.9 ± 0.8 Ma. The Kirtland Shale, at its youngest in the southwest part of the basin, just beneath the Tertiary Ojo Alamo Sandstone, is at the 335-m (1100-ft) level, the same level as the base of the *Baculites compressus* zone—thus the Kirtland must be older than 73.2 ± 0.7 Ma at that locality. The range of the oldest and youngest dated ammonites, at the 95 percent confidence level is from 71.8 (73.2 – 1.4) to 75.5 (73.9 + 1.6) Ma in the San Juan Basin (John Sutter, written communication, 1984).

It must be stated that because the spread of the three radiometric dates is only 0.7 Ma (73.9–73.2) and the possible error for these dates is also around 0.7 Ma, in effect these three dates

TABLE 2. BOUNDARIES OF SELECTED WESTERN INTERIOR AMMONITE ZONES IN THE SAN JUAN BASIN*

Western interior ammonite zone	U.S. Geological Survey Fossil locality number	Feet above or below Huerfanito Bed	Boundary level (in feet above Huerfanito Bentonite Bed) Upper	Lower	Radiometric age (Ma)
Baculites compressus	D4537	1200	?	1100	73.2
	D5120	1145			(±.7)
Didymoceras cheyennense	D4806	1100			
	D5116	1065			
	D5114	995			
	D5119	952	1100	950	
	D5094	750			
	D4800	700			
Exiteloceras jenneyi	D5118	911			
	D5117	825	950	700	73.7
					(±.7)
Didymoceras stevensoni	D389	550	700	500	
Didymoceras nebrascense	D4834	400			73.9
	D4797	260	500	200	(±.8)
Baculites scotti	D4796	-50	200	?	

*Fossil locality numbers are shown on Figures 3 and 5. Identifications of ammonites from Cobban (1973) and Cobban and others (1974). Radiometric ages (corrected) from Obradovich and Cobban (1975).

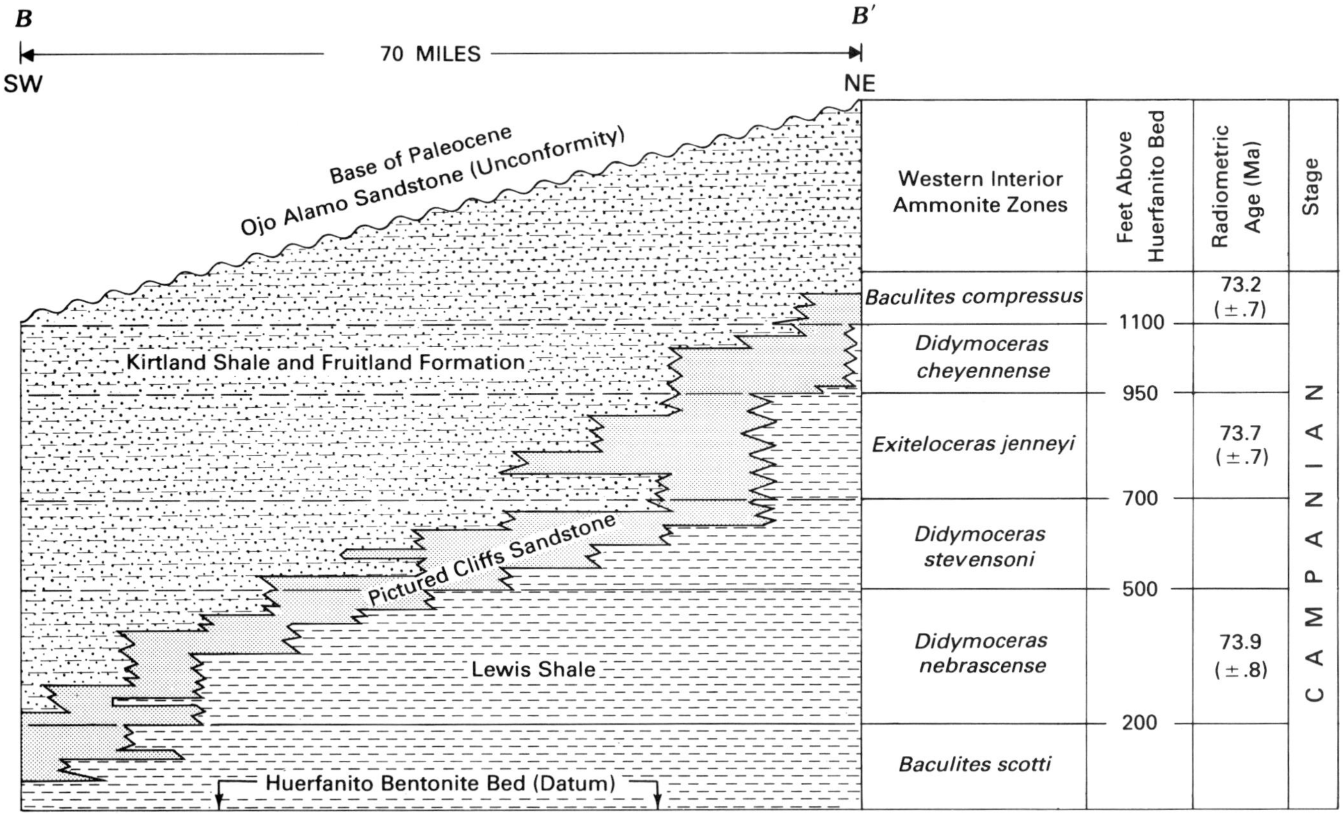

Figure 6. Northeast-trending stratigraphic cross section across the San Juan Basin. This section extends from control point 2 to control point 13 of the line-of-section B-B′ depicted on Figure 3. Placement of Western Interior ammonite zone boundaries is based on data from Cobban (1973) and Cobban and others (1974). Ages of ammonite zones are from Obradovich and Cobban (1975) corrected mathematically to the new decay constants. Stratigraphy is modified from Fassett and Hinds (1971).

basically represent only one date—the average for the three dated ammonite zones, which is 73.5 ± 0.5 Ma. We know, of course, from physical stratigraphy that the top of the Lewis Shale (base of the Pictured Cliffs) becomes progressively younger northeastward across the San Juan Basin, but because of the relative imprecision of the dates available, the use of these dates to calibrate such things as rates of deposition can only be characterized as a best estimate.

Obradovich and Cobban (1975, p. 36) placed the Campanian-Maastrichtian boundary at a level equivalent to the base of the *Baculites reesidei* zone. Because the *B. reesidei* zone is the second ammonite zone above the highest ammonite zone found in the San Juan Basin (the *B. compressus* zone), all of the rocks illustrated on Figure 6 are Campanian in age according to Obradovich and Cobban (1975). (Other workers have placed the Campanian-Maastrichtian boundary differently, some stratigraphically higher and some lower; therefore, based on some of those placements some of the rocks shown on Figure 6 could be considered lower Maastrichtian in age.)

Earlier in this report it was stated that Lindsay and others (1978) had suggested that deposition was continuous across the Cretaceous-Tertiary boundary. The dating herein of the youngest Kirtland Shale at about 73.2 ± 0.7 Ma very near the locality where the study by Lindsay and others was conducted on Ojo Alamo Arroyo (close to the southwest end of cross section B-B′ of Figures 3 and 6) demonstrates that there is a substantial hiatus there between the youngest Kirtland Shale and the overlying Tertiary Ojo Alamo Sandstone. If the Cretaceous-Tertiary boundary is at 66 Ma (Obradovich and Sutter, 1984), then the hiatus is on the order of 7 million years. If the hiatus includes some of early Paleocene time, as indicated by preliminary palynological studies (discussed below), this hiatus could represent as much as 10 million years. At a 95 percent confidence level, the age of the Kirtland Shale at the Lindsay and others (1978) locality is, at its youngest, 71.8 Ma (73.2 – 1.4), suggesting a hiatus of at least 6 million years.

The thickest Upper Cretaceous rocks of the San Juan Basin crop out in the northwest part. Figure 7 is a northwest-trending cross section that shows these thicker rocks. Because this cross section is oriented along the depositional strike of the Pictured Cliffs Sandstone (along its shoreline trend), the Pictured Cliffs is parallel to the Huerfanito Bed datum. On the left side of the section are shown a vertical scale, Western Interior ammonite zones, and ages. The time scale above the level of the *Baculites compressus* zone is an extrapolation based on the ages of the ammonite zones. The validity of this extrapolation, of course, hinges on the assumption of a constant average rate of deposition of one foot per thousand years for all of the rocks shown.

The age of the youngest Cretaceous rocks present in the northwest part of the San Juan Basin, as shown on Figure 7, is somewhat less than 72 Ma. Thus, the time gap between these youngest Upper Cretaceous rocks and the Cretaceous-Tertiary boundary (66 Ma) in the northwest part of the basin is around 6 million years. Figure 7 also dramatically shows the truncation, southeastward, of progressively older Kirtland-Fruitland rocks, the Pictured Cliffs Sandstone, and the upper Lewis Shale beneath the Tertiary Ojo Alamo Sandstone.

Because Figure 6 shows both rock thicknesses and rock ages, it provides the basic data permitting calculation of the average rate of deposition of the marine rocks discussed in this report. That rate comes out to be very close to 0.3 m (1 ft) of rock per thousand years. At that rate of deposition, the Pictured Cliffs Sandstone regression, representing a 335-m (1100-ft) stratigraphic rise across the basin, would have taken about one million years. It must be reiterated, however, that these numbers are based on a narrow range of dates (less than a million years) with relatively large errors and any conclusions regarding rates of deposition or rates of regression based on these ages are tentative, at best.

This exercise of determining the ages of the continental Kirtland and Fruitland formations by projection of dated ammonite zones is roundabout, but in the absence of precise radiometric age dates within the basin, is the only way to date these rocks at this time. I think these procedures are sound, but only the precise radiometric dating of volcanic ash beds from within the Fruitland and Kirtland formations in the San Juan Basin will provide a conclusive test of the validity of this study. As more precise radiometric age dates from ash beds within the Kirtland and Fruitland become available, they can be used to date more accurately the laterally equivalent Western Interior ammonite zones of the Lewis Shale by projection along time lines from continental to marine rocks (the reverse of the process described in this report).

OTHER EVIDENCE FOR THE AGE OF KIRTLAND-FRUITLAND ROCKS

A report by Brookins (1982) presented the ages of three bentonite beds in the lower Kirtland Shale from a site on Hunter Wash (Fig. 1) in the southwest part of the San Juan Basin. The stratigraphic level of these beds is about 153 m (500 ft) above the Huerfanito Bentonite Bed (see Fig. 6). These ages were determined by radiometric analysis of biotite and sanidine crystals from the ashes. Brookins (1982) stated: "These results are shown below and the data convincingly argue for an average date of 71.6 ± 2.7 MYBP." When corrected for the new decay constants, the average age for these lower Kirtland ash beds is 73.5 ± 2.2 Ma (John Obradovich, personal communication, 1983). The large error for these ages prevents their use to unequivocally confirm the results of the determination of the age of the Kirtland-Fruitland rocks by projection of ammonite zone ages, as discussed above. It is interesting to note, however, that the average age for these ash beds is exactly the same as the average age for the dated ammonite beds: 73.5 Ma.

Pollen and spore analysis has been conducted of rocks immediately below the Cretaceous-Tertiary boundary at a few localities in the San Juan Basin (Fassett and others, this volume). Published results of these analyses by Tschudy (1973) and by

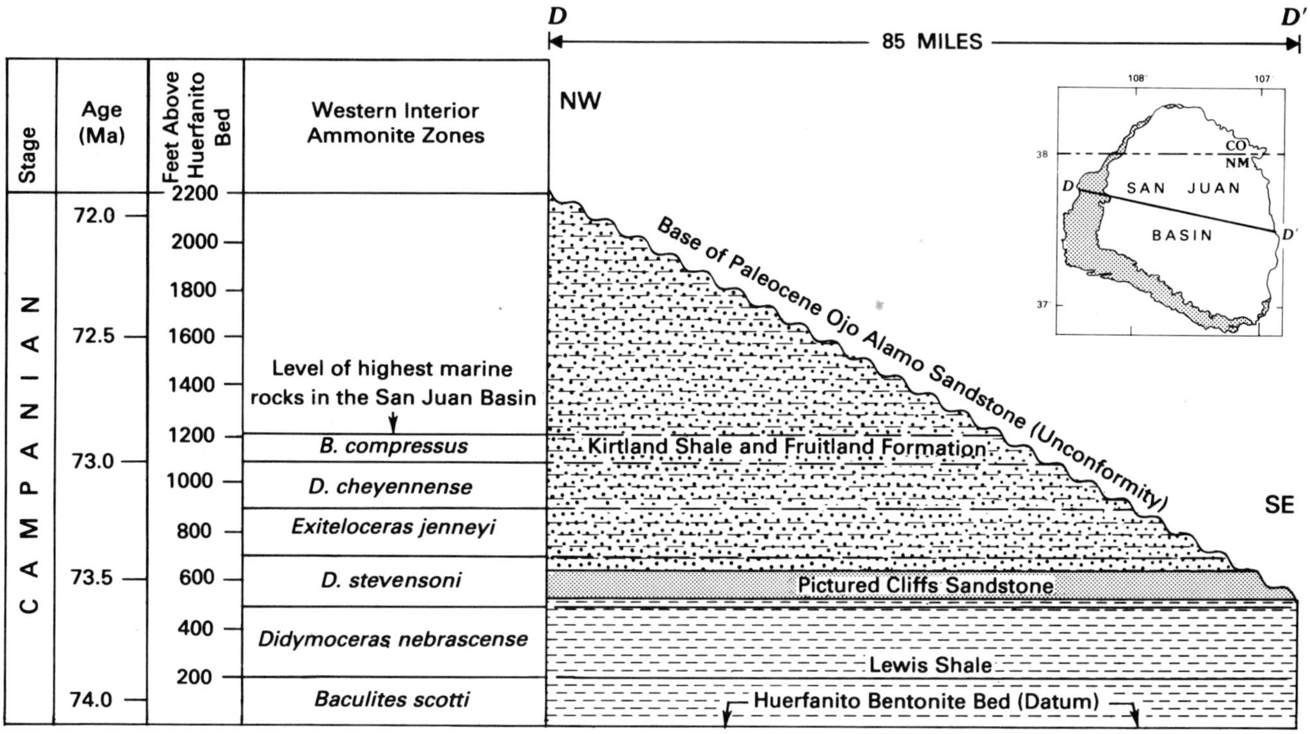

Figure 7. Stratigraphic cross section D-D′ trending southeastward across the central San Juan Basin. This cross section was drawn to parallel, as closely as possible, the 600-ft isopach line of Figure 3; thus, the top of the Pictured Cliffs Sandstone, hereon, is parallel to the Huerfanito Bentonite Bed. The Pictured Cliffs is shown diagrammatically as being the same thickness across the basin, whereas in reality its thickness along this line of section is variable because of local intertonging with the overlying Fruitland Formation and underlying Lewis Shale. Western Interior ammonite zone boundaries are based on data from Cobban (1973) and Cobban and others (1974). The age scale is based on an estimate of one foot of rock per 1000 years of deposition with the age of 73.5 Ma (the mean for the dated San Juan Basin ammonites) located at the 600-ft level above the Huerfanito Bentonite Bed (the approximate midpoint of the rocks containing the dated ammonite zones). Ammonite zone boundaries projected through continental Fruitland and Kirtland rocks are shown by dashed lines; those through the marine Lewis Shale by solid lines.

Tschudy in Fassett and Hinds (1971) indicate that uppermost Upper Cretaceous rocks are missing in the northeast-central and the southeast parts of the San Juan Basin. More recent, unpublished palynologic information (Robert S. Tschudy, written communication, 1977, 1982, 1983), further substantiates that most, if not all, of the Maastrichtian (of Tschudy, 1973) is missing in the San Juan Basin. Obviously, palynology does not yield precise ages for the Kirtland and Fruitland rocks in the San Juan Basin, but it does corroborate, in a general sense, that the age of the youngest Cretaceous rocks of the southern San Juan Basin is older than 70 Ma; this evidence certainly does not contradict the age-dating-by-projection findings of this report.

REFERENCES

Alvarez, L. W., Alvarez, W., Asaro, F., and Michel, H. V., 1980, Extraterrestrial cause for the Cretaceous-Tertiary extinction: Science, v. 208, p. 1095–1108.

Brookins, D. G., 1982, Revised K-Ar ages for the Kirtland Formation (Cretaceous), San Juan Basin, New Mexico: Isochron/West, no. 35, p. 15.

Cobban, W. A., 1973, Significant ammonite finds in uppermost Mancos Shale and overlying formations between Barker Dome, New Mexico, and Grand Junction, Colorado, *in* Fassett, J. E., ed., Cretaceous and Tertiary rocks of the southern Colorado Plateau: Four Corners Geological Society Memoir, p. 148–153.

Cobban, W. A., Landis, E. R., and Dane, C. H., 1974, Age relations of upper part of Lewis Shale on east side of San Juan Basin, New Mexico: *in* Ghost Ranch: New Mexico Geological Society 28th Field Conference Guidebook, p. 279–282.

Fassett, J. E., 1973, The saga of the Ojo Alamo Sandstone; Or, the rock stratigrapher and the paleontologist should be friends, *in* Fassett, J. E., ed., Cretaceous and Tertiary rocks of the southern Colorado Plateau: Four Corners Geological Society Memoir, p. 123–130.

——, 1979, Comment on biostratigraphy and magnetostratigraphy of Paleocene terrestrial deposits, San Juan Basin, New Mexico: Geology, v. 7, p. 69–70.

——, 1982, Dinosaurs in the San Juan Basin, New Mexico, may have survived the event that resulted in creation of an iridium-enriched zone near the Cretaceous-Tertiary boundary, *in* Geological implications of impacts of large asteroids and comets on the earth: Geological Society of America Special Paper 190, p. 435–447.

Fassett, J. E., and Hinds, J. S., 1971, Geology and fuel resources of the Fruitland Formation and Kirtland Shale of the San Juan Basin, New Mexico and Colorado: U.S. Geological Survey Professional Paper 676, 76 p.

Gill, J. R., and Cobban, W. A., 1966, The Red Bird section of the Upper Cretaceous Pierre Shale in Wyoming: U.S. Geological Survey Professional Paper 393-A, 73 p.

Lindsay, E. H., Jacobs, L. L., and Butler, R. F., 1978, Biostratigraphy and magnetostratigraphy of Paleocene terrestrial deposits, San Juan Basin, New Mexico: Geology, v. 6, p. 425–429.

Middleton, G. V., 1973, Johanes Walther's Law of the Correlation of Facies: Geological Society of America Bulletin, v. 84, p. 979–988.

Obradovich, J. D., and Cobban, W. A., 1975, A time scale for the Late Cretaceous of the Western Interior of North America: Geological Association of Canada Special Paper No. 13, p. 31–54.

Obradovich, J. D., and Sutter, J. F., 1984, What is the age of the Cretaceous-Tertiary boundary as recognized in continental deposits [abs.]: Geological Society of America Abstracts with Programs, v. 16, p. 612.

Schrader, F. C., 1905, The Durango-Gallup coal field of Colorado and New Mexico: U.S. Geological Survey Bulletin 285, p. 241–258.

Sears, J. D., Hunt, C. B., and Hendricks, T. A., 1941, Transgressive and regressive Cretaceous deposits in southern San Juan Basin, New Mexico: U.S. Geological Survey Professional Paper 193-F, p. 101–121.

Tschudy, R. S., 1973, The Gasbuggy core; A palynological appraisal, *in* Fassett, J. E., ed., Cretaceous and Tertiary rocks of the southern Colorado Plateau: Four Corners Geological Society Memoir, p. 131–143.

Walther, J., 1893–1894, Einleitung in die Geologic als historische wissenschaft: Jena, Verlag von Gustav Fischer, 3 vols., 1055 p.

Manuscript Accepted by the Society March 24, 1986

Printed in U.S.A.

Geological Society of America
Special Paper 209
1987

Dinosaurs, pollen and spores, and the age of the Ojo Alamo Sandstone, San Juan Basin, New Mexico

James E. Fassett
U.S. Geological Survey
903 National Center
Reston, Virginia 22092

Spencer G. Lucas
Department of Geology
University of New Mexico
Albuquerque, New Mexico 87131

F. Michael O'Neill
U.S. Bureau of Land Management
Taos, New Mexico 87571

ABSTRACT

The Ojo Alamo Sandstone of the San Juan Basin of New Mexico is composed of interbedded conglomeratic sandstone, sandstone, siltstone, and mudstone. Conglomerates are found in the western part of the basin; siliceous pebbles diminish in size both southward and eastward across the basin, becoming rare to nonexistent in the eastern part. There is great variation in the internal stratigraphy of the Ojo Alamo: Individual sandstone or mudstone beds thicken, thin, and pinch out laterally. The thickness of the Ojo Alamo varies from 6 m (20 ft) to more than 122 m (400 ft). The formation varies in composition from one to as many as five sandstone beds with interbeds of siltstone or mudstone.

Dinosaur bone has been found within the Ojo Alamo at several sites in the western part of the basin. Paleocene pollen has been found within the Ojo Alamo in the eastern part of the basin. To date, no Cretaceous pollen has been found at or above the stratigraphic level of dinosaur bone within the Ojo Alamo Sandstone. Near Barrel Spring, in the southwest part of the basin, both dinosaur bone and Paleocene pollen have been found. One bone, found at the top of the Ojo Alamo, was loose on the surface, and its significance is therefore equivocal. Dinosaur bone, however, has also been found in place in the upper part of the Ojo Alamo about 1.6 km (1 mi) west of Barrel Spring, at about the same stratigraphic level as Paleocene pollen from a site just east of Barrel Spring. Because there is no apparent unconformity between the highest in-place bone level and the Paleocene pollen level in this area, the Ojo Alamo dinosaurs, if not reworked, are Paleocene in age at this site and probably throughout the San Juan Basin.

INTRODUCTION

The age and the rock-stratigraphic definition of the Ojo Alamo Sandstone have been controversial for many years. The age has been changed from Cretaceous, to Tertiary(?), to Cretaceous, to Tertiary, to Cretaceous-and-Tertiary. The definition has changed from multiple sandstone beds with intervening siltstones and shales, to a single sandstone sheet, and back to multiple sandstones. Some authors have placed the contacts (especially the basal contact) of the Ojo Alamo at different levels to suit their particular perspectives at places in the San Juan Basin; different definitions have even been used at the type area itself. A detailed discussion of the history of the age and definition of the Ojo Alamo Sandstone is beyond the scope of this report, and the interested reader is directed to papers by Fassett (1973, 1982).

In this paper we discuss each locality in the San Juan Basin where either dinosaur bone or palynomorphs have been identified within the Ojo Alamo. Where unmodified, the name Ojo Alamo Sandstone in this report is used, as first defined, as a rock stratigraphic unit, mapped by Clyde Max Bauer in 1916. At each locality, however, we describe the different definitions of the Ojo Alamo used by other authors. In our conclusion, we present our interpretation of the age of the Ojo Alamo Sandstone based on all available evidence.

STRATIGRAPHIC RELATIONS

The base of the Ojo Alamo Sandstone is an unconformity; the Ojo Alamo overlies successively older rocks southeastward

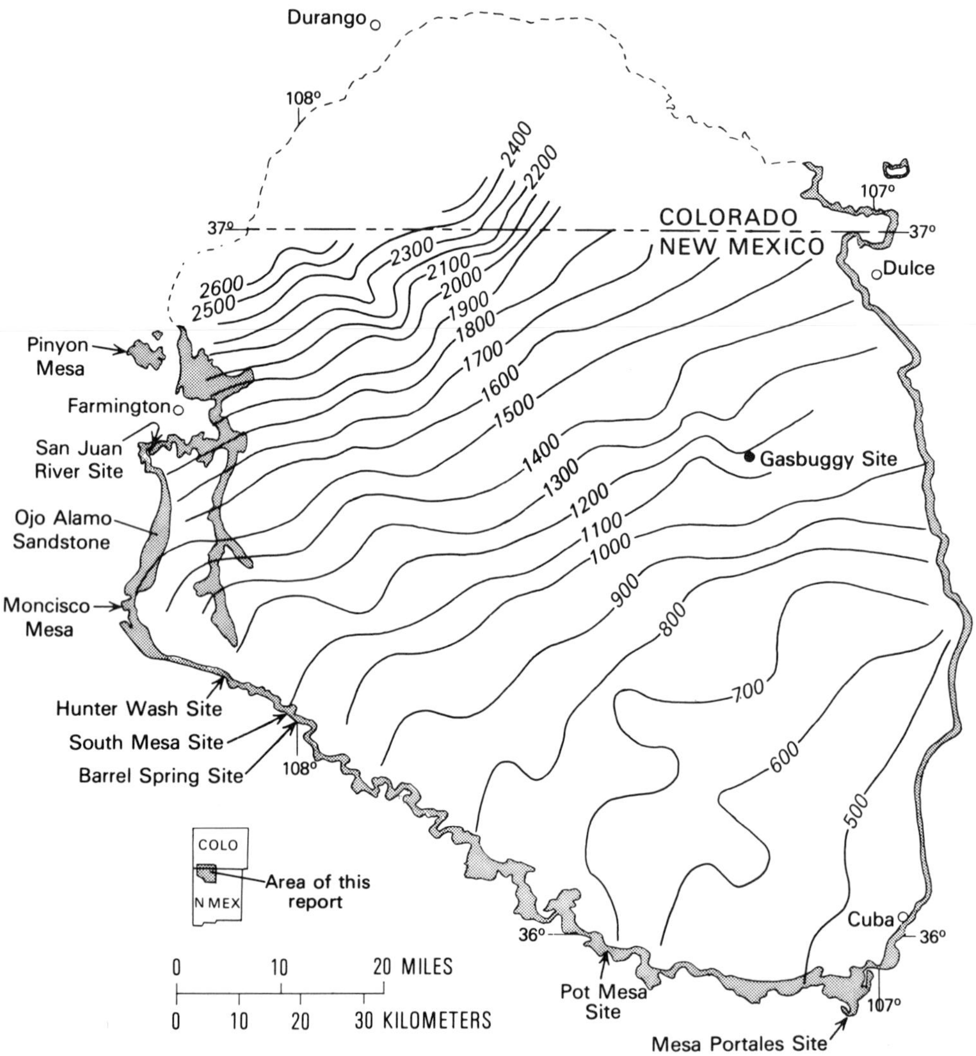

Figure 1. Outcrop map of the Ojo Alamo Sandstone (stippled) including isopachs of the interval between the Huerfanito Bentonite Bed and the base of the Ojo Alamo Sandstone. Both the geology and the isopachs are from Fassett and Hinds (1971). Fossil collection sites referred to in this report are also shown.

across the basin. The formations that underlie the Ojo Alamo, from northwest to southeast, are the McDermott Member of the Animas Formation, the Kirtland Shale, the Fruitland Formation, the Pictured Cliffs Sandstone, and on the extreme east edge of the basin, the Lewis Shale. All of these rocks are Late Createceous (Campanian to early Maastrichtian) in age. The Ojo Alamo is overlain, apparently conformably, by the upper part of the Animas Formation in the northernmost part of the basin and by the laterally equivalent Nacimiento Formation in the southern part of the basin. (For a detailed discussion of the stratigraphy and energy resources of these formations, see Fassett and Hinds, 1971).

FOSSIL LOCALITIES

Figure 1 shows the San Juan basin outlined by the outcrop of the Ojo Alamo Sandstone. Each of the sites where dinosaurs and palynological samples have been collected from the Ojo Alamo is shown on this map. Isopachs of the interval between the base of the Ojo Alamo and the Huerfanito Bentonite Bed of the Lewis Shale show the stratigraphic position of each Ojo Alamo fossil locality relative to an underlying time plane. (The Huerfanito is a time-synchronous subsurface marker bed and probably represents a volcanic ash fall into the Lewis sea in late, but not latest, Campanian time: Fassett and Hinds, 1971; Fassett, this volume).

San Juan River Site

This site (Fig. 1) represents the most recent dinosaur find in the Ojo Alamo Sandstone. It is located about 5.6 km (3.5 mi)

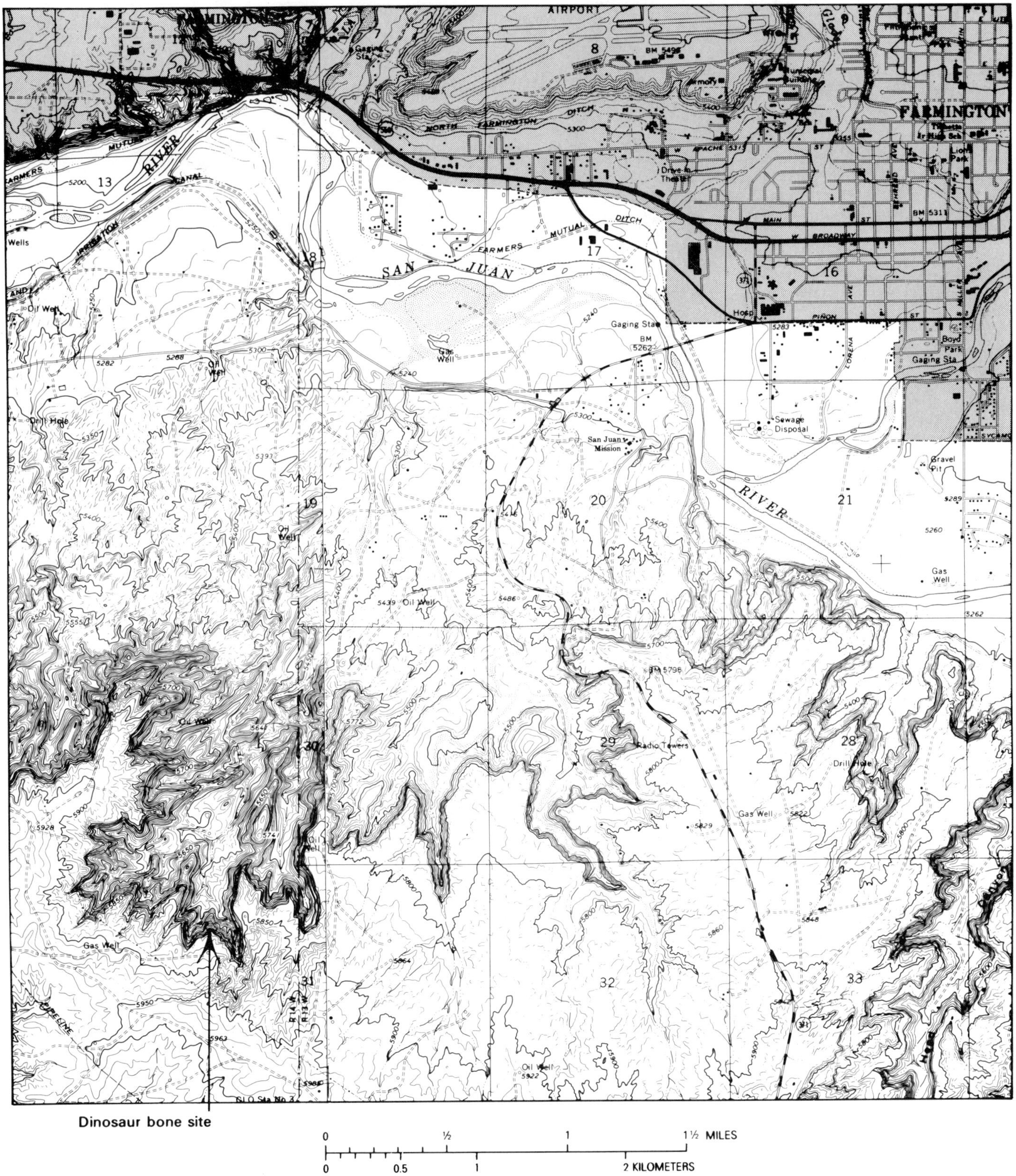

Figure 2. Topographic map showing the location of the San Juan River dinosaur bone site. Map is a composite of parts of the U.S. Geological Survey Kirtland and Farmington South 7½-minute topographic quadrangle maps.

Figure 3 (this and facing page). Photographs of the San Juan River dinosaur bone site. Photograph A shows the location of the bone site (arrow) beneath an undercut massive cliff of conglomeratic Ojo Alamo Sandstone. Vertical cliff at top is upper ledge of Ojo Alamo. Person is standing on the contact between the Ojo Alamo and underlying Kirtland Shale.

southwest of Farmington, New Mexico, south of the San Juan River. The site is on unsurveyed land on the Navajo Indian Reservation and is in the S½ Sec. 36, T.29N.,R.14W. (projected). Figure 2 is a large-scale map showing the location of the site.

The bone collected from this site is a right femur more than 112 cm (44 in) long (Fig. 3). It is from a hadrosaurian dinosaur and is virtually complete (portions of the proximal and distal ends are missing due to recent weathering). The bone is unabraded. Hadrosaurs of this size have been reported from within the Ojo Alamo Sandstone of Bauer (1916) at the Ojo Alamo type locality (Lehman, 1981); however, we consider this isolated femur to be insufficient material upon which to base a genus-level identification.

The bone site at this locality is within the lower part of the Ojo Alamo Sandstone at the base of a massive, undercut cliff of conglomeratic sandstone (Fig. 3). The bone was found in the upper part of a distinctive 1.8 to 2-m (6 to 7-ft) thick bed of green-colored conglomeratic sandstone. As Figure 3A shows, this conglomerate bed has weathered back more readily than the more massive, buff-colored, overlying conglomeratic sandstone bed. The base of the overlying buff-colored unit is a scour surface having a few inches to a foot or so of local relief. The base of the green, dinosaur-bearing, conglomerate bed is not exposed at this locality. Below this bed is a partially covered slope underlain by more easily weathered buff to white-colored conglomeratic sandstone, siltstone, and some carbonaceous mudstone. Below this unit is a rounded exposure of whitish, conglomeratic sandstone littered with iron-stained siliceous pebbles. This lower part of the Ojo Alamo is about 15 m (50 ft) thick. Figure 4 is a stratigraphic column showing the relations of these rocks and the position of the dinosaur-bone site.

The lower part of the Ojo Alamo Sandstone, the soft, slope-forming conglomeratic rocks discussed above, can be traced continuously around a small rincon (Fig. 2), southeastward, and then to the north where it becomes the lower half of a massive west-facing cliff face. Northward on this cliff face, this lower part of the Ojo Alamo is a vertical massive cliff of whitish conglomeratic sandstone. The green-colored conglomerate bed containing the

Figure 3B. Photograph B shows the bone in place in a cross-bedded conglomeratic sandstone bed of the Ojo Alamo Sandstone. Rock hammer is 28.4 cm (11.2 in) long.

dinosaur bone at the San Juan River site can also be traced continuously around these exposures to the southeast and north; it thins northward along the west-facing cliff to less than a few feet thick. At one small area, this bed completely pinches out; there the upper massive conglomeratic sandstone and the lower massive conglomeratic sandstone become one massive vertical cliff. Figure 5 is a photograph looking east at this cliff face taken from the dinosaur-bone locality.

The Ojo Alamo was first mapped and described in detail in this area by Reeside (1924). Reeside measured a stratigraphic section not very far east of this site (exact location not given). He labeled this section: "Section along San Juan River below Farmington, N. Mex." (Reeside, 1924, p. 62). In this section, he described the Ojo Alamo Sandstone as being nearly 122 m (400 ft) thick and consisting of four conglomeratic sandstone beds separated by interbeds of shale and siltstone. The part of Reeside's San Juan River stratigraphic section describing the Ojo Alamo is depicted on Figure 6. We have shown the dinosaur bone level of the San Juan River site about in the middle of the lowermost conglomeratic sandstone bed on this figure on the assumption that the lower slope-forming conglomerate beds, the green conglomeratic sandstone layer, and the overlying massive conglomeratic sandstone together constitute the lowermost, 30-m (100-ft) thick conglomeratic sandstone bed measured by Reeside in 1924 in his San Juan River section.

Reeside's (1924) section indicated that the conglomeratic Ojo Alamo Sandstone was underlain by the McDermott Formation in this area. Reeside (1924, p. 25) wrote: "At San Juan River the McDermott formation is represented by thin irregular lenses of fine purple and green tuffaceous sandstone, coarse white sandstone with clay pellets, and purple and gray shale forming 30 feet of beds just beneath the Ojo Alamo sandstone." Reeside (1924) did not include any siliceous-pebble conglomerate in his McDermott rocks south of the San Juan River.

Reeside made it clear that the appearance of conglomerates in the stratigraphic section represented a significant change in deposition in the San Juan Basin. He wrote (Reeside, 1924, p. 26): "No pebbles of any kind have been found in the formations

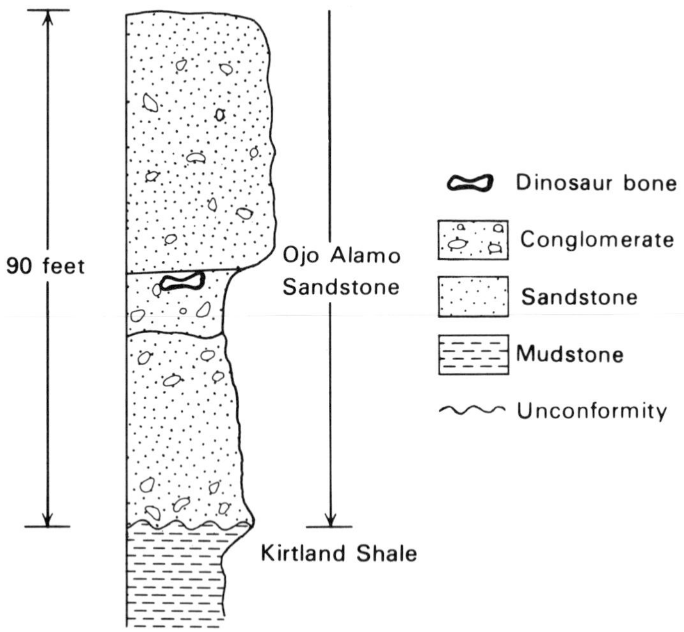

Figure 4. Stratigraphic column of the lowermost part of the Ojo Alamo Sandstone at the San Juan River dinosaur bone site. The top of the uppermost Conglomeratic Sandstone is eroded here, thus the 90 ft shown is not the total thickness of this unit.

between the Tocito sandstone lentil of the Mancos shale and the McDermott formation [north of the San Juan River]. The siliceous pebbles, therefore, appear very suddenly in the succession of strata. Their source has not been traced."

Subsequent workers have changed the definition and the mapped extent of the "McDermott formation" of Reeside (1924). Barnes and others (1954) revised the McDermott in southern Colorado from a formation overlying the Kirtland Shale to a lower member of the Animas Formation. This change in stratigraphic rank was unfortunate because it placed an unconformity representing several million years within the Animas Formation; however, further revision of the McDermott is beyond the scope of this report. Other workers, including O'Sullivan and Beikman (1963), O'Sullivan and others (1972), and Fassett and Hinds (1971), have mapped the McDermott; none of them shows the McDermott extending as far south as the San Juan River.

Baltz and others (1966, p. D9) state that the McDermott "... thins southward from its type locality in Colorado and is cut out by the unconformity at the base of the rocks assigned by Reeside to the Ojo Alamo Sandstone east of La Plata River northwest of Farmington near the center of T.30N.,R.13W. The McDermott is not present between this area and the southernmost good exposures several miles south of the San Juan River, although purplish beds occur in the upper part of the Kirtland

Figure 5. Photograph of west-facing cliff face east of the San Juan River dinosaur bone site. View is looking east from the dinosaur bone site. The rock units shown are the Kirtland Shale (Kk) and the Ojo Alamo Sandstone (Toa). The Ojo Alamo (only the lower part is here seen) consists of three subunits—A: an upper dark-colored, massive, conglomeratic sandstone bed; B: a medial interbed of silty shale and soft conglomeratic sandstone (dinosaur bed); and C: a lower light-colored, massive, conglomeratic sandstone bed. Note that the medial interbed is missing locally to the right of center of the photograph; there the upper and lower beds of conglomeratic sandstone merge into one bed.

near the road to Bisti south of the river." Therefore, south of the river at the San Juan River bone site, the conglomeratic Ojo Alamo Sandstone directly overlies fine-grained rocks now mapped as Kirtland Shale.

O'Sullivan and others (1972, p. E57) included only about the lower half of the Ojo Alamo of Reeside (1924) in their Ojo Alamo Sandstone in the San Juan River area. Figure 6 shows the Ojo Alamo of Reeside (1924) and O'Sullivan and others (1972). The Ojo Alamo of O'Sullivan and others (1972) thus consists of the two lowest conglomeratic sandstone beds of Reeside (1924), totaling as much as 67 m (220 ft).

These Ojo Alamo sandstone beds, according to O'Sullivan and others (1972, p. E57), are in places separated by a southward-and-westward-thickening mudstone interbed. They state: "Traced eastward about 3 miles from the road between Farmington and Bisti, the mudstone unit between the upper and lower conglomeratic sandstones merge to form an unbroken sequence about 200 ft thick." The Ojo Alamo, thus described, was mapped in the northwest part of the San Juan Basin by O'Sullivan and Beikman (1963). Because of the small scale of this map (1:250,000), it is not possible to locate the basal contact of the Ojo Alamo precisely enough to say positively whether or not the San Juan River bone site is within the Ojo Alamo as shown on that map; nevertheless, it appears that it is.

O'Sullivan and others (1972, p. E56) wrote: ". . . the coarse conglomeratic Ojo Alamo contrasts distinctly with the mudstone and soft sandstone of the underlying Kirtland Shale." They further noted (p. E53) that a thin layer of siliceous pebbles 11 m (35 ft) below the top of the Kirtland Shale had been found at a locality 1.6 km (1 mi) south of the San Juan River (about 4 km—2.5 mi—northeast of the San Juan River bone site) but added ". . . however, no other such pebbles were noted [in the Kirtland] elsewhere on the Navajo Indian Reservation. . . ." It thus seems clear that the San Juan River dinosaur bone site—which is on the Navajo Indian Reservation, is within a conglomeratic sandstone bed, and is some 15 m (50 ft) stratigraphically above the base of a lower conglomeratic sandstone bed—is well within the Ojo Alamo Sandstone as described and mapped by both Reeside (1924) and O'Sullivan and others (1972).

As seen on the isopachs of Figure 1, the base of the Ojo Alamo Sandstone stratigraphically is 580 m (1900 ft) above the Huerfanito Bentonite Bed at this locality. This level is about 152 m (500 ft) stratigraphically lower than the base of the Ojo Alamo at its northernmost occurrence near Pinyon Mesa (Fig. 1), where the base of the Ojo Alamo is 732 m (2,400 ft) stratigraphically above the Huerfanito Bed. This drop in the stratigraphic level of the base of the Ojo Alamo indicates that the uppermost 152 m (500 ft) of Kirtland Shale and overlying remnants of the McDermott Member of the Animas Formation (as mapped by O'Sullivan and Beikman, 1963), present beneath the Ojo Alamo at Pinyon Mesa, are missing at the San Juan River site, having been removed by erosion there, prior to deposition of the Ojo Alamo Sandstone. This isopach map thus supports the field observations of Baltz and others (1966, p. D9) that the McDermott is truncated by a pre–Ojo Alamo unconformity north of the San Juan River.

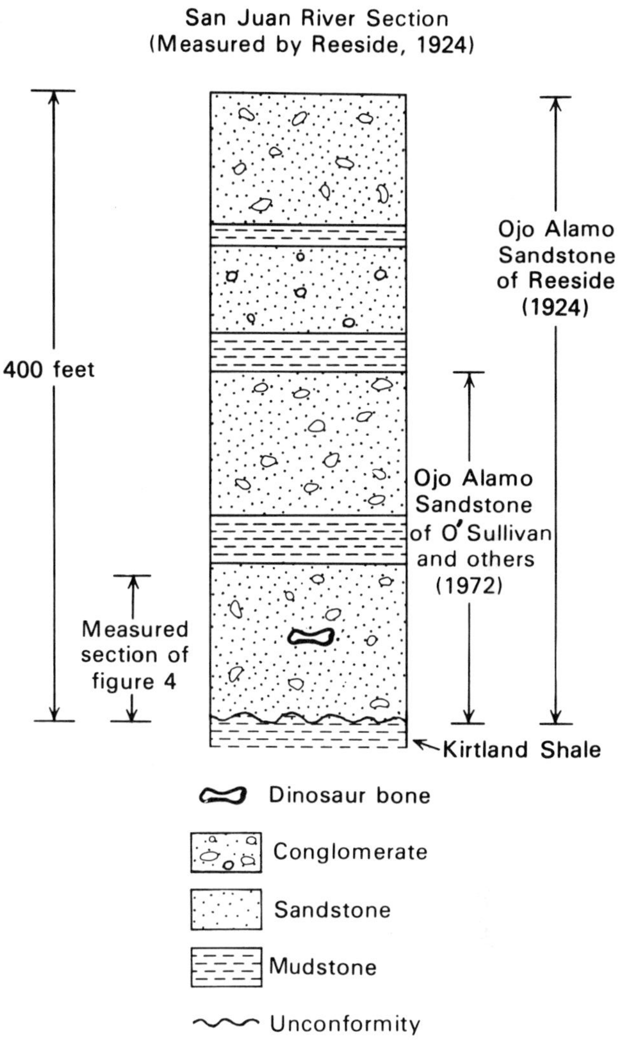

Figure 6. Stratigraphic column of the Ojo Alamo Sandstone near (probably west of) the San Juan River dinosaur bone site of Figure 4 as measured by Reeside (1924). The stratigraphic level of the dinosaur bone site is in the middle of the lowermost conglomeratic sandstone bed. The Ojo Alamo as mapped by Reeside (1924) and O'Sullivan and others (1972) is shown. Note that O'Sullivan and others (1972) did not include the two upper conglomeratic sandstone beds of Reeside's Ojo Alamo in this area.

In a paper by Fassett (this volume), the ages of the continental Fruitland Formation and Kirtland Shale are estimated based on a projection of dated ammonite zones in the Lewis Shale along time lines parallel to the Huerfanito Bentonite Bed. In that report the age of the Kirtland Shale is estimated to be a little more than 72 Ma immediately beneath the Ojo Alamo at this site. If the Cretaceous-Tertiary boundary is 66 Ma and is located at the base of the Ojo Alamo, then there is a 6-million year hiatus between the Kirtland and Ojo Alamo at this locality. If the lowermost Ojo

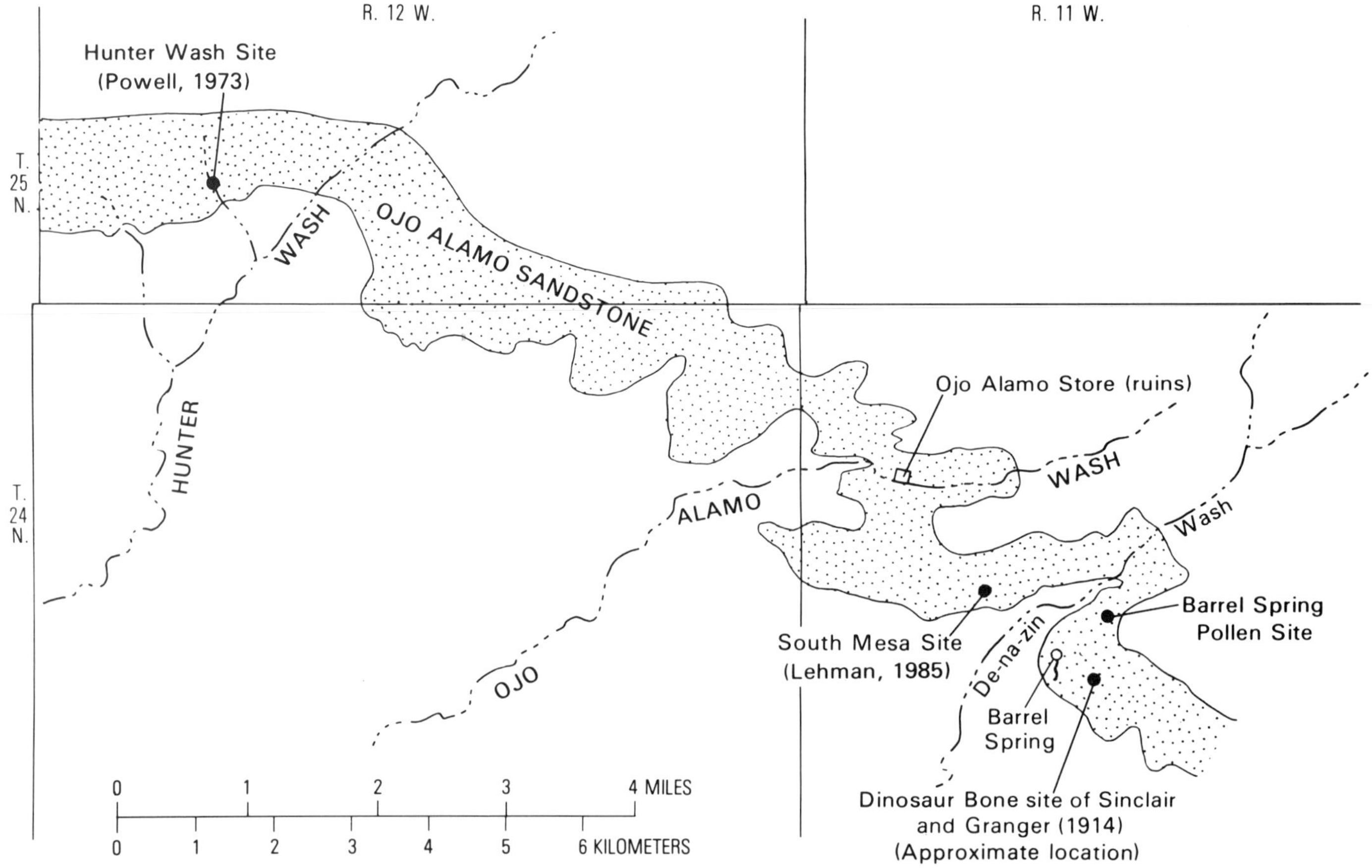

Figure 7. Map showing the outcrop of the Ojo Alamo Sandstone as defined by Bauer (1916) at the Ojo Alamo type locality. Geology is modified from Baltz and others (1916). The locations of the Hunter Wash, South Mesa, and Barrel Spring pollen and dinosaur bone collection sites are shown.

Alamo Sandstone does not represent earliest Paleocene time, as is the case in the southern San Juan Basin (on the basis of palynological analysis), then the hiatus here could be even greater.

Ojo Alamo Wash Area

General Description. The Ojo Alamo Sandstone was defined (as a rock-stratigraphic unit) and mapped in the Ojo Alamo Wash area by Bauer (1916). He measured 11 sections along about 15 km (9 mi) of Ojo Alamo outcrop in this area (Figs. 1, 7) and defined the Ojo Alamo as conglomeratic sandstones and interbeds of shale, siltstone, and sandstone. Bauer (1916, p. 276) referred to the area around Ojo Alamo Wash as "the type locality" of the Ojo Alamo, even though no specific stratigraphic section or specific site was ever designated as a type locality. It seems apparent that Bauer was referring in a general way to the entire area containing his 11 measured sections as the "type locality." In this report, we use the expression "type locality" to refer to the area around Ojo Alamo Wash where Bauer (1916, plates LXIX and LXX) measured his 11 sections and defined the rock-stratigraphic Ojo Alamo Sandstone. Bauer (1916, p. 276) stated that the Ojo Alamo ranged from 19 to 36 m (63 to 110 ft) thick at the type locality.

Baltz and others (1966) revisited the Ojo Alamo Wash area and measured 15 stratigraphic sections of the Ojo Alamo and associated rocks at and near the type locality of Bauer (1916). They restricted the Ojo Alamo in this area to the upper conglomeratic sandstone of Bauer (1916). They stated that their "restricted" Ojo Alamo ranged from 11 to 19 m (35 to 62 ft) thick in their study area. That part of Bauer's (1916) Ojo Alamo thus removed was named the Naashoibito Member of the Kirtland Shale.

Powell (1973, p. 121), who had studied the Ojo Alamo Sandstone throughout the San Juan Basin, recommended that "... the redefinition of Baltz, Ash, and Anderson (1966) of the Ojo Alamo of the type area be dropped and that the original rock stratigraphic definition of Bauer (1916) be reinstated." In addition, he suggested that the upper part of the Ojo Alamo of Bauer (1916)—the "restricted" Ojo Alamo of Baltz and others (1966)—be named the Kimbeto Member of the Ojo Alamo Sandstone and that the Naashoibito Member of the Kirtland

Shale of Baltz and others (1966) become the Naashoibito Member of the Ojo Alamo Sandstone.

Scott and others (1979) and Brown (1982a, b) mapped the geology of three 7½-minute quadrangles, including the type locality of the Ojo Alamo Sandstone of Bauer (1916) and the area mapped by Baltz and others (1966). On all three of these maps, the Ojo Alamo is described exactly as defined and mapped by Bauer in 1916. In addition, no Naashoibito Member, either of the Kirtland Shale of Baltz and others (1966) or of the Ojo Alamo Sandstone of Powell (1973), appears on any of these maps.

Beginning with Barnum Brown (1910), dinosaur bone has been identified by many workers within the type locality of the Ojo Alamo. Most of this bone has come from the middle part of the Ojo Alamo (Naashoibito Member of the Kirtland in Baltz and others [1966]; Naashoibito Member of the Ojo Alamo Sandstone of Powell [1973]). At three sites in the type locality, however, dinosaur bone has been reported from the upper conglomeratic sandstones of the Ojo Alamo (the "restricted" Ojo Alamo in Baltz and others [1966]; the Kimbeto Member of the Ojo Alamo of Powell [1973]). These three sites are discussed individually below.

Hunter Wash Site. The Hunter Wash site (Fig. 7) is located on a northern tributary of Hunter Wash in the N½NW¼ Sec. 32, T.25N., R.12W., in the northwest part of the Ojo Alamo type locality. The Ojo Alamo at this site is about 24 m (80 ft) thick and consists of "interbedded coarse sandstone, conglomerate, and shale" (Scott and others, 1979). The Ojo Alamo was first mapped in this area by Bauer (1916), was mapped in a restricted sense by Baltz and others (1966), and most recently was described and mapped by Scott and others (1979) as defined by Bauer.

Dinosaur bone was discovered at this site by Powell (1973) within the upper conglomeratic sandstone part of the Ojo Alamo. Powell (1973, p. 119) wrote:

> At this site mammal and reptile bone scraps were found in place as low as three feet above the base of the Kimbeto Member and as much as 30 feet above the base of the Kimbeto Member. Between six and ten feet above the base of the unit several dinosaur bone fragments were found, the largest of which is six inches long and three inches in diameter. The fossils are water-worn and undiagnostic, but they do establish that dinosaur remains are found in the Kimbeto Member of the Ojo Alamo Sandstone. . . .

After collecting and examining some of these dinosaur bone fragments (now catalogued as UNM [University of New Mexico] TOA-1), Lucas (1981, p. 346) offered a conclusion different from Powell's:

> I recently examined the locality described by Powell . . . and confirmed his observation by finding several *in situ* dinosaur bone fragments about 2 m above the base of the Ojo Alamo (Text-fig. 15.8f). One of these fragments, placed in the University of New Mexico Department of Geology collections (UNM TOA-1), is typical in being a well-abraded angular chunk of bone, large enough (5 cm long and 6 cm wide) to be considered dinosaurian. However, the Ojo Alamo at this locality contains much gravel and petrified wood, including a prone fossil log more than 10 m in length (Text-fig. 15.8e). This evidence of high energy of deposition supports the hypothesis that these bone fragments are reworked. . . .

(It can also be argued that the presence of large bone fragments or silicified logs in a conglomeratic, high-energy sandstone deposit indicates rapid deposition, quick burial, and little reworking; otherwise, this material would have been quickly ground to bits.) Figure 8 contains a composite stratigraphic column at this site showing the levels within the Ojo Alamo at which dinosaur bone has been found.

South Mesa Site. The South Mesa site (Fig. 7) is located west of Barrel Spring Arroyo—now named De-na-zin Arroyo on recent U.S. Geological Survey topographic quadrangle maps. The "South Mesa" name was coined by Clemens (1973, Fig. 3). He defined South Mesa as a topographic feature located in the south part of Sections 7, 8, and 9 and the north part of Sections 17 and 18, T.24N.,R.11W. above the 6400-foot contour line on the Alamo Mesa East Topographic Quadrangle map. The Ojo Alamo was first mapped here by Bauer (1916), mapped by Baltz and others (1966) in a restricted sense, and was recently mapped by Brown (1982a). Brown (1982a) described the Ojo Alamo exactly as it was described by Bauer (1916). However, even though Brown's (1982a) definition of the Ojo Alamo Sandstone is identical to Bauer's (1916) definition, Brown has not mapped the Ojo Alamo exactly as Bauer mapped it. For example, around the southwest and west sides of South Mesa and west of Barrel Spring, the lower conglomeratic part of the Ojo Alamo as mapped by Bauer (1916) has not been everywhere included in Brown's Ojo Alamo Sandstone.

Lehman (1985, p. 56) discovered an indeterminate dinosaur rib fragment and nodosaurid dermal plates 1 m (3.3 ft) above the base of the upper conglomeratic bed (the restricted Ojo Alamo in Baltz and others [1966] and the Kimbeto Member of Powell [1973]) on South Mesa in the NW¼ Sec. 17, T.24N.,R.11W (Fig. 7). Lehman (1985) noted: "These remains do not appear any more abraded than those typically found in the Naashoibito, but nothing short of articulated dinosaur remains could securely demonstrate a Cretaceous age for part of the Ojo Alamo Sandstone." A recent attempt in 1984 to relocate this occurrence by Spencer Lucas and Adrian Hunt failed. Figure 8 contains a stratigraphic column showing the levels at which dinosaur bone has been collected at South Mesa.

Barrel Spring Site. Barrel Spring (Fig. 7) is located in a small, deep, narrow canyon in the west-central part of Sec. 16, T.24N.,R.11W. on the U.S. Geological Survey Alamo Mesa East 7½-minute topographic quadrangle map (the spring is not shown on that map). The Ojo Alamo was first mapped here by Bauer (1916), mapped in a restricted sense by Baltz and others (1966), and recently mapped again by Brown (1982a). In this area, Brown's (1982a) mapping does not include the lower conglomerates of the Ojo Alamo of Bauer (1916).

Sinclair and Granger (1914, p. 301) were the first workers

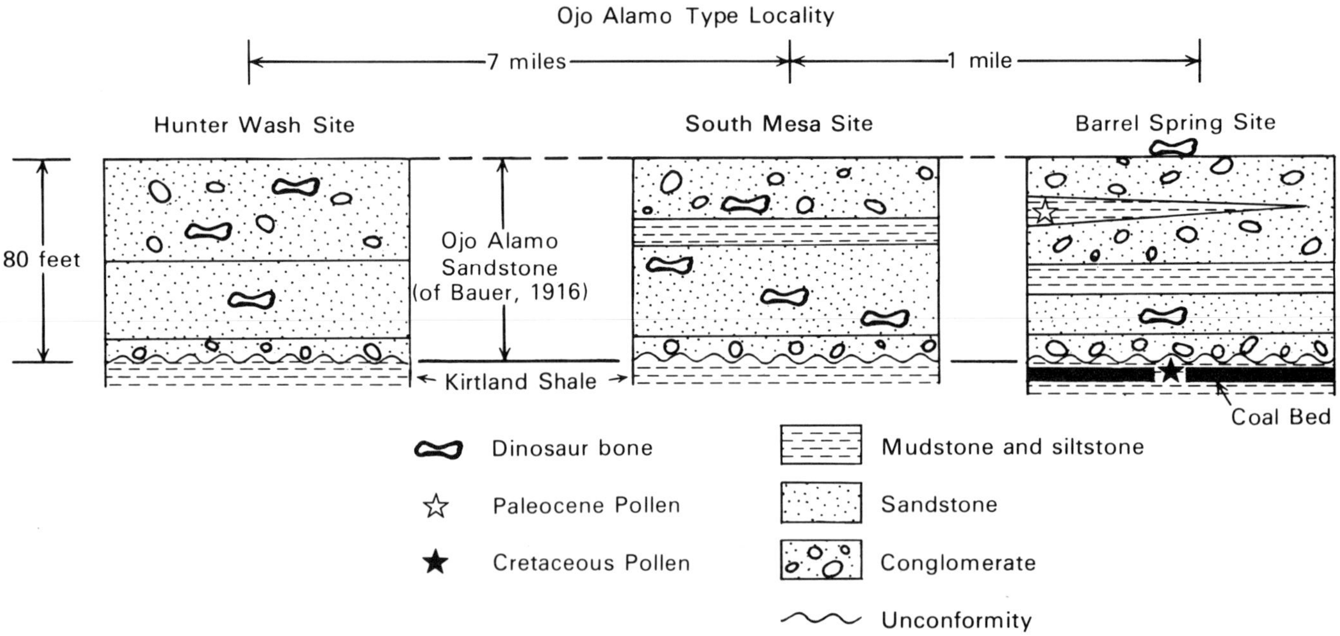

Figure 8. Stratigraphic columns of the Ojo Alamo Sandstone at the Hunter Wash, South Mesa, and Barrel Spring sites (all within the Ojo Alamo type locality of Bauer, 1916). At these sites the upper part of the Ojo Alamo Sandstone of Bauer (1916) was named the Ojo Alamo Sandstone restricted by Baltz and others (1966) and the Kimbeto Member of the Ojo Alamo by Powell (1973); the lower part was named the Naashoibito Member of the Kirtland Shale by Baltz and others (1966) and the Naashoibito Member of the Ojo Alamo Sandstone by Powell (1973). Hunter Wash site (lithology and thicknesses from Scott and others, 1979): Dinosaur bone positions are from Powell (1973) and Baltz and others (1966). South Mesa site (lithology and thicknesses from Baltz and others, 1966): Dinosaur bone positions are from Lehman (1985) and Baltz and others (1966). Barrel Spring site (lithology and thicknesses from Carl J. Orth [written communication, 1982] and Baltz and others [1966]): Dinosaur bone positions from Sinclair and Granger (1914) and Baltz and others (1966). The position of Paleocene pollen (open star symbol) is from Carl J. Orth (written communication, 1982). Solid star symbol is Upper Cretaceous pollen site. According to Robert H. Tschudy (written communication, 1982) the Paleocene pollen is late-early or early-middle Paleocene and the Cretaceous pollen is latest Campanian or earliest Maastrichtian.

to report the occurrence of dinosaur bone in the uppermost conglomeratic sandstone of the Ojo Alamo Sandstone, as follows:

> The centrum of a dinosaurian caudal vertebra was found, loose on the surface, by Mr. Jack Martin of the expedition, resting on this sandstone [Ojo Alamo Sandstone] near the edge of Split Lip Flat, in the immediate vicinity of Barrel Spring. It was not in place, but it seems hardly probable that it had been carried to the spot where it was found by the Indians, who are afraid of fossil bones. Some may question its value as an index fossil. To us it seems probable that it had weathered out of the sandstone, as did the fragments of silicified logs lying about in its vicinity. it is badly worn and lacks the arch.

This centrum, AMNH (American Museum of Natural History) 5796, was collected from Ojo Alamo outcrops, probably just south of Barrel Spring (Fig. 7). Examination of this centrum by Spencer Lucas confirms Sinclair and Granger's (1914) description of this bone as badly worn and lacking the neural arch. Although Sinclair and Granger (1914) believed that this bone indicated a Cretaceous age for the Ojo Alamo Sandstone, Baltz and others (1966, p. D15) discounted the biostratigraphic value of this fossil in the following statement:

> The badly worn centrum of the dinosaurian caudal vertebra found by Jack Martin, of Sinclair and Granger's expedition (1914, p. 301), lying loose on the restricted [uppermost conglomeratic sandstone] Ojo Alamo Sandstone near Barrel Spring cannot be considered as an index fossil; it reached the position where it was found possibly because of reworking or, more likely, because it was carried there by humans. This area had been frequented by whites as well as by Indians, and paleontologic collecting parties also had preceded Sinclair and Granger.

Figure 8 contains a stratigraphic column at Barrel Spring near the place where this dinosaur bone was collected. This column shows the level of the dinosaur bone and the level of a rock sample collected by Carl J. Orth, Los Alamos, New Mexico, from a mudstone layer about 3 m (10 ft) below the top of the Ojo Alamo. This sample was analyzed for its pollen and spore content by R. H. Tschudy of the U.S. Geological Survey, Denver, Colorado. Tschudy (written communication, 1982) gave the following report for this sample:

Sample B206 was given USGS paleobotany locality number D6391. Palynomorphs found include the following: Algal cysts; *Arecipites* cf. *A. Microreticulatus* And.; *Azolla* cf. *A schopfi* Dijkstra; *Chenopodipollis* sp.; *Cupaneidites* aff. *C. Major* Couper; *Momipites inequalis* And.; *Momipites tenuipolus* And.; *Ovoidites* sp.; *"Paliurus" triplicatus* And.; *Pandaniidite radicus* Leffingwell; *Pinus* sp.; *Polypodiisporonites* sp.; *Quercus explanata* And.; *Rectosulcites latus* And.; *Tetracolpites* 2 sp.; *Tricolpites faveolate*; *Tricolpites scabrate; Tricolporites rhomboides* And.; *Ulmipollenites* 3 and 4 pored.

This assemblage is clearly of Paleocene age. Several taxa, including *Momipites tenuipolus* were not recorded by Anderson [1960] from the Ojo Alamo, but were recorded from the overlying Nacimiento. I have not seen *M. tenuipolus* in any basal Paleocene samples from the Western Interior. This occurrence suggests that the sample is not from the basal Paleocene but rather from the upper lower or lower middle Paleocene.

Significance of Fossils of Ojo Alamo Wash Area. To summarize, dinosaur bone has been found in place at two sites, Hunter Wash and South Mesa, in the uppermost part of the Ojo Alamo Sandstone (the "restricted" Ojo Alamo Sandstone in Baltz and others, 1966) within the Ojo Alamo type locality. A loose (not in place) bone was found at the top of the uppermost Ojo Alamo near Barrel Spring. We agree with Baltz and others (1966) that the loose bone at Barrel Spring has, *in itself*, minimum stratigraphic value. The bone found at the Hunter Wash site is abraded, and probably reworked; thus its significance is questionable. The bones found at South Mesa by Lehman (1985), however, did not appear any more abraded than bone collected from the underlying middle part of the Ojo Alamo Sandstone (the upper part of the Naashoibito Member of the Kirtland Shale in Baltz and others, 1966).

Many writers have discussed the dinosaurs found within the Ojo Alamo of the type locality since Brown (1910) first described his finds there. Recently Lehman (1981) and Lucas (1981) discussed the differences between the dinosaurs of the Ojo Alamo Sandstone (Alamo Wash local fauna) and the underlying Kirtland Shale and Fruitland Formation (Hunter Wash local fauna) and came to slightly different conclusions. Lehman (1981) suggested that the Hunter Wash fauna is Campanian and the Ojo Alamo Wash fauna is Maastrichtian. Lucas (1981, p. 345), however, suggested that the Hunter Wash and the Alamo Wash local faunas were both "pre-Lancian" and that "a significant portion of Maastrichtian time is not represented by sediments preserved in the southern and western Central [San Juan] Basin." Lucas suggested that the two faunas represent animals living in different paleoecologic settings with little age difference between them. In a subsequent paper, however, Lucas and others (this volume) reassess the data and agree with Lehman's (1981) interpretation; stating that the Alamo Wash local fauna is indeed "Lancian (middle-late Maastrichtian)." They further conclude that this indicates that there cannot be as great a temporal break between rocks of the middle part of the Ojo Alamo (upper part of the Naashoibito in Baltz and others, 1966) and rocks of the upper Ojo Alamo (restricted Ojo Alamo in Baltz and others, 1966) as Baltz and others (1966) and some other workers had indicated.

The Kirtland Shale immediately underlying the Ojo Alamo Sandstone in the Ojo Alamo Wash area is 335 m (1100 ft) above the Huerfanito Bentonite Bed (see isopach map, Fig. 1). On the basis of the projection of ammonite zones described in a paper by Fassett (this volume), the 1100-foot isopach is at the level of the boundary between the *Didymoceras cheyennense* and *Baculites compressus* ammonite zones in the laterally time-equivalent marine Lewis Shale. This would make the age of the uppermost Kirtland (as defined by Bauer, 1916) at the type locality about 73.5 Ma. If there is no large temporal break within the Ojo Alamo, as suggested by Lehman (1985) and Lucas and others (this volume), then strictly on faunal evidence a hiatus of around 5.5 million years must exist between the Campanian uppermost Kirtland Shale (73.5 Ma) and the "mid-to-late Maastrichtian" middle part of the Ojo Alamo Sandstone (say, 68 Ma). Furthermore, this hiatus must be located at the base of the lower conglomeratic sandstone bed of the Ojo Alamo Sandstone.

Even more interesting: If the palynologic analysis of the pollen from the Ojo Alamo Sandstone near Barrel Spring by Tschudy is correct—that is, if the upper part of the Ojo Alamo Sandstone is late early or early middle Paleocene in age (let's assume 63 Ma) and the uppermost Kirtland is 73.5 Ma—then there must be a 10.5-million-year hiatus somewhere within or, more likely, at the base of the Ojo Alamo Sandstone. If the entire hiatus is at the base of the Ojo Alamo, then the "Lancian" dinosaurs within the Ojo Alamo are really post-Lancian or Paleocene (but not earliest Paleocene) in age!

Pot Mesa Site

The Pot Mesa site (Fig. 1) is located in the NE¼ NE¼ NE¼ Sec. 16, T.20N.,R.6W. on the U.S. Geological Survey Star Lake 7½-minute topographic quadrangle map. Figure 9 shows the geology of this area as mapped by Scott and others (1980) and shows the exact location of the Pot Mesa dinosaur bone site. The Ojo Alamo in this area was first mapped by Dane (1936) as a single sandstone bed (the upper sandstone bed of the Ojo Alamo of Fig. 9). Scott and others (1980) later mapped two beds of sandstone (Dane's bed and a lower sandstone bed) as Ojo Alamo in this same area, with the lower sandstone bed terminating eastward in the north-central part of Sec. 16, T.20N.,R.6W. (see Fig. 9). They described the Ojo Alamo in this quadrangle as follows:

Interbedded, light-gray, yellowish-brown, or dark-brown, medium- to coarse-grained sandstone that contains abundant petrified wood in the lower part and spherical sandstone and ironstone concretions as much as 1.2 m (4 ft) in diameter. Lower part marked by thin lenses of pebble conglomerate. Pebbles generally are less than 2.5 cm (1 in.) in diameter, very well rounded, bright-red to purple quartzite, gray quartz, and black to yellow chert. Thickness is about 27 m (90 ft).

In field checking this area, Fassett discovered that the Ojo Alamo does indeed consist of two sandstone beds but that the

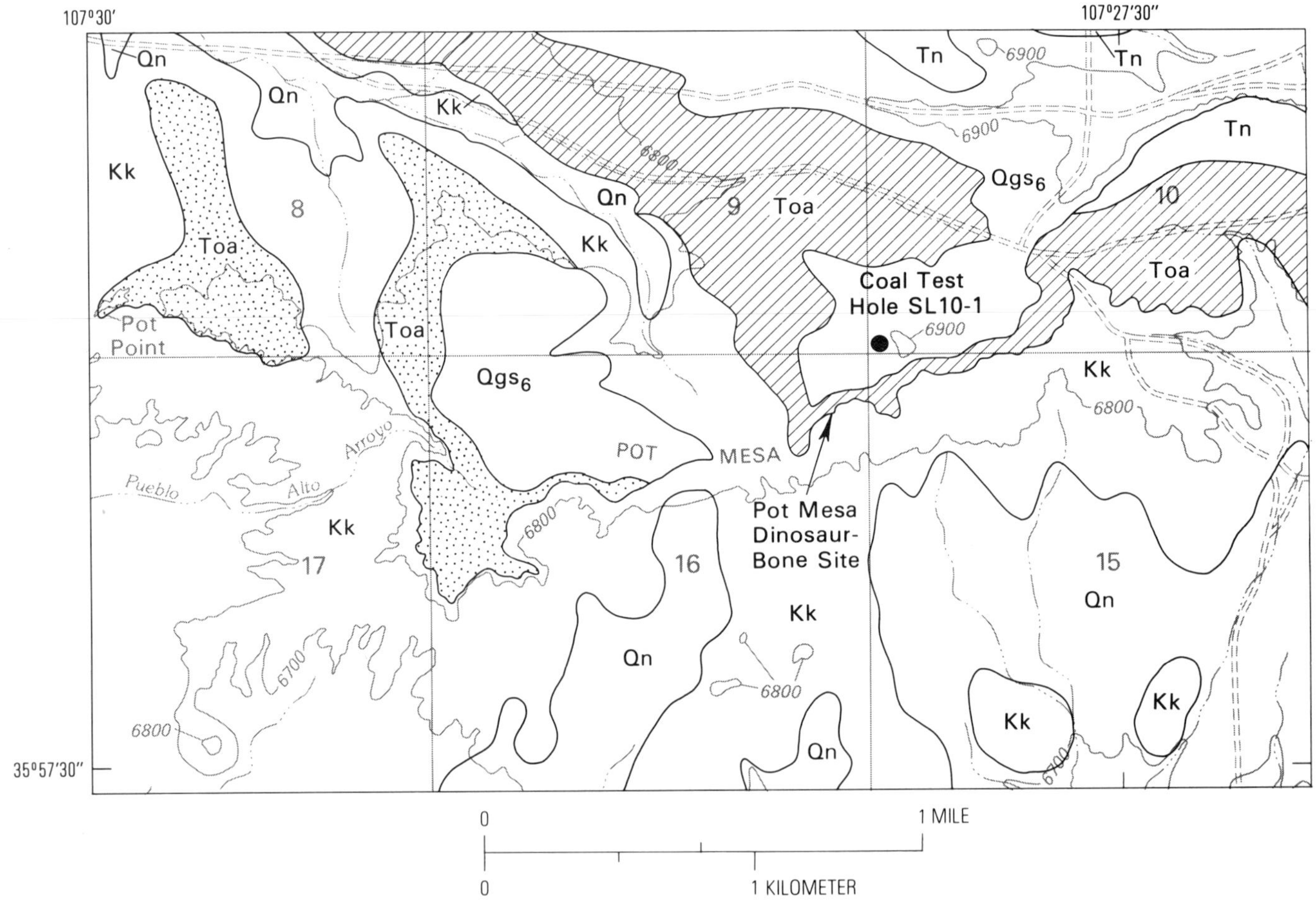

EXPLANATION

Toa — Upper bed of Ojo Alamo Sandstone

Toa — Lower bed of Ojo Alamo Sandstone

Figure 9. Geologic map of the Pot Mesa area in T.20N., R.6W. Geology from Scott and others (1980). The locations of the Pot Mesa dinosaur bone site (see photograph, Fig. 11) and coal drill hole SL 10-1 (Fig. 10) are shown. The map units shown are Kk, Kirtland Shale (Upper Cretaceous); Toa, Ojo Alamo Sandstone (Paleocene); Tn, Nacimiento Formation (Paleocene); Qgs_6, gravelly sand (Pleistocene); Qn, Naha alluvium of Hack (1941) (Holocene). We have patterned the upper and lower sandstone beds of the Ojo Alamo Sandstone differently on this map to clarify their relationships.

lower sandstone bed is present eastward on the outcrop beyond the place where Scott and others (1980) show it pinching out in the north-central part of Sec. 16. Fassett would include the mudstone interbed between the upper and lower sandstone beds in this area as part of the Ojo Alamo Sandstone, not as a tongue of the Kirtland Shale as mapped by Scott and others (1980) and shown on Figure 9. This lower sandstone bed is exposed intermittently (where not covered by alluvium) across the northern part of Sec. 15, but no attempt was made to remap this area for this report.

A coal test hole was drilled in the SW¼ SW¼ SW¼ Sec. 10 (Fig. 10) in 1976. The upper part of this hole was cored, and two Ojo Alamo sandstone beds were described from this core (Jentgen and Fassett, 1977). A geophysical log from this drill hole (Fig. 10) clearly shows the presence of the two sandstone beds that constitute the Ojo Alamo interval at this location.

Dinosaur bone is abundant in this area within the Ojo Alamo Sandstone and in the underlying Kirtland Shale. The only bone collected at this site was a large, badly weathered and fragmented dinosaur limb(?) bone (Fig. 11). The bone (UNM TOA—2) was discovered by Fassett in 1982 and collected by Lucas in 1983. Although the bone had disintegrated due to weathering when photographed, its general shape and size could still be seen. It was bulbous at the exposed end and tapered into

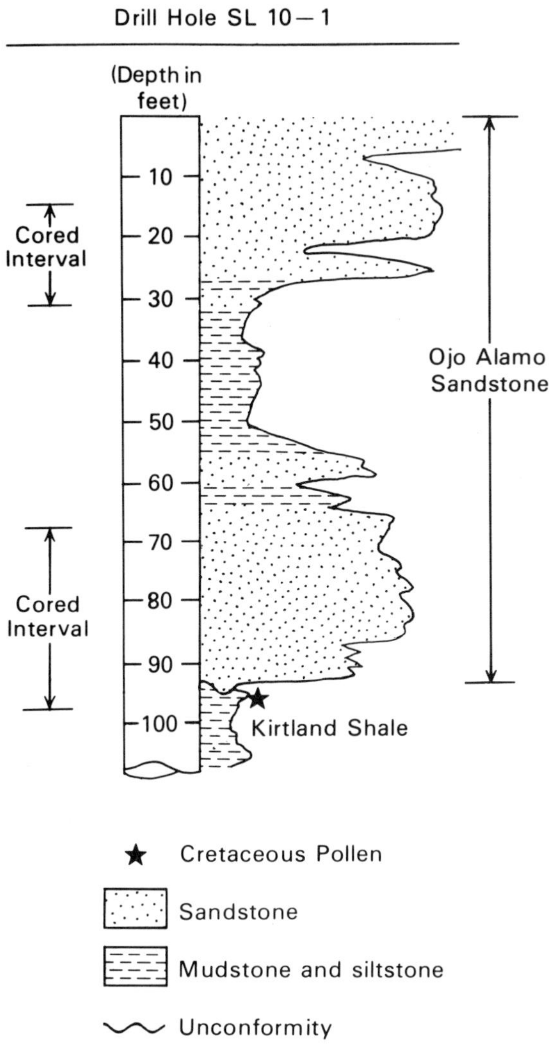

Figure 10. Electric log of the upper part of coal test hole SL 10-1 (Fig. 9), showing the upper and lower sandstone beds of the Ojo Alamo Sandstone. The lithology of the Ojo Alamo in this drill hole is described in detail in Jentgen and Fassett (1977). Robert H. Tschudy analyzed a sample of drill core from this well from the uppermost Kirtland Shale from a depth of 29 m (94 ft) (solid star symbol) and reported (written communication, 1981) that the pollen assemblage indicated an age as old as late Campanian.

Figure 11. Photograph of large, fragmented dinosaur bone at the Pot Mesa site (Fig. 9). Bone is weathering out of the base of the upper bed of the Ojo Alamo Sandstone. Bone has a bulbous end (to right of hammer point) and tapers into the bedrock at top of photo. Hammer is about 0.3 m (1 ft) long and is scaled in tenths of a foot. Abundant fragments of this bone litter the slope below the main mass of the bone. The cliff face above this site is about 4.9 m (16 ft) high here. The slope in the foreground is a mudstone to siltstone interbed within the Ojo Alamo Sandstone.

the outcrop. Abundant small fragments from this bone were scattered down-slope from it. The lower part of the bone was encapsulated in a hard, brown, fine-to-medium-grained sandstone rind.

This bone came from the base of the uppermost Ojo Alamo sandstone bed (Fig. 12). Because the bone was fragmented, it was difficult to determine whether or not it had been reworked. However, the outer cortical surface of the bone is not abraded; if the bone had been reworked, we would expect to find significant abrasion of this surface, as is the case with the bone fragments described by Powell (1973) at the Hunter Wash site. Ceratopsian frill fragments were also found a few feet higher in the upper sandstone bed of the Ojo Alamo near this site (Sloan, personal communication, 1984).

This site is at the 700-ft isopach on Fig. 1 (at a stratigraphic level 213 m [700 ft] above the Huerfanito Bed). This is the level of the boundary of the *Didymoceras stevensoni* and the *Exiteloceras jenneyi* ammonite zones (Fassett, this volume). The age of this boundary is slightly older than 73.7 Ma.

Mesa Portales Site

Mesa Portales is located in the southeast part of the San Juan Basin, about 16 km (10 mi) southwest of Cuba, New Mexico (Fig. 1). The Ojo Alamo on Mesa Portales consists of from one to several sandstone beds separated by sandstone and siltstone interbeds (Fig. 13); the thickness of the Ojo Alamo on the mesa ranges from 12 to 88 m (40 to 290 ft) (Fassett, 1966). This site is located on a south-facing cliff in the SE¼ SE¼ SW¼ Sec. 9, T.19N.,R.2W. The area was discussed in detail in Fassett and Hinds (1971), and a geologic section and photograph (Figs. 10, 12) of the Ojo Alamo at this locality are included in that report. The Ojo Alamo contains no conglomerate in this part of the San Juan Basin.

The rocks of the Mesa Portales area were first mapped in detail by Dane (1936), who mapped the Ojo Alamo as consisting of from one to several sandstone beds with intervening mudstone and siltstone interbeds. Baltz (1962) mapped the Ojo Alamo on Mesa Portales as only a single sandstone sheet and labeled the lower sandstone beds of Dane's (1936) Ojo Alamo "unit B of the

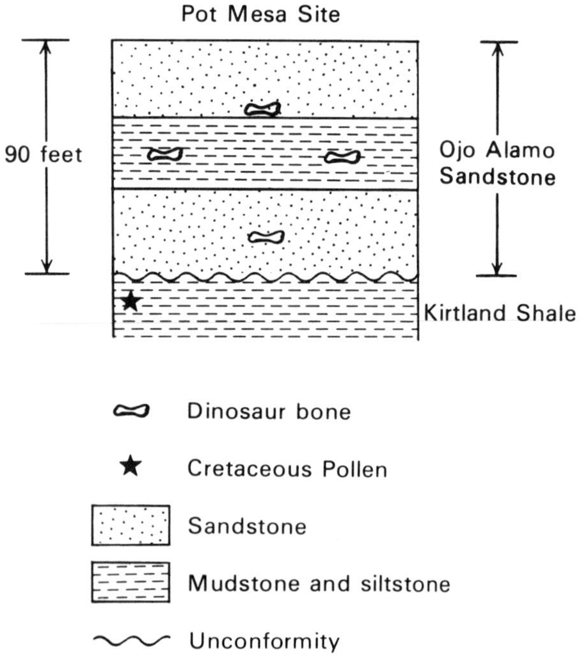

Figure 12. Stratigraphic column of the Ojo Alamo Sandstone at the Pot Mesa site. Lithology and thicknesses are based on field observations by Fassett and on the electric log (Fig. 10) and lithologic descriptions of the Ojo Alamo from drill hole SL 10-1 in Jentgen and Fassett (1977). Positions of dinosaur bone found within the Ojo Alamo Sandstone at this locality are shown. Solid star shows the position of pollen recovered from rocks in this area; pollen was determined to be as old as late Campanian by Robert H. Tschudy (written communication, 1981).

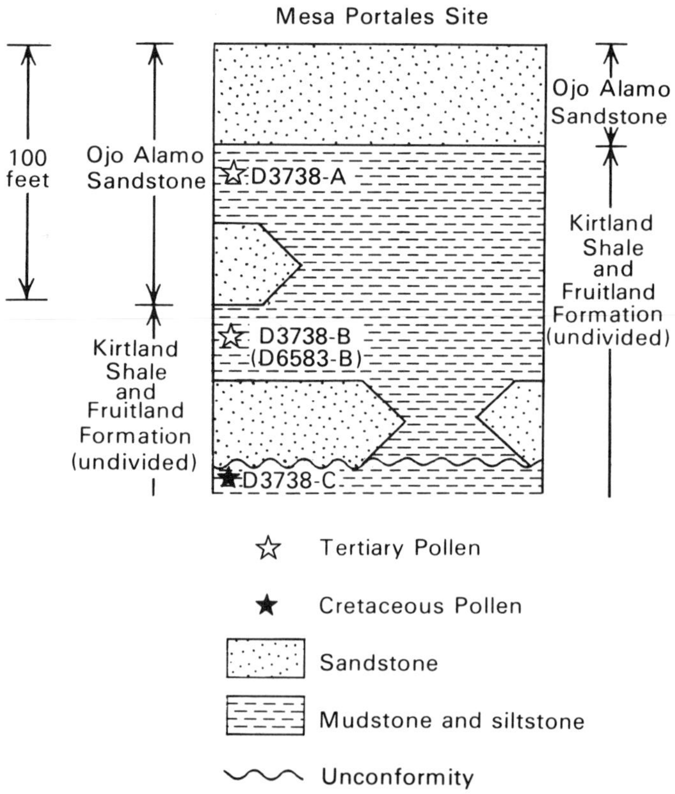

Figure 13. Stratigraphic column of the Ojo Alamo Sandstone at the Mesa Portales site. (Because the thicknesses of the units shown here change markedly over short distances in this area and because the lateral scale is condensed, this column is more diagrammatic than the other stratigraphic columns shown in this report.) Lithology and thicknesses from Fassett (1966) and Fassett and Hinds (1971). Star symbols show positions of pollen sample levels: U.S.G.S. paleobotany localities D 3738-A, B, and C reported on by Tschudy in Fassett and Hinds (1971) and D 6583-B reported on by Tschudy in 1983 (written communication). The Ojo Alamo is shown on this column as mapped by Fassett (1966).

undivided Fruitland Formation and Kirtland Shale." Fassett (1966) mapped the Ojo Alamo on Mesa Portales using the definition of Dane (1936) of one to several sandstone beds with interbeds of siltstone and mudstone. (Baltz, in 1967, republished his 1962 map and report.)

Figure 13 is a stratigraphic column of the Ojo Alamo Sandstone and underlying rocks at the Mesa Portales locality. The Ojo Alamo Sandstone as here shown is the formation as it was mapped in this area by Dane (1936) and by Fassett (1966). At this site the unit consists of two massive sandstone beds separated by a silty to sandy mudstone interbed. The Ojo Alamo is underlain here by a mudstone and siltstone bed about 6 m (20 ft) thick; this bed is in turn underlain by a sandstone bed that varies in thickness in this area from 0 to 6 m (0 to 20 ft) or so in thickness. This lowermost discontinuous sandstone bed is lithologically similar to the two overlying massive, cliff-forming sandstones and is probably genetically related to them. Fassett and Hinds (1971, fig. 10) show the Cretaceous-Tertiary boundary, queried, at the base of this discontinuous sandstone bed. Their placement of the boundary at that level was based on pollen and spore data (discussed below).

No dinosaur bone has as yet been discovered within the Ojo Alamo at Mesa Portales. Pollen and spore assemblages, however, have been identified from this site (Fig. 13) by Robert H. Tschudy, U.S. Geological Survey, Denver, Colorado, on two different occasions. In 1971, Tschudy (in Fassett and Hinds, 1971) listed and discussed palynomorphs collected from within and just beneath the Ojo Alamo at Mesa Portales and concluded that the pollen from these two levels was Paleocene in age. Tschudy examined additional rock samples collected in 1983 from the same levels as the samples reported on in Fassett and Hinds (1971) and again reported (written communication, 1983) that the pollen from these rocks was Paleocene. One of the 1983 samples, D6583-B, is equivalent to sample number D3738-B of Fassett and Hinds (1971). This sample contained *Momipites tenuipolus.* Tschudy wrote: "The presence of *Momipites tenuipolus* strongly suggests a Paleocene age. We have not observed this taxon in other than Paleocene rocks. Furthermore, we have not found this taxon in basal Paleocene rocks. We suggest that this sample, though poor, is possibly from the lower middle Paleocene."

It seems clear, based on analyses of palynomorphs from within and below the Ojo Alamo Sandstone at Mesa Portales, that all of the rocks mapped there as Ojo Alamo by all workers (and as much as 9 m—30 ft—or so of underlying strata) are Paleocene in age.

The base of the Ojo Alamo Sandstone at the Mesa Portales site is located below the 500-ft isopach line of Figure 1. The youngest Cretaceous rocks at this site are therefore within the uppermost part of the *Didymoceras nebrascense* Western Interior ammonite zone (Fassett, this volume) and thus have an age of about 73.9 Ma. If the oldest Paleocene rocks at this site are early middle Paleocene, then the Cretaceous-Tertiary hiatus at this site could be as great as 10 to 11 million years (8 million years missing from the uppermost Upper Cretaceous and 2 to 3 million years from the lowermost Paleocene).

Gasbuggy Site

The Gasbuggy site is located in the northeast part of the basin (Fig. 1) in the NE¼ SW¼ Sec. 36, T.29N.,R.4W. This site is drill-hole GB-1 of Project Gasbuggy. In this hole, strata from 1047 to 1316 m (3436 to 4316 ft) below the surface were continuously cored. The formations cored, from top down, were the Nacimiento Formation, Ojo Alamo Sandstone, Fruitland Formation, Pictured Cliffs Sandstone, and Lewis Shale. The Ojo Alamo Sandstone in this core is from 1061 to 1222 m (3480 to 3680 ft), a thickness of 61 m (200 ft), and consists of mostly massive sandstone beds with some few thin siltstone and mudstone interbeds. Figure 14 is a representation of the Ojo Alamo based on a description of the core from this hole by Fassett (1968a, b).

Tschudy (1973) reported on the pollen and spores identified from samples from the Gasbuggy core. The sample levels from the Ojo Alamo in this core are shown on Figure 14. All of the pollen from samples of Ojo Alamo in this core was determined by Tschudy (1973) to be Paleocene in age. One of the samples was especially interesting; Tschudy (1973, p. 133) wrote: "The sample at 3515.6 feet contained *Maceopolipollenites tenuipolus,* a fossil that elsewhere in the Rocky Mountain region is limited to the Paleocene but is not present in the lowermost Paleocene."

The highest (youngest) Cretaceous rocks present in this core are 366 m (1200 ft) above the Huerfanito Bentonite Bed (on the 1200-ft isopach line on Fig. 1). This level falls within the projected *Baculites compressus* ammonite zone (Fassett, this volume) with an age of about 73 Ma. The hiatus between rocks of Cretaceous and Paleocene age in this core hole thus may be on the order of 10 million years.

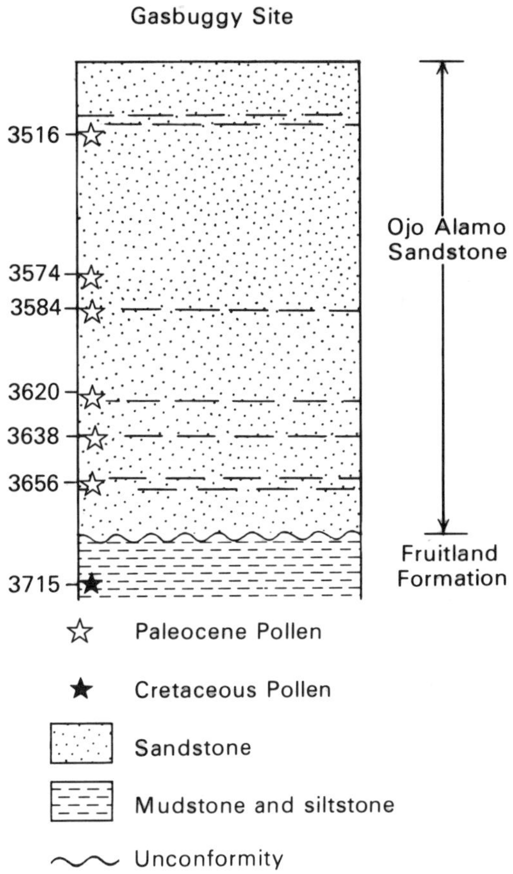

Figure 14. Stratigraphic column of the Ojo Alamo Sandstone at the Gasbuggy site. Lithology and thicknesses from Fassett (1968 a and b). Drilled depths on left represent positions where samples were collected from drill core for pollen and spore analysis. Pollen analysis from this site is in Tschudy (1973); he wrote that the Paleocene pollen is late-early or early-middle Paleocene and that the Cretaceous pollen is Campanian in age.

SUMMARY

Physical Characteristics and Definition of Ojo Alamo Sandstone

In this report we have examined the Ojo Alamo Sandstone at seven localities in the San Juan Basin. Based on this limited study, it is apparent that the thickness, composition, and component parts of the Ojo Alamo vary greatly from place to place. Figure 15 shows a stratigraphic column of the Ojo Alamo at each of the sites discussed in this report. The overall thickness of the Ojo Alamo varies from 9 to nearly 122 m (30 to 400 ft); its rock types grade from siliceous-pebble conglomerate on the west side of the basin to coarse-grained, gritty sandstone on the east side; the mix of interbeds of sandstone beds with silty-mudstone interbeds varies from one to four sandstones (zero to three silty-mudstones). The thicknesses of individual sandstone beds range from a few feet to more than 610 m (200 ft).

When tracing Ojo Alamo beds in local areas, such as Mesa Portales, it is common to see a given bed of sandstone pinch out or merge with an overlying or underlying sandstone bed or to see a channel of an overlying sandstone bed cut down into an underlying sandstone bed (Fassett, 1966; Fassett and Hinds, 1971). In a

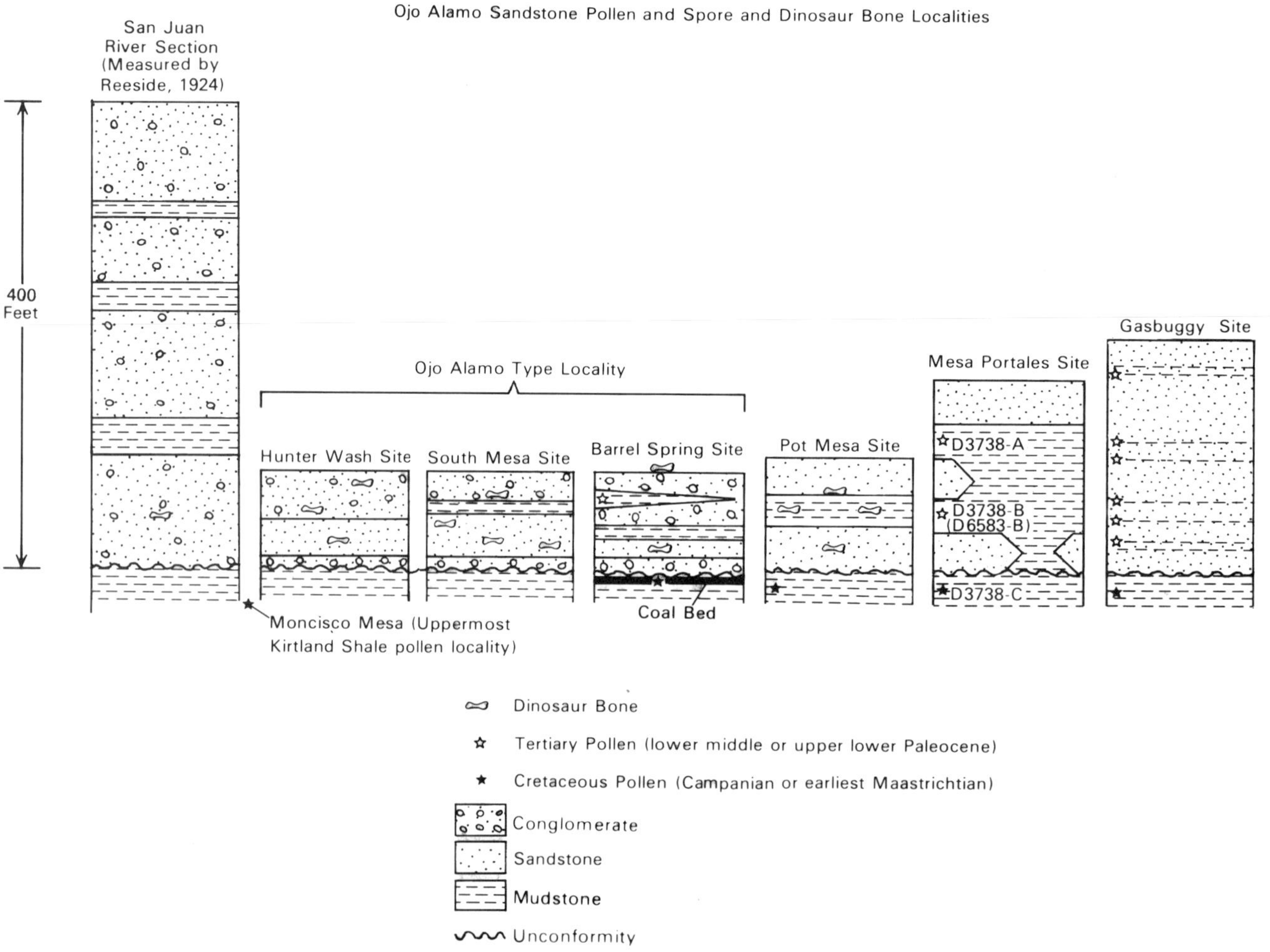

Figure 15. Comparison of the stratigraphic columns at the seven fossil collection sites discussed in this report. The columns are aligned on the base of the Ojo Alamo Sandstone (except for the Mesa Portales column, which is aligned on the Cretaceous-Tertiary boundary); individual beds do not necessarily correlate from one site to another. Distances between columns are not to scale. The bone symbol shows the positions of dinosaur bone collected from the Ojo Alamo Sandstone; the star symbol shows positions of palynomorphs identified from the Ojo Alamo, Kirtland Shale, or Fruitland Formation. Site locations are shown on Figure 1. The Kirtland Shale pollen locality labeled Moncisco Mesa was collected by Carl J. Orth; Tschudy analyzed the pollen and stated (written communication, 1982) that it represents a Campanian or early Maastrichtian age.

word, the internal stratigraphy of the Ojo Alamo sandstone is variable.

Fassett and Hinds (1971, p. 28–29) described the Ojo Alamo, basin-wide as follows:

> The Ojo Alamo throughout most of the basin consists of overlapping massive sandstone beds interbedded with shale. The sandstone beds are sheetlike and discontinuous. Sheets terminate by merging with underlying or overlying sandstone sheets or by wedging out into overlying or underlying shale beds. Sandstone sheets are, for the most part, parallel; intervening shale beds maintain a relatively constant thickness. Large-scale channeling at the base of the sandstone beds is common. Channels of 50 or more feet cut into the underlying Kirtland or the Fruitland have been measured by Fassett (1966) and Hinds (1966). In some areas, channels at the base of an upper sandstone sheet have cut through the intervening shale bed and into an underlying sandstone sheet.
>
> The thickness of the Ojo Alamo varies according to the number of sandstone beds that constitute it at any given locality. Thicknesses range from less than 20 feet to more than 400 feet throughout the basin; thicknesses of about 50–150 feet are the most common.

Because of the extreme variability of the Ojo Alamo from place to place in the basin, we suggest that the formal naming of individual sandstone beds or mudstone interbeds within the Ojo Alamo be avoided. The redefining of the Ojo Alamo at the type locality by Baltz and others (1966) is a case in point. There the upper part of Bauer's (1916) Ojo Alamo was named the "restricted" Ojo Alamo, and the rest of the unit was renamed the Naashoibito Member of the Kirtland Shale. Some subsequent workers have not used this terminology, for example, Clemens (1973), Fassett (1973), Powell (1973), Scott and others (1979), and Brown (1982a, b).

A few workers have misinterpreted the intent of the Ojo Alamo revision in Baltz and others (1966) at the Ojo Alamo Wash area by stating that the Naashoibito Member of the Kirtland Shale (Baltz and others, 1966) is present outside of its type locality. O'Sullivan and others (1972, p. E56), however, wrote that "Westward, the Ojo Alamo Sandstone and underlying beds are poorly exposed for about 6 miles, from the head of Hunter Wash [northwest end of the type locality] to the outcrops at the south end of Moncisco Mesa. The Ojo Alamo Sandstone rests on the upper shale member of the Kirtland Shale at Moncisco Mesa, and the Naashoibito Member apparently is absent." Lehman (1981, p. 194) also questioned the existence of the Naashoibito outside of its type area; he wrote: "However, lack of exposures on the divides between the major wash systems north and south of the Hunter Wash–De-na-zin Wash confluence [Ojo Alamo type locality] make it difficult, if not impossible, to physically trace the Naashoibito, into other areas. Moreover, the absence in other areas of the distinctly colored shale and the soft sandstone characteristic of the Naashoibito, as well as the lack of any persistent conglomeratic sandstone, suggests that the Naashoibito may be restricted to the Ojo Alamo region."

We believe that the name Naashoibito should be used only in the Ojo Alamo Wash area. Furthermore, one of us (Fassett) supports Powell's (1973) recommendation that the Naashoibito be a member of the Ojo Alamo rather than a member of the Kirtland Shale as designated by Baltz and others (1966). Powell's (1973) terminology would retain Bauer's (1916) definition of the Ojo Alamo, a definition that is as valid today as it was in 1916. A return to Bauer's (1916) definition for the Ojo Alamo Sandstone in the San Juan Basin would greatly reduce confusion in future discussions of this rock unit.

Age of the Ojo Alamo Sandstone

The age of the Ojo Alamo Sandstone is still a problem. As Figure 15 shows, the formation contains Paleocene pollen in the eastern part of the San Juan Basin and widespread dinosaur bone in the western part. At only one locality, the Ojo Alamo Wash area, has dinosaur bone been found at or above the level of Paleocene pollen. At the Barrel Spring site a dinosaur bone was collected from the top of the Ojo Alamo; however, since this bone was found loose on the surface (Sinclair and Granger, 1914), its significance is questionable. Less than 1.6 km (1 mi) west of Barrel Spring, in-place dinosaur bone has been found on South Mesa at about the same stratigraphic level as the Barrel Spring Paleocene pollen site (Fig. 15) in the upper part of the Ojo Alamo Sandstone. If there is no significant unconformity between the stratigraphic levels of the Barrel Spring pollen and the South Mesa dinosaurs, of which there is no indication on the outcrop in this area, and if the dinosaur bone is not reworked, then the South Mesa dinosaurs are late early or early middle Paleocene in age.

If there are no significant unconformities at any stratigraphic level within the Ojo Alamo Sandstone, as Lehman (1981), Lucas and others (this volume), and Fassett (1985) have suggested for the type locality (which includes the Barrel Spring and South Mesa sites), then there is no question that both Paleocene pollen and dinosaur bone do coexist in the Ojo Alamo Sandstone in the San Juan Basin. If this condition does exist, which of these fossil forms should be *the* Cretaceous-Tertiary determinant? In other words, do we have Tertiary dinosaurs or Cretaceous "Paleocene" pollen?

In an earlier report, Fassett (1982) reviewed this problem and reasoned as follows: (1) In the Raton Basin, the Cretaceous-Tertiary pollen boundary coincides with an iridium-enriched clay layer; (2) an iridium-enriched clay layer has been discovered coincident with the marine Cretaceous-Tertiary mass-extinction level at many sites around the world; and (3) the Cretaceous-Tertiary pollen break in the San Juan Basin is at the base of the Ojo Alamo Sandstone, at least in the eastern part of the basin. Fassett (1982) therefore concluded that in the San Juan Basin, dinosaur bone found within the Ojo Alamo Sandstone, based on available evidence, is Paleocene in age. This report describes new dinosaur bone and palynomorph localities within the Ojo Alamo Sandstone and further strengthens the conclusion of Fassett (1982) that the dinosaur bone of the Ojo Alamo Sandstone of the San Juan Basin is Paleocene in age.

REFERENCES CITED

Anderson, R. Y., 1960, Cretaceous-Tertiary palynology, eastern side of the San Juan Basin, New Mexico: New Mexico Bureau of Mines and Mineral Resources Memoir 6, 59 p.

Baltz, E. H., Jr., 1962, Stratigraphy and geologic structure of uppermost Cretaceous and Tertiary rocks of the east-central part of the San Juan Basin, New Mexico: U.S. Geological Survey Open-File report, 278 p.

—— , 1967, Stratigraphy and regional tectonic implications of part of Upper Cretaceous and Tertiary rocks, east-central San Juan basin, New Mexico: U.S. Geological Survey Professional Paper 552, 101 p.

Baltz, E. H., Ash, S. R., and Anderson, R. Y., 1966, History of nomenclature and stratigraphy of rocks adjacent to the Cretaceous-Tertiary boundary, western San Juan Basin, New Mexico: U.S. Geological Survey Professional Paper 524-D, p. 1–23.

Barnes, H., Baltz, E. H., Jr., and Hayes, P. T., 1954, Geology and fuel resources of the Red Mesa area, La Plata and Montezuma Counties, Colorado: U.S. Geological Survey Oil and Gas Investigations Map OM-149.

Bauer, C. M., 1916, Stratigraphy of a part of the Chaco River valley: U.S. Geological Survey Professional Paper 98-P, p. 271–278 [1917].

Brown, B., 1910, The Cretaceous Ojo Alamo beds of New Mexico, with description of the new dinosaur genus *Kritosaurus:* Bulletin of the American Museum of Natural History, v. 28, p. 267–274.

Brown, J. L., 1982a, Geologic map of the Alamo Mesa East quadrangle, San Juan County, New Mexico: U.S. Geological Survey Miscellaneous Field Studies Map MF-1497.

—— , 1982b, Geologic map of the Huerfano Trading Post SW quadrangle, San Juan County, New Mexico: U.S. Geological Survey Miscellaneous Field

Studies Map MF-1498.

Clemens, W. A., 1973, The roles of fossil vertebrates in interpretation of Late Cretaceous stratigraphy of the San Juan Basin, New Mexico, *in* Fassett, J. E., ed., Four Corners Geological Society Memoir, p. 154–167.

Dane, C. H., 1936, The La Ventana–Chacra Mesa coal field, pt. 3 *of* Geology and fuel resources of the southern part of the San Juan Basin, New Mexico: U.S. Geological Survey Bulletin 860-C, p. 81–161 [1937].

Fassett, J. E., 1966, Geologic map of the Mesa Portales quadrangle, Sandoval County, New Mexico: U.S. Geological Survey Geologic Quadrangle Map GQ-590.

——, 1968a, Core description from GB-1 (Gasbuggy 1) in the northeastern part of the San Juan Basin, Rio Arriba County, New Mexico: U.S. Geological Survey Open-File Report, 93 p.

——, 1968b, Summary of geologic data obtained from borehole GB-1, Project Gasbuggy, *in* San Juan–San Miguel–La Plata region: New Mexico Geological Society 19th Field Conference Guidebook, p. 24–27.

——, 1973, The saga of the Ojo Alamo Sandstone; or, the rock stratigrapher and the paleontologist should be friends, *in* Fassett, J. E., ed., Cretaceous and Tertiary rocks of the southern Colorado Plateau: Four Corners Geological Society Memoir, p. 123–130.

——, 1982, Dinosaurs in the San Juan Basin, New Mexico, may have survived the event that resulted in creation of an iridium-enriched zone near the Cretaceous-Tertiary boundary, *in* Geological Society of America Special Paper 190, p. 435–447.

——, 1985, Early Tertiary paleogeography and paleotectonics of the San Juan Basin area, New Mexico and Colorado, *in* Flores, R. M., and Kaplan, S. S., eds., Cenozoic paleogeography of the west-central United States: Rocky Mountain Section, Society of Economic Paleontologists and Mineralogists, p. 317–334.

Fassett, J. E., and Hinds, J. S., 1971, Geology and fuel resources of the Fruitland Formation and Kirtland Shale of the San Juan Basin, New Mexico and Colorado: U.S. Geological Survey Professional Paper 676, 76 p.

Hack, J. T., 1941, Dunes of the western Navajo country: Geographical Review, v. 31, no. 2, p. 240–263.

Hinds, J. S., 1966, Geologic map of the Johnson Trading Post quadrangle, Sandoval County, New Mexico: U.S. Geological Survey Geologic Quadrangle Map GQ-591.

Jentgen, R. W., and Fassett, J. E., 1977, Sundance-Bisti–Star Lake 1976 coal drilling in McKinley and San Juan Counties, northwest New Mexico: U.S. Geological Survey Open-File Report 77-369, 80 p., 25 figs.

Lehman, T. M., 1981, The Alamo Wash local fauna: A new look at the old Ojo Alamo fauna, *in* Lucas, S. G., Rigby, J. K., Jr., and Kues, B. S., eds., Advances in San Juan Basin paleontology: Albuquerque, University of New Mexico Press, p. 189–221.

——, 1985, Depositional environments of the Naashoibito Member of the Kirtland Shale, Upper Cretaceous, San Juan Basin, New Mexico: New Mexico Bureau of Mines and Mineral Resources Circular 195, p. 55–79.

Lucas, S. G., 1981, Dinosaur communities of the San Juan Basin: A case for lateral variations in the composition of Late Cretaceous dinosaur communities, *in* Lucas, S. G., Rigby, J. K., Jr., and Kues, B. S., eds., Advances in San Juan Basin paleontology: Albuquerque, University of New Mexico Press, p. 337–393.

O'Sullivan, R. B., and Beikman, H. M., 1963, Geology, structure, and uranium deposits of the Shiprock Quadrangle, New Mexico and Colorado: U.S. Geological Survey Miscellaneous Geologic Investigations Map I-345.

O'Sullivan, R. B., Repenning, C. A., Beaumont, E. C., and Page, H. G., 1972, Stratigraphy of the Cretaceous rocks and the Tertiary Ojo Alamo Sandstone, Navajo and Hopi Indian Reservations, Arizona, New Mexico, and Utah: U.S. Geological Survey Professional Paper 521-E, 65 p.

Powell, J. S., 1973, Paleontology and sedimentation of the Kimbeto Member of the Ojo Alamo Sandstone, *in* Fassett, J. E., ed., Cretaceous and Tertiary rocks of the southern Colorado Plateau: Four Corners Geological Society Memoir, p. 111–122.

Reeside, J. B., 1924, Upper Cretaceous and Tertiary formations of the western part of the San Juan Basin of Colorado and New Mexico: U.S. Geological Survey Professional Paper 134, 70 p.

Scott, G. R., O'Sullivan, R. B., and Mytton, J. W., 1979, Reconnaissance geologic map of the Alamo Mesa West Quadrangle, San Juan County, New Mexico: U.S. Geological Survey Miscellaneous Field Studies Map MF-1074.

Scott, G. R., Mytton, J. W., and Schneider, G. B., 1980, Geologic map of the Star Lake Quadrangle, McKinley County, New Mexico: U.S. Geological Survey Miscellaneous Field Studies Map MF-1248.

Sinclair, W. J., and Granger, W., 1914, Paleocene deposits of the San Juan Basin, New Mexico: Bulletin of the American Museum of Natural History, v. 33, p. 297–316.

Tschudy, R. H., 1973, The Gasbuggy core—a palynological appraisal, *in* Fassett, J. E., ed., Cretaceous and Tertiary rocks of the southern Colorado Plateau: Four Corners Geological Society Memoir, p. 131–143.

Manuscript Accepted by the Society March 24, 1986

Printed in U.S.A.

Geological Society of America
Special Paper 209
1987

Dinosaurs, the age of the Fruitland and Kirtland Formations, and the Cretaceous-Tertiary boundary in the San Juan Basin, New Mexico

Spencer G. Lucas
Department of Geology
University of New Mexico
Albuquerque, New Mexico 87131

Niall J. Mateer
Department of Geology
McMurry College
Abilene, Texas 79697

Adrian P. Hunt
Department of Geology
University of New Mexico
Albuquerque, New Mexico 87131

F. Michael O'Neill
U.S. Bureau of Land Management
Taos, New Mexico 87571

ABSTRACT

Dinosaurs from the Upper Cretaceous Fruitland Formation and Kirtland Shale in the San Juan Basin, New Mexico, pertain to the following taxa: Ornithomimidae, cf. *Ornithomimus edmonticus,* cf. *Struthiomimus altus,* Dromaeosauridae, *Albertosaurus* sp., cf. *Tyrannosaurus rex, Alamosaurus sanjuanensis,* ?Pachycephalosauridae, Ankylosauria, Ankylosauridae, Nodosauridae, ?*Euoplocephalus* sp., ?*Panoplosaurus* sp., Hadrosauridae, *Kritosaurus navajovius, Parasaurolophus tubicen, P. cyrtocristatus,* Ceratopsidae, cf. *Chasmosaurus* sp., *Pentaceratops sternbergii, P. fenestratus,* and *Torosaurus* cf. *T. utahensis.* The dinosaur fauna of the Fruitland Formation is temporally equivalent to the dinosaur faunas of the Judith River (Montana) and Oldman (Alberta) Formations and is of late Campanian (Judithian) age. This correlation is based primarily on the absence in the Fruitland Formation of dinosaurs typical of post-Judithian dinosaur faunas elsewhere in western North America. The dinosaur fauna of the Kirtland Shale below the Naashoibito Member is virtually identical to that of the Fruitland Formation. Based on stratigraphic relationships, the Kirtland Shale must be younger than the Fruitland Formation and may be as young as Edmontonian (latest Campanian–early Maastrichtian) below the Naashoibito Member. The Naashoibito Member of the Kirtland Shale contains cf. *Tyrannosaurus rex, Alamosaurus sanjuanensis,* and *Torosaurus* cf. *T. utahensis,* taxa indicative of a Lancian (middle-late Maastrichtian) age. Therefore, *Kritosaurus* from the Naashoibito Member represents the youngest known occurrence of this genus. The Lancian age of the Naashoibito Member indicates that the unconformity at the base of the Ojo Alamo Sandstone is not of as great a temporal magnitude as most previous workers had believed. Thus, there is a nearly complete record of the Cretaceous-Tertiary transition in the west-central San Juan Basin, indicated by dinosaur-based correlation of the Fruitland and Kirtland Formations. This correlation is consistent with most other evidence, except magnetostratigraphy, that has been used to determine the age of the Fruitland and Kirtland Formations.

INTRODUCTION

Since 1863 an understanding of dinosaur fossils has been important in the placement of the Cretaceous-Tertiary boundary. In that year James Dwight Dana, in his classic *Manual of Geology,* made the following assertion: "The Saurian type in the later Jurassic rises also to the grade of Dinosaurs, the highest in rank and among the largest of Reptiles; and these all disappear before the close of the Cretaceous period" (Dana, 1863, p. 498). More than 30 years later, the French geologist Grossouvre (1897) placed the Cretaceous-Tertiary boundary above the last occurrence of ammonites, belemnites, inoceramids, rudistids, mosasaurs, and dinosaurs in western Europe. Grossouvre thereby laid the basis for the long accepted, though sometimes questioned, idea that there was a synchronous extinction of marine invertebrates and vertebrates and the terrestrial dinosaurs at the end of the Cretaceous.

In continental rocks, such as those that encompass the Cretaceous-Tertiary boundary in the San Juan Basin in northwestern New Mexico, the stratigraphically highest dinosaur occurrence has long been taken to indicate the upper limit of the Cretaceous. However, in the San Juan Basin dinosaurs have been of additional importance to the placement of the Cretaceous-Tertiary boundary: They have been used to correlate the Upper Cretaceous continental sediments of the San Juan Basin, the Fruitland Formation, and the Kirtland Shale with other Upper Cretaceous continental sediments in western North America. In this way, the use of dinosaurs contributed to the evaluation of the temporal significance of the unconformity at the base of the Ojo Alamo Sandstone (sensu Baltz and others, 1966) that most workers agree lies at, or near, the Cretaceous-Tertiary boundary in the San Juan Basin.

The purpose of this paper is to pursue this second use of dinosaurs in the placement of the Cretaceous-Tertiary boundary in the San Juan Basin by re-examining the age of the Fruitland and Kirtland Formations on the basis of their dinosaur faunas. In order to do so, we first review the dinosaurs known from the Fruitland and Kirtland Formations, then summarize the age determinations they indicate, and finally, briefly compare these age determinations with those based on other evidence. In this paper the following abbreviations are used: AMNH—American Museum of Natural History, New York; FMNH—Field Museum of Natural History, Chicago; MNA—Museum of Northern Arizona, Flagstaff; OSM—Stovall Museum of Science and History, University of Oklahoma, Norman; PMU—Paleontologiska Museet, Uppsala University, Uppsala, Sweden; UALP—University of Arizona Laboratory of Paleontology, Tucson; UKVP—University of Kansas, Lawrence; UNM—University of New Mexico, Albuquerque; USNM—National Museum of Natural History, Smithsonian Institution, Washington, D.C.

DINOSAURS FROM THE FRUITLAND AND KIRTLAND FORMATIONS

Introduction

All dinosaur fossils from the Fruitland and Kirtland Formations discussed here are from localities in the west-central San Juan Basin, New Mexico (Fig. 1B; Appendix 1). The brief review of the dinosaurs from these localities presented here aims at re-identification, establishment of stratigraphic provenance, and documentation of new (and some old) occurrences, where needed.

Theropoda

Ornithomimidae. Kues and others (1977) first reported the presence of ornithomimids in the Fruitland and Kirtland Formations, but none of these specimens were described. These ornithomimid specimens, all in the UNM collection, are isolated and fragmentary bones that defy precise identification. They do, however, indicate the presence of ornithomimids throughout the Fruitland and Kirtland formations. These specimens include:

1. UNM B-115B, distal end of a metatarsal from UNM locality 77-92 (Fruitland Formation)
2. UNM B-157, distal end of a phalanx from UNM locality 77-90 (Fruitland Formation)
3. UNM B-167, distal end of a metatarsal from UNM locality 77-127 (Fruitland Formation)
4. UNM B-215, foot-bone fragments from UNM locality 77-187 (Fruitland Formation)
5. UNM B-433C, distal end of a phalanx from UNM locality 77-404 (lower shale member of Kirtland Shale)
6. UNM B-479A, foot bone fragment from UNM locality 77-464 (Fruitland Formation)
7. UNM B-741A, claw from UNM locality 77-779 (middle part of Kirtland Shale)
8. UNM FKK-019, partial claw from UNM locality 77-675 (Naashoibito Member of Kirtland Shale)

Cf.* Ornithomimus *Marsh, 1890. UNM B—476 from the Fruitland Formation (UNM locality 77-460) is the claw of digit II of the left manus (Fig. 2E–F). This claw is 7.8+ cm long, nearly as large as that of *Struthiomimus altus* (Osborn, 1916, Fig. 3; Russell, 1972, Table 3), but lacks the marked curvature characteristic of claws of *Struthiomimus.* In this feature, UNM B-476 more closely resembles *Ornithomimus edmontonicus* (Parks, 1933, Pl. 3, Fig. 1), although it is somewhat larger than previously described specimens of *O. edmontonicus.*

Cf.* Struthiomimus *Osborn, 1916. Some of the ornithomimid specimens in the UNM collection are relatively large and compare favorably with specimens of *Struthiomimus altus* Osborn, 1916.

1. UNM B-499B from the lower shale member of the Kirtland Shale (UNM locality 77-421) is the centrum of a dorsal vertebra (Fig. 2G–I). This centrum is 6.3 cm long and thus is within the size range of *S. altus* (Russell, 1972, Table 3).

2. UNM B-590 from the Farmington Sandstone Member of the Kirtland Shale (UNM locality 77-591) is the distal portion of a left fourth metatarsal (Fig. 2C–D) that is very similar to, though

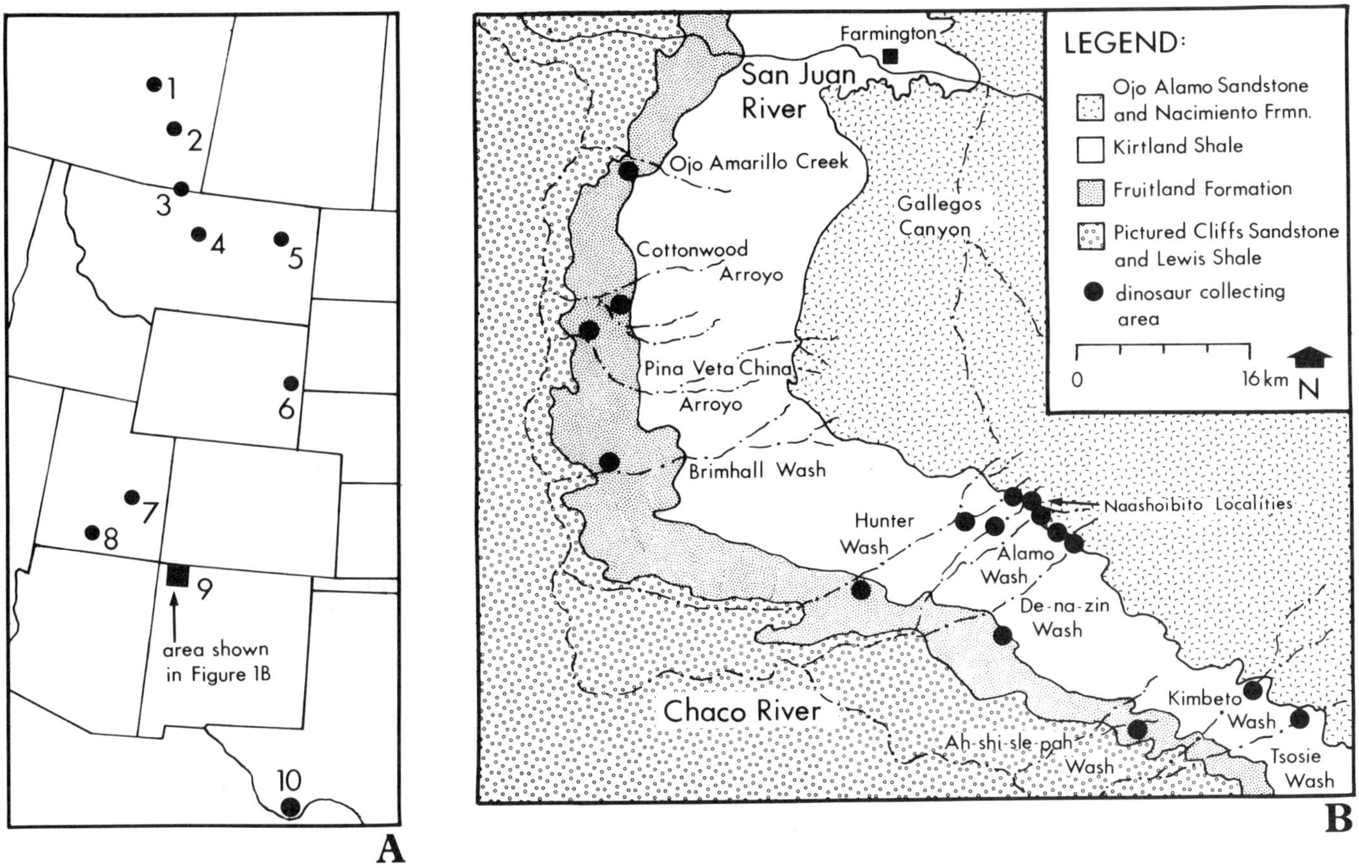

Figure 1. Location map. A: Map of part of western North America showing the location of dinosaur-bearing Upper Cretaceous units discussed in the text. Dinosaur-bearing units are: 1—Edmonton Formation (traditional usage), Alberta; 2—"Oldman" Formation, Alberta; 3—Milk River Formation, Alberta; 4—Judith River Formation, Montana; 5—Hell Creek Formation, Montana; 6—Lance Formation, Wyoming; 7—North Horn Formation, Utah; 8—Kaiparowits Formation, Utah; 9—Fruitland and Kirtland Formations, New Mexico; 10—Aguja and Javelina Formations, Texas. B: Map of part of the west-central San Juan Basin, New Mexico, showing dinosaur collecting areas in the Fruitland and Kirtland Formations (see Appendix 1 for precise locations of all dinosaur localities).

slightly smaller than, that of AMNH 5339, a pes of *S. altus* (Osborn, 1916, Fig. 11).

3. UNM B-745 from the lower part of the Kirtland Shale (UNM locality 77-784) is the proximal phalanx of digit III of the left pes (Fig. 2A–B). It is 7.5 cm long and thus is about the same size as that bone in *S. altus* (Russell, 1972, Table 3).

Dromaeosauridae. A large number of small, isolated teeth that are bladelike and serrated on one or both edges have been found at various levels in the Fruitland and Kirtland Formations. Several names—Coeluridae, Dromaeosauridae, Saurornithoididae, *Paronychodon*—have been applied to these teeth (Kues and others, 1977; Armstrong-Ziegler, 1978, 1980; Lehman, 1981). We assign all of these teeth to the Dromaeosauridae (sensu Ostrom, 1976; also see Ostrom, 1981, Table 2) because of their close resemblance to the teeth of various members of this family, especially *Velociraptor* and *Saurornithoides* (Osborn, 1924; Barsbold, 1974). We do not use the name *Paronychodon* Cope, 1876 because we consider it to be little more than a form-genus for small theropod teeth and therefore of as little use as the name *Deinodon* Leidy, 1856 for the identification of large theropod teeth.

Typical of the dromaeosaur teeth from the San Juan Basin are UNM B-568 (Fig. 2J) and UNM B-311B (Fig. 2K). These teeth have 24–26 serrations per 5 mm on their posterior edges, do not bear anterior serrations, and are quite similar to teeth of *Velociraptor* (Osborn, 1924, Fig. 1).

Albertosaurus *Osborn, 1905.* Cope (1885), Brown (1910; in Sinclair and Granger, 1914), Gilmore (1916, 1919, 1935), Armstrong-Ziegler (1978, 1980), Lehman (1981), and Rowe and others (1981) have mentioned, described, and (or) illustrated remains of large theropods from the Fruitland and Kirtland Formations. Virtually all of these remains are isolated teeth or fragmentary postcrania, and a variety of names—*Laelaps, Deinodon, Gorgosaurus, Albertosaurus*—have been applied to them.

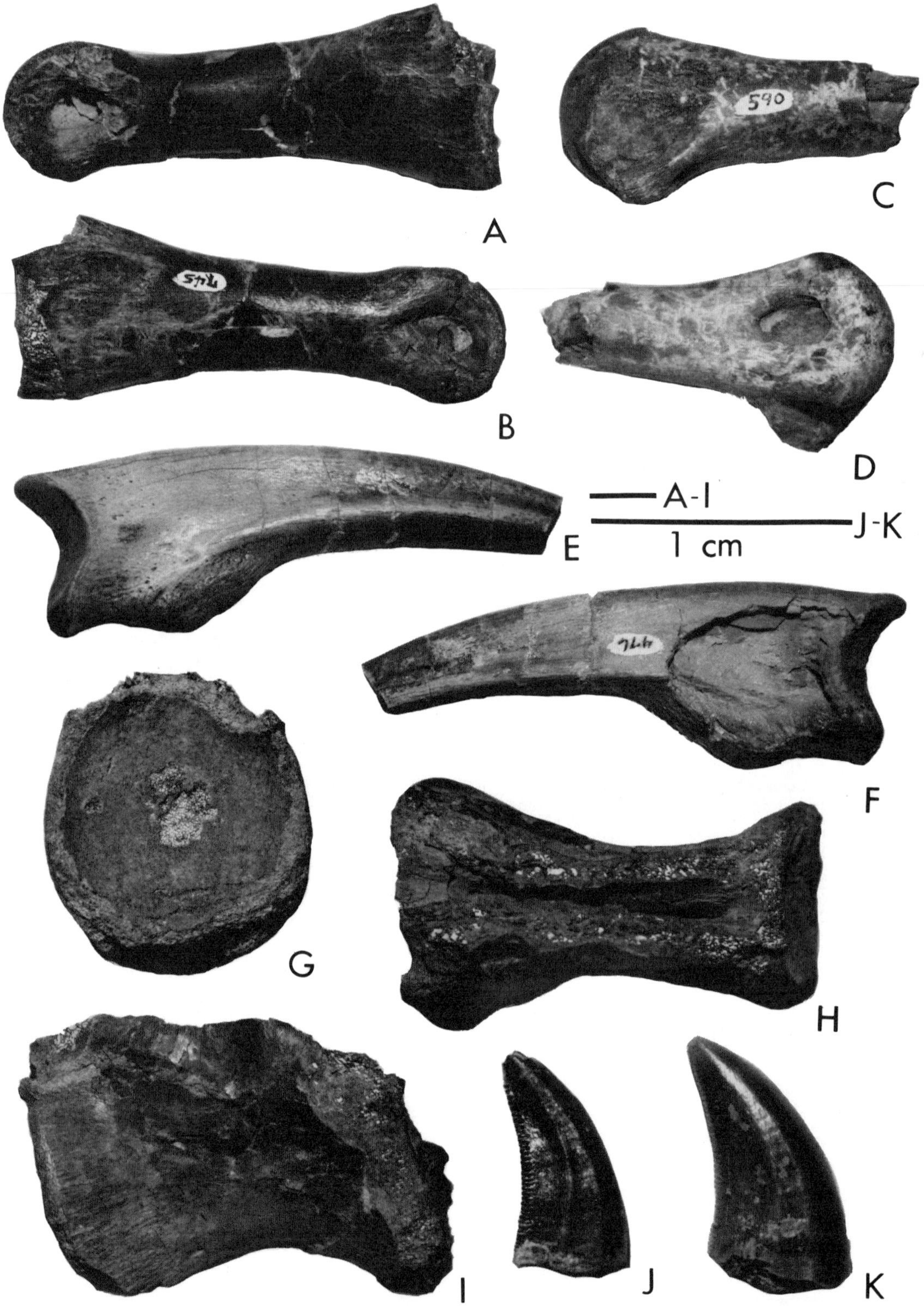

Figure 2. Cf. *Struthiomimus altus,* cf. *Ornithomimus edmontonicus,* and Dromaeosauridae from the Fruitland and Kirtland formations. A, B: cf. *S. altus,* UNM B-745, proximal phalanx of digit III of left pes, external (A) and internal (B) views; C, D: cf. *S. altus,* UNM B-590, distal portion of left fourth metatarsal, external (C) and internal (D) views; E, F: cf. *O. edmontonicus,* UNM B-476, incomplete claw of digit II of left manus, internal (E) and external (F) views; G, H, I: cf. *S. altus,* UNM B-499B, centrum of dorsal vertebra, anterior (G), dorsal (H), and left lateral (I) views; J, K: Dromaeosauridae, teeth, external views of UNM B-568 (J) and UNM B-311B (K).

Strictly speaking, most of these specimens are not generically diagnostic, but those that are seem to pertain to *Albertosaurus* (sensu Russell, 1970), including most of the isolated teeth. Examples of these teeth include UNM FKK-077, 078, 079, and 080 (Fig. 3A–D, F) from the Naashoibito Member of the Kirtland Shale. These teeth have serration counts of 9–11 per 5 mm and are within the size range of those of *Albertosaurus* (Lambe, 1917; Ostrom, 1969; Russell, 1970). Other specimens that almost certainly pertain to *Albertosaurus* include OSM 40-V-1-7, a partial skeleton from the lower shale member of the Kirtland Shale or uppermost Fruitland Formation (Langston in Kues and others, 1977, p. 376); USNM 8346, a left dentary from the upper shale member of the Kirtland Shale (Gilmore, 1916, Pl. 73, Fig. 1); and a right metatarsal from the Naashoibito Member of the Kirtland Shale reported by Lehman (1981, text-Fig. 9.6).

Cf.* Tyrannosaurus *Osborn, 1905. UNM FKK-076 (Fig. 3E, G) is an extremely large theropod tooth (12.8 cm long) that has a serration count of six per 5 mm and thus closely conforms to teeth of *Tyrannosaurus rex* Osborn, 1905 (Osborn, 1912; Ostrom, 1969, Table 2; Russell, 1970). This tooth may therefore pertain to *T. rex,* but we are unwilling to make a definite generic assignment, as were Lucas and Mateer (1983) and O'Neill (1983). As Gilmore (1916, p. 288) expressed it, "the provisional reference of detached [theropod] teeth to a genus . . . is in the highest degree conjectural. . . ."

Sauropoda

Alamosaurus *Gilmore, 1922.* In the San Juan Basin, specimens of *Alamosaurus* have been found only in the Naashoibito Member of the Kirtland Shale. Most of these specimens, which are assigned to *A. sanjuanensis* Gilmore, 1922, the type and only species of the genus, have been described and (or) illustrated by Gilmore (1921, 1922, 1946), Mateer (1976, 1981), Kues and others (1980), and Lehman (1981). All of the specimens of *Alamosaurus* from the Naashoibito Member represent several individuals and are as follows:

1. USNM 10486, left scapula, holotype of *A. sanjuanensis* (Gilmore, 1922, Pl. 1), UNM locality V-143.
2. USNM 10487, right ischium, paratype of *A. sanjuanensis* (Gilmore, 1922, Pl. 2), UNM locality V-143.
3. PMU. R. 280, sacral vertebrae 1-3, topotype of *A. sanjuanensis* (Mateer, 1976, Fig. 2; 1981, text-Fig. 3.2), UNM locality V-143.
4. PMU. R. 281, right ilium, topotype of *A. sanjuanensis* (Mateer, 1976, Fig. 1; 1981, text-Fig. 3.2), UNM locality V-143.
5. UNM B-522, distal caudal centrum (Fig. 4B), UNM locality 77-505.
6. UNM B-656, partial tooth (Kues and others, 1980, text-Figs. 3–4), UNM locality 77-675.
7. UNM B-709, right juvenile scapula (Lehman, 1981, text-Fig. 9.9). UNM locality 77-748. Originally identified as ?Hypsilophondontidae by Lehman (1981, p. 203) but subsequently recognized as sauropod by Lehman (1985).
8. UNM FKK-003A-G, seven partial teeth (Kues and others, 1980, text-Fig. 4), UNM locality 77-675.
9. UNM FKK-029, incomplete tooth, topotype of *A. sanjuanensis* (Kues and others, 1980, text-Fig. 5), UNM locality V-143.
10. UNM FKK-033, partial right femur, topotype of *A. sanjuanensis* (Fig. 4A), UNM locality V-143. This femur was briefly mentioned by Lehman (1981, p. 203); it is over 150 cm long, both of its ends are incomplete, and it is anteroposteriorly crushed.
11. UNM FKK-034, nearly complete tooth (Kues and others, 1980, text-Fig. 2), from just north of UNM locality V-143.
12. USNM 15658, distal caudal centrum and neural spine, topotype (?) of *A. sanjuanensis,* UNM locality V-143? Gilmore (1946, p. 31) briefly mentioned but did not illustrate this specimen.

Ornithopoda

?Pachycephalosauridae. Armstrong-Ziegler (1978, table 1; 1980, p. 27, Pl. 3C) tentatively assigned an isolated tooth (MNA Pl. 1642) from the upper part of the Fruitland Formation to the Pachycephalosauridae. The only other report of a pachycephalosaur from the San Juan Basin, by Kues and others (1977; also see Lucas, 1981, Table 15.1), was based on an isolated tooth here considered to pertain to an ankylosaur (see discussion that follows).

Ankylosauria. The few specimens of ankylosaurs from the Fruitland and Kirtland Formations are so incomplete that no definite generic assignments can be made. Gilmore (1916, 1919, 1935) and Lehman (1981) have described and (or) illustrated most of the known specimens, which can be placed into five categories:

1. Ankylosauria: The UNM collection contains isolated dermal plates from various levels in the Fruitland and Kirtland Formations.

2. Ankylosauridae: UNM FKK-084 (Fig. 4F–G) is an incomplete left humerus from UNM locality V-141 in the Naashoibito Member of the Kirtland Shale. The humerus is relatively short and stout and has a relatively small deltopectoral crest. We tentatively identify it as the humerus of a juvenile ankylosaurid (Coombs, 1978, p. 156, text-Fig. 10).

3. Nodosauridae: USNM 8610 (Gilmore, 1919, Fig. 33) and 8611 are dermal plates of a nodosaurid from the Naashoibito Member of the Kirtland Shale (Lehman, 1981). UNM B-732 from UNM locality 77-770 is an isolated tooth (Fig. 4D–E) previously identified as ?Pachycephalosauridae (Kues and others, 1977; Lucas, 1981, Table 15.1). This tooth closely resembles a maxillary tooth of *Panoplosaurus* (Gilmore, 1930, Fig. 4) in having a rugose lingual cingulum and nine denticles, the three central denticles being highly elevated. Therefore, we refer this tooth to the Nodosauridae.

4. ?*Euoplocephalus:* A right humerus, USNM 8360, from the upper shale member of the Kirtland Shale, originally de-

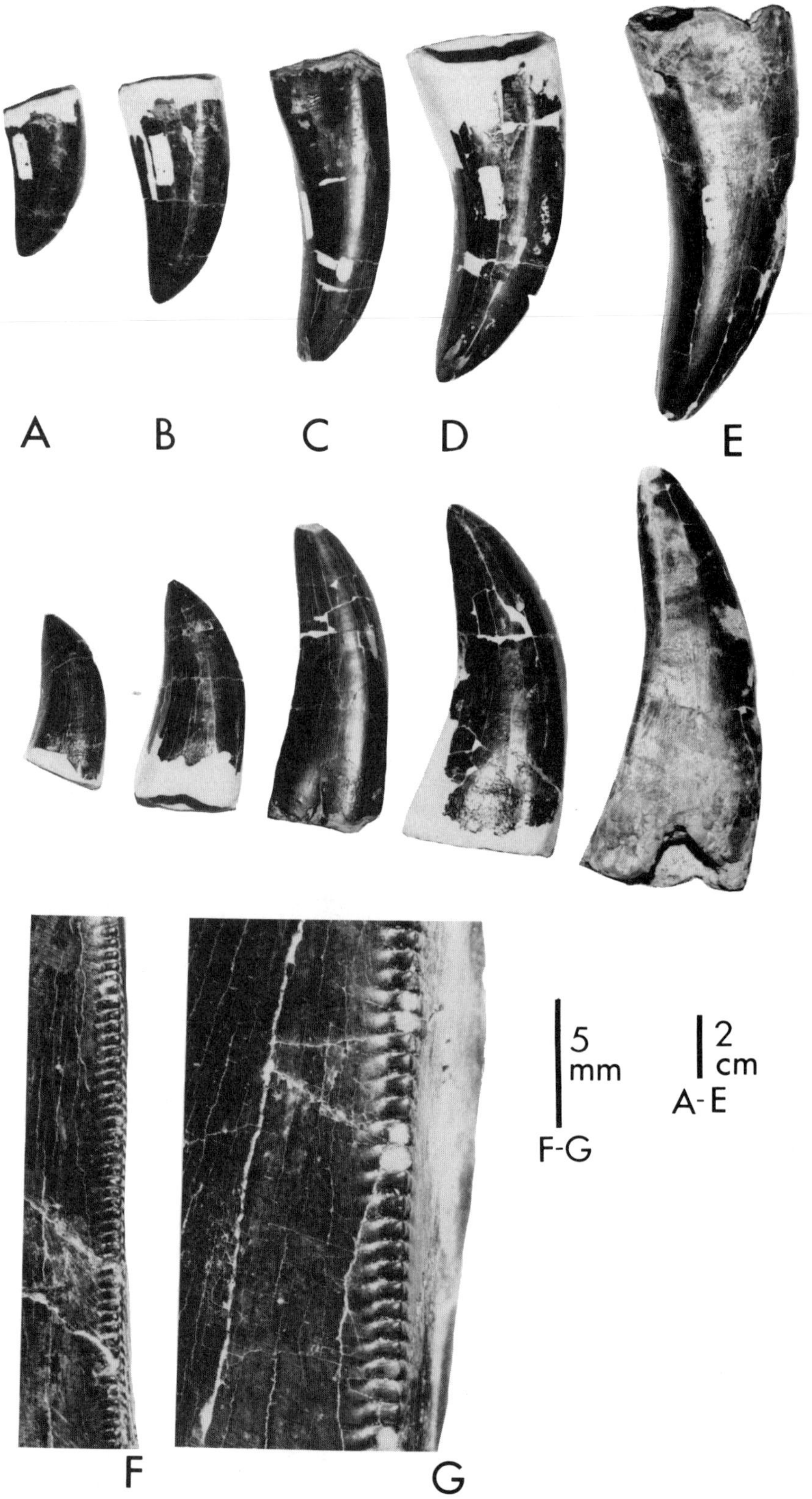

Figure 3. *Albertosaurus* sp. and cf. *Tyrannosaurus rex* from the Kirtland Formation (Naashoibito Member). A, B, C, D: *Albertosaurus* sp., teeth, internal (above) and external (below) views of UNM FKK-080 (A), –079 (B), –078 (C) and –077 (D); E: cf. *T. rex,* tooth, internal (above) and external (below) views of UNM FKK-076; F: *Albertosaurus* sp., detail of serrations of UNM FKK-080; G: cf. *T. rex,* detail of serrations of UNM FKK-076.

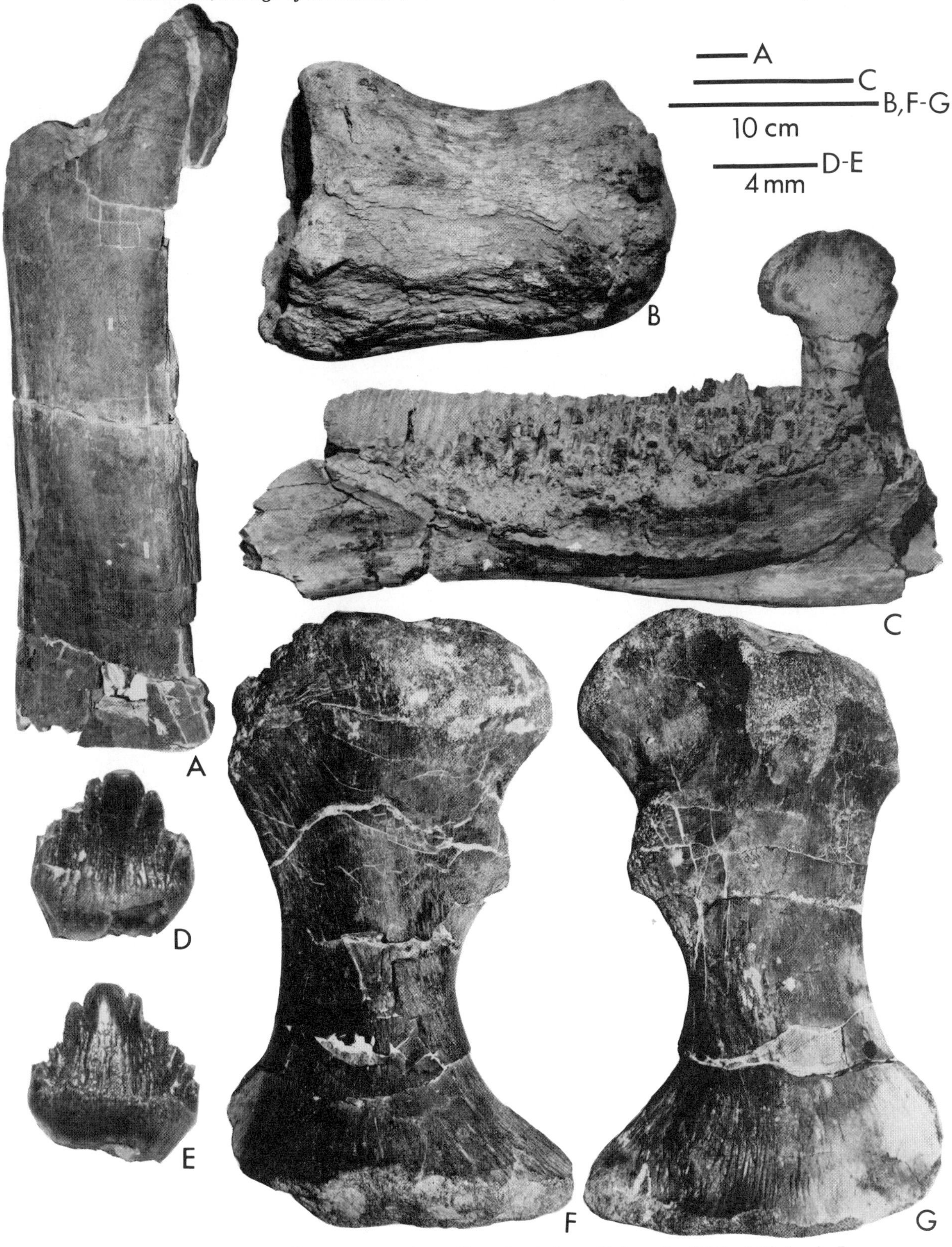

Figure 4. *Alamosaurus, Kritosaurus,* nodosaurid, and ankylosaurid from the Kirtland Shale. A, B: *Alamosaurus sanjuanensis,* anterior view of UNM FKK-033, right partial femur (A) and ventral view of UNMB B-522, distal caudal centrum (B); C: *Kritosaurus navajovius,* internal view of UNM FKK-085, right lower jaw; D, E: Nodosauridae, internal (D) and external (E) views of UNM B-732, maxillary (?) tooth; F, G: Ankylosauridae, dorsal (F) and ventral (G) views of UNM FKK-084, juvenile left humerus.

scribed by Gilmore (1916, 1935) was tentatively referred to *Euoplocephalus* by Lehman (1981, p. 213).

5. ?*Panoplosaurus:* Lehman (1981, text-Fig. 9.14) illustrated USNM 8571, a left scapula (also see Gilmore, 1919, Pl. 119, Fig. 2) from the Naashoibito Member of the Kirtland Shale and tentatively referred it to *Panoplosaurus.*

Ornithopoda

Hadrosauridae. Gilmore (1916) reported numerous generically indeterminate specimens of hadrosaurs, mostly from the Naashoibito Member of the Kirtland Shale. There also are many fragmentary and (or) isolated hadrosaur bones in the UNM collection from various stratigraphic levels in the Fruitland and Kirtland Formations. In addition, Lehman (1981, p. 204, text-Fig. 9.10) briefly described a humerus and tibia from the Naashoibito Member that are within the size range of the large hadrosaurs known from the "El Gallo" Formation in Baja California, Mexico (Morris, 1972, 1981).

Kritosaurus *Brown, 1910.* Specimens of *Kritosaurus*[1] from the Fruitland and Kirtland Formations have been described and (or) illustrated by Brown (1910), Gilmore (1916, 1919, 1935), and Mateer (1981). One species, *K. navajovius* Brown, 1910, has been recognized. Specimens of *Kritosaurus* are known from the Fruitland Formation (for example, MNA Pl. 1626, isolated teeth [Armstrong-Ziegler, 1980, p. 28, pl. 3F-G] and UNM B-151, an ischium from UNM locality 77-71) and from the upper shale member of the Kirtland Shale (for example, AMNH 5799, the holotype skull and lower jaws of *K. navajovius*). USNM 8629, a partial skull and vertebrae of *Kritosaurus* (Gilmore, 1935) from Kimbeto Wash, comes from a horizon in the Kirtland Shale equivalent to either the lower shale or Farmington Sandstone members farther north. Occurrences of *Kritosaurus* in the Naashoibito Member of the Kirtland Shale include AMNH 5797, part of a maxillary and lower jaw (Sinclair and Granger, 1914), and UNM FKK-085, a right lower jaw (Fig. 4C) from UNM locality V-142 assigned to the genus because of its close resemblance to the lower jaw of AMNH 5797 (Brown, 1910, Pl. 29). The occurrence of *Kritosaurus* throughout the Fruitland and Kirtland Formations thus seems reasonably well established.

Parasaurolophus *Parks, 1922.* Specimens of *Parasaurolophus* from the San Juan Basin have been described and (or) illustrated by Wiman (1931), Gilmore (1935), Ostrom (1961, 1963) and Mateer (1981). Two species have been named, *P. tubicen* Wiman, 1931 and *P. cyrtocristatus* Ostrom, 1961. *Parasaurolophus* is present in the Fruitland Formation (for example, FMNH P 2793, the holotype skull and skeleton of *P. cyrtocristatus,* and MNA P1. 529, a partial maxilla), the lower shale member of the Kirtland Shale (for example, USNM 13492, a partial skeleton), and the Naashoibito Member of the Kirtland Shale (for example, PMU. R. 222, the holotype partial skull of *P. tubicen*). The inability to definitely assign hadrosaur specimens from the upper shale and Farmington Sandstone members of the Kirtland Shale to *Parasaurolophus* is almost certainly due to the incompleteness of hadrosaur specimens from these units; it does not necessarily mean that the genus was not present in the San Juan Basin during this interval (see Lucas, 1981).

Ceratopsia

Ceratopsidae. Brown (1910, p. 296, Fig. 1) and Gilmore (1916, p. 285–287, Figs. 29–30; 1919, p. 64–65, pl. 25) identified fragmentary ceratopsian specimens from the San Juan Basin as *Monoclonius*? (Fruitland Formation, Naashoibito Member of the Kirtland Shale) and *Ceratops* sp. (lower shale member of the Kirtland Shale). There also are a large number of fragmentary and (or) isolated ceratopsian bones from various horizons in the Fruitland and Kirtland Formations, especially in the UNM collection.

Lehman (1981, p. 207) refuted Brown's (1910) assignment of AMNH 5798, a partial horn-core from the Naashoibito Member of the Kirtland Shale, to *Monoclonius*? sp., and we agree with him that this specimen is generically indeterminate. We also view USNM 7347, fragmentary postcrania from the Fruitland Formation referred by Gilmore (1916) to *Monoclonius*? sp., and USNM 8604, the median bar of a frill from the lower shale member of the Kirtland Shale referred by Gilmore (1919) to *Ceratops* sp., as generically indeterminate (Gilmore [1935, p. 160] reached the same conclusion).

cf.* Chasmosaurus *Lambe, 1914. Gilmore (1935, p. 164, Fig. 6) referred a left supraorbital horn-core to *Chasmosaurus* sp. This horn-core was collected "5 miles west of Brimhall's store" according to Gilmore (1935) and thus almost certainly came from the lower shale member of the Kirtland Shale. The horn-core is short, blunt-tipped, and slightly recurved; it closely resembles horn-cores of *Chasmosaurus belli* and *C. brevirostris* (Lull, 1933, Pls. 6–7). Therefore, we follow Gilmore (1935) by identifying this specimen as cf. *Chasmosaurus* sp.

Pentaceratops *Osborn, 1923.* *Pentaceratops* Osborn, 1923 is endemic to the San Juan Basin (Gilmore's [1933] provisional assignment to *Pentaceratops* of a specimen from the Mongolian People's Republic has been discounted by Rowe and others [1981, p. 31]). Specimens of *Pentaceratops* from the San Juan Basin have been described and (or) illustrated by Osborn (1923), Wiman (1930), Gilmore (1935), Lehman (1981), Mateer (1981), and Rowe and others (1981). Two species have been named, *P. sternbergii* Osborn, 1923 (the type species) and *P. fenestratus* Wiman, 1930. Most workers (but see Mateer, 1981) have questioned the validity of Wiman's species (for example, Gilmore, 1935; Lull, 1933; Rowe and others, 1981).

Pentaceratops is present in the Fruitland Formation (for example, MNA Pl. 1747, the skull and partial skeleton described by Rowe and others [1981]) and in the lower shale member of

[1]Although Baird and Horner's (1977) conclusion that *Kritosaurus* is a synonym of *Hadrosaurus* has been accepted by many workers (for example, Lehman, 1981; Dodson, 1983; Weishampel and Weishampel, 1983), we continue to use the name *Kritosaurus* until this conclusion is adequately documented.

the Kirtland Shale (for example, AMNH 6325, the holotype skull of *P. sternbergii,* UALP 13342, skull and UKVP 16100, skull). The specimens referred to *Pentaceratops* by Gilmore (1935, p. 163), USNM 12002 and 12743, also probably came from the lower shale member (locality *fide* Gilmore [1935] is the SW¼ T24N, R13W, San Juan County, New Mexico). Wiman's (1930) specimens of *Pentaceratops,* PMU. R. 200 (the holotype skull of *P. fenestratus*) and PMU. R. 268 (postcranial skeleton), came from the Kirtland Shale below the Naashoibito Member, but from which member of the Kirtland they were collected is uncertain (Mateer, 1981). Specimens of *Pentaceratops* from the Naashoibito Member of the Kirtland Shale include OSM 40-IX-1-41 through 40-IX-44-41, a skull and postcranial skeleton (Langston in Kues and others, 1977, p. 377); UNM FKK-081, a partial frill (Fig. 5F) and postcranial skeleton (left scapula: Fig. 5E) from UNM locality V-144; and USNM 12741, a partial skull (Lehman, 1981, p. 208). Although specimens of *Pentaceratops* have not been defnitely identified from the Farmington Sandstone and upper shale members of the Kirtland Shale, its presence in the lower shale and Naashoibito members suggests continued presence of the genus throughout Kirtland time.

Torosaurus ***Marsh, 1891.*** Specimens from the San Juan Basin that we and Lehman (1981, 1985) assign to *Torosaurus* come from the Naashoibito Member of the Kirtland Shale. These specimens are UNM B-628, a partial left squamosal (Fig. 5B–C; Lehman, 1981, text-Fig. 9.12); UNM FKK-013, a partial left parietal (Fig. 5D; Lehman, 1981, text-Fig. 9.13); and UNM FKK-082, a partial right horn-core (Fig. 5A). Lehman (1981) originally identified UNM B-628 and FKK-013 as aff. *Torosaurus utahensis* but later (Lehman, 1985) identified these specimens as *T.* cf. *T. utahensis.* We agree with this later identification and also identify UNM FKK-082 as *T.* cf. *T. utahensis* for the following reasons:

1. The squamosal (UNM B—628) is relatively large, broadly triangular in outline, and lacking either significant undulations or epoccipitals on its ventral border. It thus closely resembles USNM 15583, the holotype skull of *T. utahensis* (Gilmore, 1946, Fig. 12). However, UNM B-628 is not well enough preserved to allow determination of whether it bears a small epoccipital along its ventral border as does USNM 15583.

2. The partial parietal (UNM FKK-013) has a broadly triangular cross section and a low and slightly sinuous central ridge on its dorsal surface. In these features it closely resembles parietals of *T. utahensis* (Gilmore, 1946, Pl. 14; Lawson, 1976, text-Fig. 5) but differs from the skull of *T. latus* described by Colbert and Bump (1947, Pl. 3).

3. The partial supraorbital horn core is relatively large and very slightly curved outward; it has an ovate cross section. In these features it closely resembles *Torosaurus* among the Ceratopsidae (Lambe, 1915, p. 17; Lull, 1933, Pl. 15).

There may be other specimens of *Torosaurus* known from the Naashoibito Member of the Kirtland Shale (Lehman, 1985), but we believe that they are too fragmentary to permit certain identification.

AGE OF THE FRUITLAND AND KIRTLAND FORMATIONS BASED ON DINOSAURS

Introduction

In this section, we answer the following question: What do dinosaur fossils, themselves, indicate are the ages of the Fruitland and Kirtland Formations? To answer this question, we compare the dinosaurs of the Fruitland and Kirtland Formations with dinosaurs known elsewhere from Aquilan (early-middle Campanian), Judithian (late Campanian), Edmontonian (latest Campanian–early Maastrichtian), and Lancian (middle-late Maastrichtian) strata ("age" terms of Russell, 1975; also see Clemens and others, 1979). Dinosaur faunas used for comparison are Milk River (Aquilan); "Oldman" and Judith River (Judithian); Aguja (Judithian?); lower part of Edmonton of traditional usage (Edmontonian); and Hell Creek, Lance, North Horn, and Javelina (Lancian) (see Russell, 1975; Clemens and others, 1979; Lehman, 1982; Dodson, 1983; Rice and Shurr, 1983). Finally, the dinosaurs from the Fruitland and Kirtland Formations considered here come from only part of the San Juan Basin: a narrow belt in the west-central basin between Tsosie Wash (south) and Ojo Amarillo Creek (north) (Fig. 1B). Therefore, the dinosaur-based age determinations proposed here apply only to strata of the Fruitland and Kirtland Formations in this area; time transgression of the Fruitland and Kirtland Formations may make them older or younger outside of this area.

Age of the Fruitland Formation

Dinosaurs present in the Fruitland Formation are Ornithomimidae, cf. *Struthiomimus altus,* Dromaeosauridae, *Albertosaurus* sp., ?Pachycephalosauridae, Ankylosauria, Hadrosauridae, *Kritosaurus navajovius, Parasaurolophus cyrtocristatus,* Ceratopsidae, and *Pentaceratops sternbergii* (Fig. 6). Only three of these taxa can be used at the generic level to correlate the Fruitland Formation with other dinosaur-bearing units in western North America: *Kritosaurus, Parasaurolophus,* and *Pentaceratops.* The other dinosaur taxa known from the Fruitland Formation occur in Judithian through Lancian strata elsewhere in western North America (Fig. 6) and thus are of little use in providing a refined age determination.

Outside of the San Juan Basin, *Parasaurolophus* and *Kritosaurus* are known only from pre-Lancian strata, with one exception, the occurrence of *Parasaurolophus* in the Lancian Kaiparowits Formation (see discussion below). Although *Pentaceratops* is endemic to the San Juan Basin, Lull (1933) and Colbert (1948) considered it to be of Edmontonian age based on their proposed phylogenies of the Ceratopsidae. Pending a long overdue revision of the ceratopsids and reanalysis of their phylogeny, this conclusion must be considered tentative at best. The absence in the Fruitland Formation of dinosaurs characteristic of Lancian time (for example, *Tyrannosaurus, Alamosaurus, Torosaurus, Triceratops*) is the most persuasive argument that the

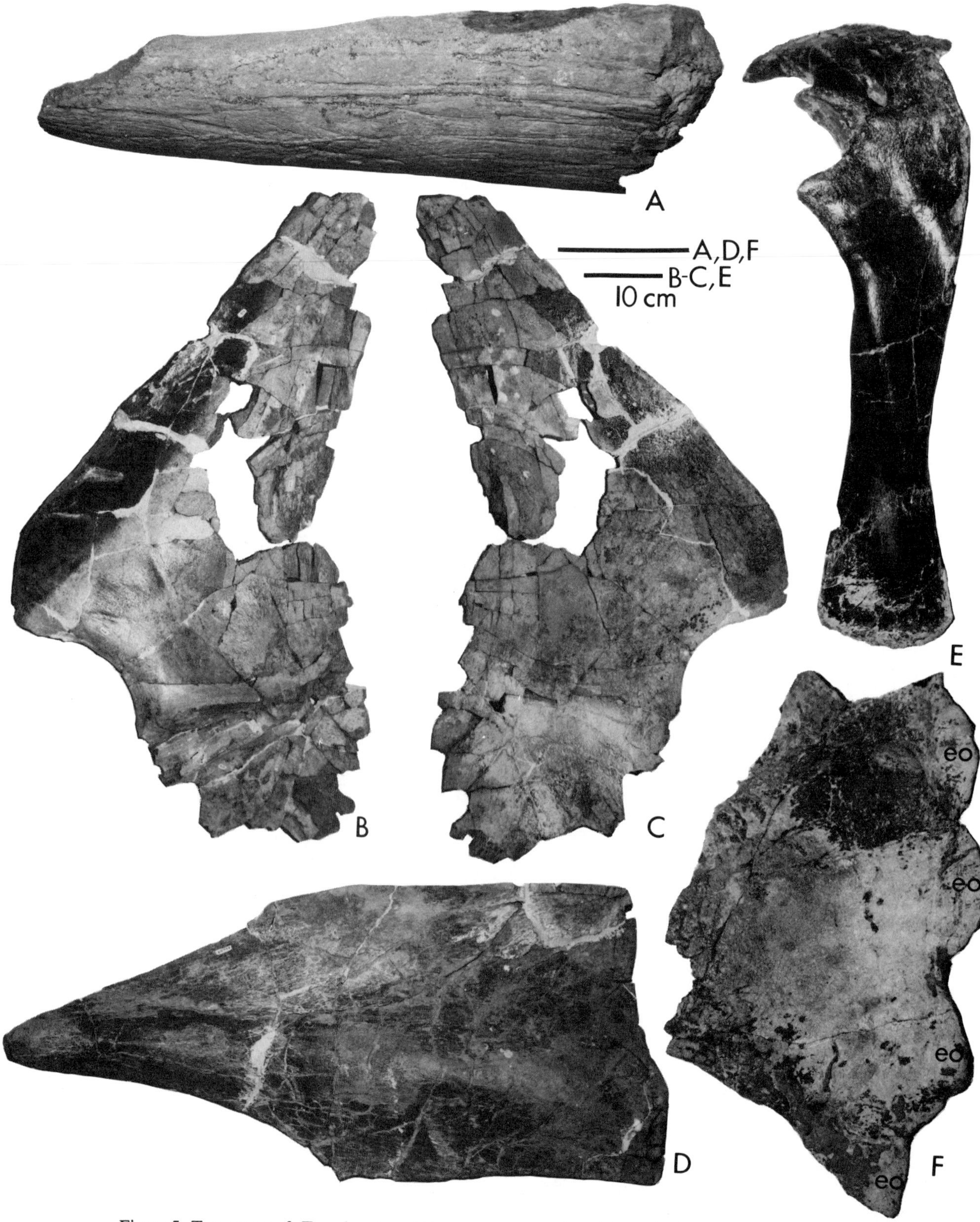

Figure 5. *Torosaurus* cf. *T. utahensis* and *Pentaceratops sternbergii* from the Kirtland Shale (Naashoibito Member). A, B, C, D: *T.* cf. *T. utahensis,* anterior view (A) of partial right supraorbital horn-core (UNM FKK-082), internal (B) and external (C) views of partial left squamosal (UNM B-628), and dorsal view (D) of partial left parietal (UNM FKK-013); E, F: *P sternbergii,* external view (E) of left scapula and dorsal view (F) of part of left squamosal (eo = epoccipital), both UNM FKK-081.

		Kirtland Shale												
TAXA: / UNITS:	Fruitland Formation	lower shale member	Farmington Sandstone Member	upper shale member	Naashoibito Member	"Oldman Formation"	Judith River Formation	Edmonton Formation (lower part)	Milk River Formation	Aguja Formation	Javelina Formation	North Horn Formation	Hell Creek Formation	Lance Formation
Ornithomimidae	●	●	●	●	●	●	●	●	●	●			●	●
Ornithomimus	cf					●		●					●	●
Struthiomimus		cf	cf			●		●						
Dromaeosauridae	●	●	●	●	●	●	●						●	●
Albertosaurus	●	●	●	●	●	●	●	●	●	?		?		●
Tyrannosaurus					cf						●		●	●
Alamosaurus					●						●	●		
Pachycephalosauridae	?					●	●	●	●	?			●	
Ankylosauria	●	●	●	●	●	●	●	●	●	●			●	●
Ankylosauridae					●	●		●					●	●
Nodosauridae				●	●	●	●	●	●	●			●	●
Euoplocephalus				?		●	●	●						
Panoplosaurus					?	●		●		●				
Hadrosauridae	●	●	●	●	●	●	●	●	●	●	●	●	●	●
Kritosaurus	●	●	●	●	●		●		cf	●				
Parasaurolophus	●	●			●	●								
Ceratopsidae	●	●	●	●	●	●	●	●	●	●	●	●	●	●
Chasmosaurus		cf				●				●				
Pentaceratops	●	●			●									
Torosaurus					●						●	●	●	●

● - taxon present

Figure 6. Dinosaur taxa known from the Fruitland and Kirtland Formations and their distribution in the Fruitland and Kirtland and selected dinosaur-bearing Upper Cretaceous units in western North America. See Figure 1A for location of these dinosaur-bearing units. Data on dinosaur occurrences in units outside the San Juan Basin are from Russell (1930, 1964), Gilmore (1946), Coombs (1971), Lawson (1972), Sahni (1972), Carpenter (1982), Lehman (1982), and Dodson (1983).

Fruitland is pre-Lancian. However, beyond this all that can be said on the basis of the dinosaur fauna is that the Fruitland probably is Judithian or might be as young as Edmontonian.

Age of the Lower Shale, Farmington Sandstone, and Upper Shale Members of the Kirtland Shale

The dinosaur fauna of the Kirtland Shale below the Naashoibito Member is virtually identical to that of the Fruitland Formation (Fig. 6). It was mainly for this reason that the vertebrate fauna of the Fruitland Formation and Kirtland Shale up through the upper shale member were included in one local fauna (the Hunter Wash local fauna) by Clemens (1973), Powell (1973), Lehman (1981), and Lucas (1981). The differences between the Fruitland Formation dinosaur faunas and those of the Kirtland Shale below the Naashoibito Member are few: the possible occurrences of *Chasmosaurus* and *Euoplocephalus* and the occurrence of nodosaurid ankylosaurs in the Kirtland Shale (Fig. 6). These taxa, however, are consistent with either a Judithian or Edmontonian age assignment. Of course, the Kirtland Shale, based on its stratigraphic relationship to the underlying Fruitland Formation, must be younger than the Fruitland in any localized area (Fassett and Hinds, 1971). For the same reasons previously stated for the dinosaur fauna of the Fruitland Formation, the dinosaurs of the lower shale, Farmington Sandstone, and upper shale members of the Kirtland Shale are pre-Lancian. They may be Edmontonian, but additional dinosaur taxa from the Kirtland Shale will be needed to demonstrate this conclusively.

Age of the Naashoibito Member of the Kirtland Shale

Dinosaurs present in the Naashoibito Member of the Kirtland Shale are Ornithomimidae, Dromaeosauridae, *Albertosaurus* sp., cf. *Tyrannosaurus rex, Alamosaurus sanjuanensis,* Ankylosauria, Ankylosauridae, Nodosauridae, ? *Panoplosaurus* sp., Hadrosauridae, *Kritosaurus navajovius, Parasaurolophus tubicen,* Ceratopsidae, *Pentaceratops sternbergii,* and *Torosaurus* cf. *T. utahensis* (Fig. 6). Except for cf. *T. rex, A. sanjuanensis,* Ankylosauridae, ?*Panoplosaurus,* and *T.* cf. *T. utahensis,* all of these taxa are present in either the underlying members of the Kirtland Shale or the Fruitland Formation, or both. *Panoplosaurus* is known from pre-Lancian strata outside of the San Juan Basin (Coombs, 1971), but its occurrence in the Naashoibito Member is questionable and probably should not be accorded any biostratigraphic significance. The same can be said for the possible ankylosaurid in the Naashoibito Member; ankylosaurids are found in Judithian-Lancian strata elsewhere (Coombs, 1971).

The presence of *Alamosaurus* and *Torosaurus* and possible presence of *Tyrannosaurus* in the Naashoibito Member of the Kirtland Shale provide very strong evidence that the Naashoibito is of Lancian age. Earlier arguments that *Alamosaurus* is a formgenus of North American Late Cretaceous sauropods, and therefore of little biostratigraphic value (Lucas and Rigby, 1979; Lucas, 1981; Lucas and Mateer, 1983), can be devalued. Although few skeletal elements are shared by the specimens of *Alamosaurus* from the Naashoibito Member, North Horn Formation, and Javelina Formation, ischia of the paratype of *A. sanjuanensis* from the Naashoibito (Gilmore, 1922) and from the North Horn Formation (Gilmore, 1946) and ilia of a topotype of *A. sanjuanensis* from the Naashoibito Member (Mateer, 1976, 1981) and from the Javelina Formation (Lawson, 1972) are identical. Since the North Horn and Javelina dinosaurs are of Lancian age, the occurrence of *Alamosaurus* in the Naashoibito supports a Lancian age assignment, even if it is a facies-fossil restricted to inland environments (Lucas, 1981) or to southern localities in western North America (Sloan, 1970).

Outside of the San Juan Basin, *Torosaurus* is known only from Lancian strata (Hell Creek, Lance, North Horn, and Javelina Formations). Its occurrence in the Naashoibito Member also supports a Lancian age assignment. The possible occurrence in the Naashoibito of *Tyrannosaurus,* a typical Lancian genus, also supports assignment of a Lancian age to the Naashoibito.

The presence in the Naashoibito Member of *Kritosaurus,* a genus known elsewhere from pre-Lancian strata (Weishampel and Weishampel, 1983), has deterred some workers (for example, Baltz and others, 1966; Lucas, 1981) from assigning a Lancian age to the Naashoibito. However, in view of the presence of *Alamosaurus, Torosaurus,* and cf. *Tyrannosaurus* in the Naashoibito, we think it is most reasonable to regard the occurrence of *Kritosaurus* in this unit as an extension of its temporal range upward into the Lancian. *Parasaurolophus* from the Naashoibito Member also has been considered to be a taxon indicative of a pre-Lancian age (Baltz and other, 1966; Lucas, 1981). However, *Parasaurolophus,* besides its pre-Lancian occurrences outside of the San Juan Basin, also is known from the Kaiparowits Formation in south-central Utah (Weishampel and Jensen, 1979) that Lohrengel (1969) considered to be of Lancian age based on palynology. Therefore, we do not consider the occurrence of *Parasaurolophus* in the Naashoibito Member to be a deterrent to a Lancian age assignment.

All of the remaining dinosaurs in the Naashoibito Member, with the exception of the endemic ceratopsian *Pentaceratops,* are known outside of the San Juan Basin from the Judithian-Lancian strata (Fig. 6) and therefore are of no use in a refined age determination. However, neither do they provide evidence against the Lancian age assignment for the Naashoibito Member argued for earlier.

COMPARISONS WITH OTHER CORRELATIONS

Introduction

In this section, we briefly compare the dinosaur-based correlation of the Fruitland and Kirtland Formations with correlations based on other groups of fossils as well as those based on radiometric dates and magnetostratigraphy.

Other Biostratigraphic Correlations

The megafossil plants of the Fruitland and Kirtland Formations were summarized by Tidwell and others (1981), who essentially reviewed the work of Knowlton (1916). These plants are clearly of Late Cretaceous (Campanian-Maastrichtian) age, but until they are thoroughly restudied we believe no more precise age determination should be based on them. The same can be said of the nonmarine invertebrates from the Fruitland and Kirtland Formations that have not been studied since Stanton (1916).

Based on the occurrence of late Campanian (*Baculites compressus* zone) ammonites at the top of the Lewis Shale and in the Pictured Cliffs Sandstone in the north- and east-central San Juan Basin (Cobban, 1973; Cobban and others, 1974), Clemens (1973) and Lucas and Schoch (1982) inferred that the overlying Fruitland Formation in the west-central San Juan Basin (Fig. 1B) is of late Campanian age. This age assignment is consistent with a Judithian age assignment for the Fruitland Formation based on its dinosaur fauna.

Nondinosaurian lower vertebrates from the Fruitland and Kirtland Formations are part of a recurring complex of lower vertebrates present in most Late Cretaceous continental deposits in western North America (Lucas, 1981). They thus seem to be of little use in arriving at a precise age determination. Most fossil mammals from the Fruitland and Kirtland formations were collected from a narrow stratigraphic interval that encompasses the uppermost Fruitland Formation and lowermost Kirtland Shale. These mammals probably are of Edmontonian age (Clemens, 1973; Clemens and others, 1979; Rigby and Wolberg, 1983). Dinosaur-based assignment of a Judithian-Edmontonian age to the Fruitland through upper shale member of the Kirtland Shale sequence is consistent with the mammalian age determination. Lehman (1985) reported the presence of *Essondon* cf. *E. browni* in the Naashoibito Member. This mammal is known from the Lance and Hell Creek Formations (Clemens and others, 1979), and its presence thus supports the dinosaur-based assignment of a Lancian age to the Naashoibito Member.

The only biostratigraphic data that apparently conflict with the dinosaur-based correlation of the Fruitland and Kirtland Formations are palynological. Most palynological age assignments for the Fruitland Formation and Kirtland Shale below the Naashoibito Member indicate a late Campanian–early Maastrichtian age (e.g., Tschudy, 1973; Newman, 1982, 1983; Manfrino, 1984). However, Zavada (1976) assigned the Fruitland Formation a Maastrichtian age based on his palynological studies. In addition, Newman (1982, 1983) has presented palynological evidence that indicates an early Maastrichtian age for the Farmington Sandstone, upper shale, and Naashoibito Members of the Kirtland Shale. Clearly, there are disagreements among the palynologists about the age of the Fruitland and Kirtland Formations, and some, but not all, of their conclusions are not in accord with the dinosaur-based correlation. As vertebrate paleontologists, we are unable to adequately analyze the palynological data from the Fruitland and Kirtland Formations, especially since few of these data have been adequately published. Instead, we merely point out that there are some discrepancies between the dinosaur-based and the palynological correlations of the Fruitland and Kirtland Formations that need to be resolved.

Radiometric Dates

Lindsay and others (1981) reported a latest Maastrichtian–earliest Paleocene fission track date of 64.6 ± 3.0 Ma from an ash bed in the Farmington Sandstone Member of the Kirtland Shale. Not only is this date inconsistent with all biostratigraphic correlations (Lucas and Schoch, 1982), but K-Ar dates reported by Brookins and Rigby (1983) from the lower part of the Kirtland Shale have an average of about 72.7–73.7 Ma (corrected for new decay constants). Brookins and Rigby's age determination is very close to the Campanian-Maastrichtian boundary defined by Obradovich and Cobban (1975) and thus is consistent with the dinosaur-based correlation of the Fruitland and Kirtland Formations.

Magnetostratigraphic Correlations

The magnetic-polarity stratigraphy of Butler and others (1977) and Lindsay and others (1978, 1981) indicates that the entire Fruitland-Kirtland sequence is of late Maastrichtian (Lancian) age. Clearly, this disagrees with the dinosaur-based correlation that assigns an earlier age to the Fruitland Formation and majority of the Kirtland Shale. Previous attempts (by Alvarez and Vann [1979], Fassett [1979], Lucas and Rigby [1979], and Lucas and Schoch [1982]) to reconcile the magnetostratigraphic correlations with biostratigraphic correlations of the Fruitland and Kirtland Formations have relied heavily on the argument that there is a major temporal gap at the base of the Ojo Alamo Sandstone. The belief of these workers that this unconformity represents much of Maastrichtian time (i.e., at least all of the Lancian) has led them to conclude that the uppermost Cretaceous is not represented by sediments in the west-central San Juan Basin and that the magnetic-polarity sequence presented by Butler and others (1977) and Lindsay and others (1978, 1981) consequently is missing one or more magnetozones. Thus, revised correlation of this polarity sequence is possible, bringing it into agreement with biostratigraphic correlations (for example, see Lucas and Schoch, 1982, Fig. 3).

Dinosaurs, however, indicate that the Naashoibito Member of the Kirtland Shale immediately below the unconformity at the base of the Ojo Alamo Sandstone is of Lancian age. Therefore, this unconformity cannot represent much of Maastrichtian time. Although this removes a mainstay of the argument against the magnetostratigraphic correlations of Butler and others (1977) and Lindsay and others (1978, 1981), we still do not accept the magnetostratigraphic correlations, because there evidently is a normal overprint carried by hematite in the samples that were used to construct the magnetic-polarity zonation for the Upper Cretaceous of the San Juan Basin.

Lindsay and others (1982) revealed this overprint in samples from the Naashoibito Member of the Kirtland Shale. They did so in a test of these samples aimed at determining whether significant amounts of hematite exist in the sediment. The following sentences from Lindsay and others (1982, p. 452) are very revealing when juxtaposed by removing intervening text:

> If normal polarity sites exhibited higher σ values [calculated from IRM acquisition tests], it would be possible that the normal polarity zones are the result of overprints carried by hematite. . . .However, he [Butler] found a mean σ value from the Ojo Alamo Sandstone [Naashoibito of this paper], where [normal polarity] magnetozone γ + occurs, is 0.242, significantly higher than the mean σ value for all San Juan Basin sites (0.120). . . .The available data do not favor an overprint origin for magnetozone γ +, although we are continuing to investigate additional techniques, including thermal demagnetization, for critical evaluation of this magnetozone.

We, however, do believe that the data Lindsay and others (1982) presented indicate a normal overprint of their γ + magnetozone and thus render the magnetic-polarity zonation for the Naashoibito Member reported by Butler and others (1977) and Lindsay and others (1978, 1981) incorrect. Additional paleomagnetic studies will be necessary to obtain an accurate polarity zonation for the Upper Cretaceous of the San Juan Basin that can then be used for correlation.

CONCLUSION: DINOSAURS AND THE CRETACEOUS-TERTIARY BOUNDARY IN THE SAN JUAN BASIN

The stratigraphically highest dinosaur bones in the San Juan Basin are in the Ojo Alamo Sandstone (sensu Baltz and others,

1966), but all of these bones probably are reworked (Lehman, 1981, 1985; Lucas, 1981). Thus, the highest in situ dinosaur bones in the San Juan Basin are those found in the uppermost Kirtland Shale (Naashoibito Member in part of the west-central basin). Since most, or all, of the Ojo Alamo Sandstone is Tertiary (Baltz and others, 1966), by a traditional definition the Cretaceous-Tertiary boundary lies at or very near the base of the Ojo Alamo in the west-central San Juan Basin. The Lancian dinosaur fauna from the Naashoibito Member of the Kirtland Shale indicates that in the west-central basin the unconformity at the base of the Ojo Alamo Sandstone, thought by many workers to represent a major hiatus (most or all of Maastrichtian time), no longer appears to have been of such great temporal magnitude. This unconformity, which has long plagued the placement of the Cretaceous-Tertiary boundary in the San Juan Basin, definitely exists; its nature has been documented by detailed physical stratigraphy. However, the evaluation of the temporal significance of this unconformity is a geochronological problem, and the biostratigraphic evidence provided by dinosaurs indicates that the interval of time it represents has often been greatly overestimated.

ACKNOWLEDGMENTS

First and foremost, we thank Brad Petersen and David Thomas for their collecting efforts in the Naashoibito Member of the Kirtland Shale, which have brought many new and important specimens to light. We also thank T. Lehman for access to his unpublished data on San Juan Basin and Big Bend dinosaurs. Over the years, we have studied San Juan Basin dinosaurs in numerous collections, both in the United States and Sweden, and we are grateful to all who have aided in these studies. Comments on an earlier version of this paper by P. Dodson and D. A. Russell improved its content and clarity.

APPENDIX 1—LOCALITY LIST

The following list includes the location and stratigraphic position of all dinosaur localities in the Fruitland and Kirtland Formations mentioned in the text of this paper. It also includes the same information, to the degree of detail available, for the type localities of *Alamosaurus sanjuanensis, Kritosaurus navajovius, Parasaurolophus tubicen, P. cyrtocristatus, Pentaceratops sternbergii,* and *P. fenestratus.*

LOCALITY	LOCATION	STRATIGRAPHIC UNIT
UNM 77-71	SW¼, SE¼, Sec. 1, T.23N., R.13W.	Fruitland Formation
UNM 77-90	SE¼, SE¼, Sec. 6, T.23N., R.13W.	Fruitland Formation
UNM 77-92	NW¼, SW¼, Sec. 32, T.24N., R.13W.	Fruitland Formation
UNM 77-127	SE¼, SE¼, Sec. 33, T.24N., R.13W.	Fruitland Formation
UNM 77-187	NE¼, NE¼, Sec. 34, T.23N., R.13W.	Fruitland Formation
UNM 77-404	NE¼, NW¼, Sec. 13, T.24N., R.13W	Kirtland (lower shale)
UNM 77-421	NE¼, SW¼, Sec. 23, T.24N., R.13W.	Kirtland (lower shale)
UNM 77-460	SW¼, NE¼, Sec. 32, T.24N., R.13W.	Fruitland Formation
UNM 77-464	SW¼, NW¼, Sec. 33, T.24N., R.13W.	Fruitland Formation
UNM 77-505	NE¼, NW¼, Sec. 3, T.24N., R.12W.	Kirtland (Naashoibito)
UNM 77-591	NE¼, SW¼, Sec. 27, T.24N., R.12W.	Kirtland (Farmington SS)
UNM 77-675	NE¼, NW¼, Sec. 18, T.24N., R.11W.	Kirtland (Naashoibito)
UNM 77-770	NW¼, NE¼, Sec. 1, T.22N., R.11W.	Kirtland ("middle part")
UNM 77-779	NW¼, SE¼, Sec. 7, T.22N., R.10W.	Kirtland ("middle part")
UNM 77-784	SW¼, SW¼, Sec. 8, T.22N., R.10W.	Kirtland ("lower part")
UNM V-141	NW¼, NW¼, Sec. 3, T.24N., R.12W.	Kirtland (Naashoibito)
UNM V-142	NW¼, SE¼, Sec. 2, T.24N., R.12W.	Kirtland (Naashoibito)
UNM V-143	NW¼, SW¼, Sec. 17, T.24N., R.11W.	Kirtland (Naashoibito)
UNM V-144	SE¼, SW¼, Sec. 2, T.24N., R.12W.	Kirtland (Naashoibito)

TYPE LOCALITY OF:	LOCATION	STRATIGRAPHIC UNIT
A. sanjuanensis[1,2]	NW¼, SW¼, Sec. 17, T.24N., R.11W.	Kirtland (Naashoibito)
K. navajovius[3]	T.11 or T.12N., R.11 or R.12W.	Kirtland (upper shale)
Parasaurolophus tubicen[4]	Secs. 17 or 20?, T.24N., R.11W.	Kirtland (Naashoibito)
P. cyrtocristatus[5]	T.22N., R.12W.	Fruitland Formation
Pentaceratops sternbergii[6]	T.23N., R.12W.	Fruitland Formation
P fenestratus[7]	T.22N., R9 or R.10W.	Kirtland[8]

NOTES: [1] Locality from Kues and others (1980); [2] The same locality as UNM V-143; [3] Locality from Lehman (1981); [4] Locality from Mateer (1981); [5] Locality from Ostrom (1961); [6] Locality from Osborn (1923); [7] Locality from Mateer (1981); [8] Below Naashoibito Member.

REFERENCES CITED

Alvarez, W., and Vann, D. W., 1979, Comment *on* 'Biostratigraphy and magnetic polarity stratigraphy of Paleocene terrestrial deposits, San Juan Basin, New Mexico': Geology, v. 7, p. 66–67.

Armstrong-Ziegler, J. G., 1978, An aniliid snake and associated vertebrates from the Campanian of New Mexico: Journal of Paleontology, v. 52, p. 480–483.

——, 1980, Amphibia and Reptilia from the Campanian of New Mexico: Fieldiana Geology, new series, v. 4, p. 1–39.

Baird, D., and Horner, J., 1977, A fresh look at the dinosaurs of New Jersey and Delaware [abs.]: New Jersey Academy of Science Bulletin, v. 22, p. 50.

Baltz, E. H., Ash, S. R., and Anderson, R. Y., 1966, History of nomenclature and stratigraphy of rocks adjacent to the Cretaceous-Tertiary boundary western San Juan Basin, New Mexico: U.S. Geological Survey Professional Paper 524-D, p. 1–23.

Barsbold, R., 1974, Saurornithoididae, a new family of small theropod dinosaurs from central Asia and North America: Palaeontologia Polonica, v. 30, p. 5–22.

Brookins, D. G., and Rigby, J. K., Jr., 1983, Geochronlogic and geochemical study of volcanic ashes from the Kirtland Formation (Cretaceous), San Juan Basin, New Mexico: Geological Society of America Abstracts with Programs, v. 15, p. 307.

Brown, B., 1910, The Cretaceous Ojo Alamo beds of New Mexico, with description of the new dinosaur genus *Kritosaurus:* Bulletin of the American Museum of Natural History, v. 28, p. 267–274.

Butler, R. F., Lindsay, E. H., Jacobs, L. L., and Johnson, N. M., 1977, Magnetostratigraphy of the Cretaceous-Tertiary boundary in the San Juan Basin, New Mexico: Nature, v. 267, p. 318–323.

Carpenter, K., 1982, Baby dinosaurs from the Late Cretaceous Lance and Hell Creek formations and a description of a new species of theropod: Contributions to Geology, University of Wyoming, v. 20, p. 123–134.

Clemens, W. A., 1973, The roles of fossils vertebrates in interpretation of Late Cretaceous stratigraphy of the San Juan Basin, New Mexico: Four Corners Geological Society Memoir, p. 154–167.

Clemens, W. A., Lillegraven, J. A., Lindsay, E. H., and Simpson, G. G., 1979, Where, when and what; A survey of known Mesozoic mammal distribution, *in* Lillegraven, J. A., Kielan-Jaworowska, Z., and Clemens, W. A., eds., Mesozoic mammals: The first two-thirds of mammalian history: Berkeley, University of California Press, p. 7–58.

Cobban, W. A., 1973, Significant ammonite finds in uppermost Mancos Shale and overlying formations between Barker Dome, New Mexico, and Grand Junction, Colorado: Four Corners Geological Society Memoir, p. 148–153.

Cobban, W. A., Landis, E. R., and Dane, C. H., 1974, Age relations of upper part of Lewis Shale on east side of San Juan Basin, New Mexico: New Mexico

Geological Society Guidebook 25, p. 279–282.

Colbert, E. H., 1948, Evolution of the horned dinosaurs: Evolution, v. 2, p. 145–163.

Colbert, E. H., and Bump, J. D., 1947, A skull of *Torosaurus* from South Dakota and a revision of the genus: Proceedings of the Academy of Natural Sciences of Philadelphia, v. 99, p. 93–106.

Coombs, W. P., Jr., 1971, The Ankylosauria [Ph.D. thesis]: Columbia University, 465 p.

—— , 1978, The families of the ornithischian dinosaur order Ankylosauria: Palaeontology, v. 21, p. 143–170.

Cope, E. D., 1885, The relations of the Puerco and Laramie deposits: American Naturalist, v. 19, p. 985–986.

Dana, J. D., 1863, Manual of geology; Treating of the principles of the science with special reference to American geological history, etc.: New York, American Press.

Dodson, P., 1983, A faunal review of the Judith River (Oldman) Formation, Dinosaur Provincial Park, Alberta: The Mosasaur, v. 1, p. 89–118.

Fassett, J. E., 1979, Comment on 'Biostratigraphy and magnetostratigraphy of Paleocene terrestrial deposits, San Juan Basin, New Mexico: Geology, v. 7, p. 69–70.

Fassett, J. E., and Hinds, J. S., 1971, Geology and fuel resources of the Fruitland Formation and Kirtland Shale of the San Juan Basin, New Mexico and Colorado: U.S. Geological Survey Professional Paper 676, p. 1–76.

Gilmore, C. W., 1916, Vertebrate faunas of the Ojo Alamo, Kirtland, and Fruitland formations: U.S. Geological Survey Professional Paper 98-Q, p. 279–308.

—— , 1919, Reptilian faunas of the Torrejon, Puerco, and underlying Upper Cretaceous formations of San Juan County, New Mexico: U.S. Geological Survey Professional Paper 119, p. 1–71.

—— , 1921, Discovery of sauropod dinosaur remains in the Upper Cretaceous of New Mexico: Science, v. 54, p. 274.

—— , 1922, A new sauropod dinosaur from the Ojo Alamo Formation of New Mexico: Smithsonian Miscellaneous Collections, v. 72, no. 14, p. 1–9.

—— , 1930, On dinosaurian reptiles from the Two Medicine Formation of Montana: Proceedings of the U.S. National Museum, v. 77, article 77, p. 1–39.

—— , 1933, Two new dinosaurian reptiles from Mongolia with notes on fragmentary specimens: American Museum Novitates, no. 679, p. 1–20.

—— , 1935, On the Reptilia of the Kirtland Formation of New Mexico, with descriptions of new species of fossil turtles: Proceedings of the U.S. National Museum, v. 83, p. 159–188.

—— , 1946, Reptilian fauna of the North Horn Formation of central Utah: U.S. Geological Survey Professional Paper 210-C, p. 29–53.

Grossouvre, A. de., 1897, Sur la limite du Crétacé et du Tertiare: Bulletin Societe Geologique de France, third series, v. 25, p. 57–81.

Knowlton, F. H., 1916, Flora of the Fruitland and Kirtland formations: U.S. Geological Survey Professional Paper 98, p. 327–333.

Kues, B. S., and others, 1977, Paleontological survey, resource assessment, and mitigation plan for the Bisti–Star Lake area, northwestern New Mexico: Unpublished report to U.S. Bureau of Land Management, Albuquerque Office, p. 1–1525.

Kues, B. S., Lehman, T., and Rigby, J. K., Jr., 1980, The teeth of *Alamosaurus sanjuanensis,* a Late Cretaceous sauropod: Journal of Paleontology, v. 54, p. 864–869.

Lambe, L. M., 1915, On *Eoceratops canadensis,* gen. nov., with remarks on other genera of Cretaceous horned dinosaurs: Canadian Geological Survey Museum Bulletin, no. 12, p. 1–49.

—— , 1917, The Cretaceous theropodous dinosaur *Gorgosaurus*: Canadian Geological Survey Memoir, no. 100, p. 1–84.

Lawson, D. A., 1972, Paleoecology of the Tornillo Formation, Big Bend National Park, Brewster, Co., Texas [M.S. thesis]: University of Texas at Austin, 182 p.

—— , 1976, *Tyrannosaurus* and *Torosaurus,* Maestrichtian dinosaurs from Trans-Pecos, Texas: Journal of Paleontology, v. 50, p. 158–164.

Lehman, T. M., 1981, The Alamo Wash local fauna: A new look at the old Ojo Alamo fauna, *in* Lucas, S. G., Rigby, J. K., Jr., and Kues, B. S., eds., Advances in San Juan Basin paleontology: Albuquerque, University of New Mexico Press, p. 189–221.

—— , 1982, A ceratopsian bonebed from the Aguja Formation (Upper Cretaceous), Big Bend National Park, Texas [M.S. thesis]: University of Texas at Austin, 210 p.

—— , 1985, Depositional environments of the Naashoibito Member of the Kirtland Shale, Upper Cretaceous, San Juan Basin, New Mexico: New Mexico Bureau of Mines and Mineral Resources Circular 195, p. 55–79.

Lindsay, E. H., Jacobs, L. L., and Butler, R. F., 1978, Biostratigraphy and magnetostratigraphy of Paleocene terrestrial deposits, San Juan Basin, New Mexico: Geology, v. 6, p. 425–429.

Lindsay, E. H., Butler, R. F., and Johnson, N. M., 1981, Magnetic polarity zonation and biostratigraphy of Late Cretaceous and Paleocene continental deposits, San Juan Basin, New Mexico: American Journal of Science, v. 281, p. 390–435.

—— , 1982, Testing of magnetostratigraphy in Late Cretaceous and early Tertiary deposits, San Juan Basin, New Mexico: Geological Society of America Special Paper 190, p. 449–453.

Lohrengel, C. F., II, 1969, Palynology of the Kaiparowits Formation, Garfield County, Utah: Brigham Young University Geology Studies, v. 16, no. 3, p. 61–186.

Lucas, S. G., 1981, Dinosaur communities of the San Juan Basin: A case for lateral variations in the composition of Late Cretaceous dinosaur communities, *in* Lucas, S. G., Rigby, J. K., Jr., and Kues, B. S., eds., Advances in San Juan Basin paleontology: Albuquerque, University of New Mexico Press, p. 337–393.

Lucas, S. G., and Mateer, N. J., 1983, Dinosaurs and the Cretaceous-Tertiary boundary in the San Juan Basin, New Mexico: Geological Society of America Abstracts with Programs, v. 15, p. 306.

Lucas, S. G., and Rigby, J. K., Jr., 1979, Comment *on* 'Biostratigraphy and magnetostratigraphy of Paleocene terrestrial deposits, San Juan Basin, New Mexico': Geology, v. 7, p. 323–325.

Lucas, S. G., and Schoch, R. M., 1982, Discussion magnetic polarity zonation and biostratigraphy of Late Cretaceous and Paleocene continental deposits, San Juan Basin, New Mexico: American Journal of Science, v. 282, p. 920–927.

Lull, R. S., 1933, A revision of the Ceratopsia: Memoirs of the Peabody Museum of Natural History, v. 3, p. 1–175.

Manfrino, C., 1984, Stratigraphy and palynology of the upper Lewis Shale, Pictured Cliffs Sandstone, and lower Fruitland Formation (Upper Cretaceous) near Durango, Colorado: The Mountain Geologist, v. 21, p. 115–132.

Mateer, N. J., 1976, New topotypes of *Alamosaurus sanjuanensis* Gilmore (Reptilia: Sauropoda): Bulletin of the Geological Institutions of the University of Uppsala, new series, v. 6, p. 93–95.

—— , 1981, The reptilian megafauna from the Kirtland Shale (Late Cretaceous) of the San Juan Basin, New Mexico, *in* Lucas, S. G., Rigby, J. K., Jr., and Kues, B. S., eds., Advances in San Juan Basin paleontology: Albuquerque, University of New Mexico Press, p. 49–75.

Morris, W. J., 1972, A giant hadrosaurian dinosaur from Baja California: Journal of Paleontology, v. 46, p. 777–779.

—— , 1981, A new species of hadrosaurian dinosaur from the Upper Cretaceous of Baja California—?*Lambeosaurus laticaudus:* Journal of Paleontology, v. 55, p. 453–462.

Newman, K. R., 1982, Biostratigraphy of Fruitland, Kirtland, and Animas formations (Cretaceous and Paleocene), northern San Juan Basin, Colorado: Geological Society of America Abstracts with Programs, v. 14, p. 344.

—— , 1983, Biostratigraphic correlation of K-T rocks from northwestern Colorado to San Juan Basin, Colorado and New Mexico: Geological Society of America Abstracts with Programs, v. 15, p. 307.

Obradovich, J. D., and Cobban, W. A., 1975, A time-scale for the Late Cretaceous of the western interior of North America: Geological Association of Canada Special Paper 13, p. 31–54.

O'Neill, F. M., 1983, New finds from the upper Kirtland Shale, San Juan Basin,

New Mexico: Geological Society of America Abstracts with Programs, v. 15, p. 306.
Osborn, H. F., 1912, Crania of *Tyrannosaurus* and *Allosaurus:* American Museum of Natural History Memoir, new series, v. 1, p. 1–30.
—— , 1916, Skeletal adaptations of *Ornitholestes, Struthiomimus, Tyrannosaurus:* Bulletin of the American Museum of Natural History, v. 35, p. 733–771.
—— , 1923, A new genus and species of Ceratopsia from New Mexico, *Pentaceratops sternbergii*: American Museum Novitates, no. 93, p. 1–3.
—— , 1924, Three new Therapoda, *Protoceratops* zone, central Mongolia: American Museum Novitates, no. 144, p. 1–12.
Ostrom, J. H., 1961, A new species of hadrosaurian dinosaur from the Cretaceous of New Mexico: Journal of Paleontology, v. 35, p. 575–577.
—— , 1963, *Parasaurolophus cyrtocristatus,* a crested hadrosaurian dinosaur from New Mexico: Fieldiana Geology, v. 14, p. 143–168.
—— , 1969, Osteology of *Deinonychus antirrhopus,* an unusual theropod from the Lower Cretaceous of Montana: Peabody Museum of Natural History Bulletin, no. 30, p. 1–165.
—— , 1976, *Archaeopteryx* and the origin of birds: Biological Journal of the Linnaean Society of London, v. 8, p. 91–182.
—— , 1981, *Procompsognathus*—theropod or thecodont?: Palaeontographica Abteilung A, v. 175, p. 179–185.
Parks, W. A., 1933, New species of dinosaurs and turtles from the Upper Cretaceous formations of Alberta: University of Toronto Studies, Geological Series, v. 34, p. 3–33.
Powell, J. S., 1973, Paleontology and sedimentation models of the Kimbeto Member of the Ojo Alamo Sandstone: Four Corners Geological Society Memoir, p. 111–122.
Rice, D. D., and Shurr, G. W., 1983, Patterns of sedimentation and paleogeography across the western interior seaway during time of deposition of Upper Cretaceous Eagle Sandstone and equivalent rocks, northern Great Plains, *in* Reynolds, M. W., and Dolly, E. D., eds., Mesozoic paleogeography of west-central United States: Denver, Rocky Mountain Society of Economic Paleontologists and Mineralogists, p. 337–358.
Rigby, J. K., Jr., and Wolberg, D. L., 1983, Late Cretaceous mammals from the fossil forest area, San Juan Basin, New Mexico: Geological Society of America Abstracts with Programs, v. 15, p. 307.
Rowe, T., Colbert, E. H., and Nations, J. D., 1981, The occurrence of *Pentaceratops* (Ornithischia: Ceratopsia) with a description of its frill, *in* Lucas, S. G., Rigby, J. K., Jr., and Kues, B. S., eds., Advances in San Juan Basin paleontology: Albuquerque, University of New Mexico Press, p. 29–48.
Russell, D. A., 1970, Tyrannosaurs from the Late Cretaceous of western Canada: National Museum of Natural Sciences, Publications in Palaeontology, no. 1, p. 1–34.
—— , 1972, Ostrich dinosaurs from the Late Cretaceous of western Canada: Canadian Journal of Earth Sciences, v. 9, p. 375–402.
Russell, L. S., 1930, Upper Cretaceous dinosaur faunas of North America: Proceedings of the American Philosophical Society, v. 69, p. 133–159.
—— , 1964, Cretaceous non-marine faunas of northwestern North America: Royal Ontario Museum Life Science Contributions, no. 61, p. 1–24.
—— , 1975, Mammalian faunal succession in the Cretaceous System of western North America: Geological Association of Canada Special Paper 13, p. 139–161.
Sahni, A., 1972, The vertebrate fauna of the Judith River Formation, Montana: Bulletin of the American Museum of Natural History, v. 147, p. 321–412.
Sinclair, W. J., and Granger, W., 1914, Paleocene deposits of the San Juan Basin, New Mexico: Bulletin of the American Museum of Natural History, v. 33, p. 297–316.
Sloan, R. E., 1970, Cretaceous and Paleocene terrestrial communities of western North America: Proceedings of the North American Paleontological Convention I, p. 427–453.
Stanton, T. W., 1916, Nonmarine Cretaceous invertebrates of the San Juan Basin: U.S. Geological Survey Professional Paper 98, p. 309–326.
Tidwell, W. D., Ash, S. R., and Parker, L. R., 1981, Cretaceous and Tertiary floras of the San Juan Basin, *in* Lucas, S. G., Rigby, J. K., Jr., and Kues, B. S., eds., Advances in San Juan Basin paleontology: Albuquerque, University of New Mexico Press, p. 307–332.
Tschudy, R. H., 1973, The Gasbuggy core; A palynological appraisal: Four Corners Geological Society Memoir, p. 131–143.
Weishampel, D. B., and Jensen, J. A., 1979, *Parasaurolophus* (Reptilia: Hadrosauridae) from Utah: Journal of Paleontology, v. 53, p. 1422–1427.
Weishampel, D. B., and Weishampel, J. B., 1983, Annotated localities of ornithopod dinosaurs: Implications to Mesozoic paleobiogeography: The Mosasaur, v. 1, p. 43–87.
Wiman, C., 1930, Über Ceratopsia aus der Oberen Kreide in New Mexico: Nova Acta Regiae Societatis Scientiarum Upsaliensis, fourth series, v. 7, no. 2, p. 1–19.
—— , 1931, *Parasaurolophus tubicen* n. sp. aus der Kreide in New Mexico: Nova Acta Regiae Societatis Scientiarum Upsaliensis, fourth series, v. 7, no. 5, p. 1–11.
Zavada, M. S., 1976, Palynology of the Upper Cretaceous Fruitland Formation, San Juan Basin, New Mexico [M.S. thesis]: Arizona State University, 157 p.

Manuscript Accepted by the Society March 24, 1986

Printed in U.S.A.

Geological Society of America
Special Paper 209
1987

The Therian mammalian fauna (Campanian) of Quarry 1, Fossil Forest study area, San Juan Basin, New Mexico

J. Keith Rigby, Jr.
Department of Earth Sciences
University of Notre Dame
Notre Dame, Indiana 46556-1020

Donald L. Wolberg
New Mexico Bureau of Mines and Mineral Resources
Socorro, New Mexico 87801

ABSTRACT

Cretaceous mammals have been recovered from the lowermost Kirtland Shale (Quarry 1 of the Fossil Forest study area). Of the 13 therian mammal taxa described, 6 are new (*Alphadon parapraesagus,* n. sp., *Ectocentrocristus foxi* n. gen. and sp., *Pediomys fassetti* n. sp., *Aquiladelphis paraminor,* n. sp., *Gypsonictops clemensi* n. sp., and *Cimolestes lucasi* n. sp.). Faunal comparison with other known Cretaceous sites show that Fossil Forest mammals are most closely related to faunas described by Sahni (1972) from the Judith River Formation and by Fox (1977, 1979a, 1979b, 1979c, and 1981) from the Oldman Formation (Judith River equivalent). The age of the lowermost part of the Kirtland Shale must be near the Campanian-Maastrichtian boundary based on mammalian evidence. The long normal magnetochrons (30 and 31 of Lindsay and others [1981]) must, therefore, be renumbered. Two alternatives for renumbering are discussed; the fossil mammal evidence favors renumbering the magnetochrons as anomaly 33. This interpretation requires that one or more unconformities be present between the lower part of the Kirtland Shale and overlying formations on the south flank of the San Juan Basin. This results in the loss of approximately 6.4 Ma of rock record that probably includes the Cretaceous-Tertiary boundary event.

INTRODUCTION

The Fossil Forest study area is located in sections 13, 14, 23, and 24, T.23N., R.12W., in southeast-central San Juan County, New Mexico (Fig. 1). The area is within the Navajo Section of the Colorado Plateau (Hawley and Love, 1981) and is managed by the U.S. Bureau of Land Management (BLM).

Fossil mammals collected from Quarry 1 of the Fossil Forest study area are a particularly valuable biostratigraphic tool in determining the age of the Kirtland and Fruitland formations of the San Juan Basin. The fauna from Quarry 1 is diverse and represents the largest collection of Cretaceous fossil mammals obtained to date from New Mexico. These fossils provide much information important to the long-standing problem of the age of Cretaceous continental deposits of the San Juan Basin.

The term *Fossil Forest* has been applied to the study area because of the presence of a large concentration of fossilized stumps in growth position. C. Robison, A. Hunt, and D. Wolberg have mapped more than 40 stumps in situ and have determined the orientation of 11 logs in a single stump field. Additionally, Hunt (1983) has determined that 23 other stumps and 1 log occur at the same stratigraphic position just to the south. BLM files document the occurrence of several hundred stumps within the 3.9-sq km (1.5-sq mi) study area. The stumps are generally less than 2 m (6.4 ft) in diameter and less than 1 m (3.2 ft) in height, and most logs are less than 10 m (32 ft) in length. C. Robison has suggested that the forest was largely composed of taxodiaceous trees and that the flora was one of low diversity. Hunt (1983) provides a detailed critical evaluation of possible scenarios leading to the destruction of the forest; it is probable that a single

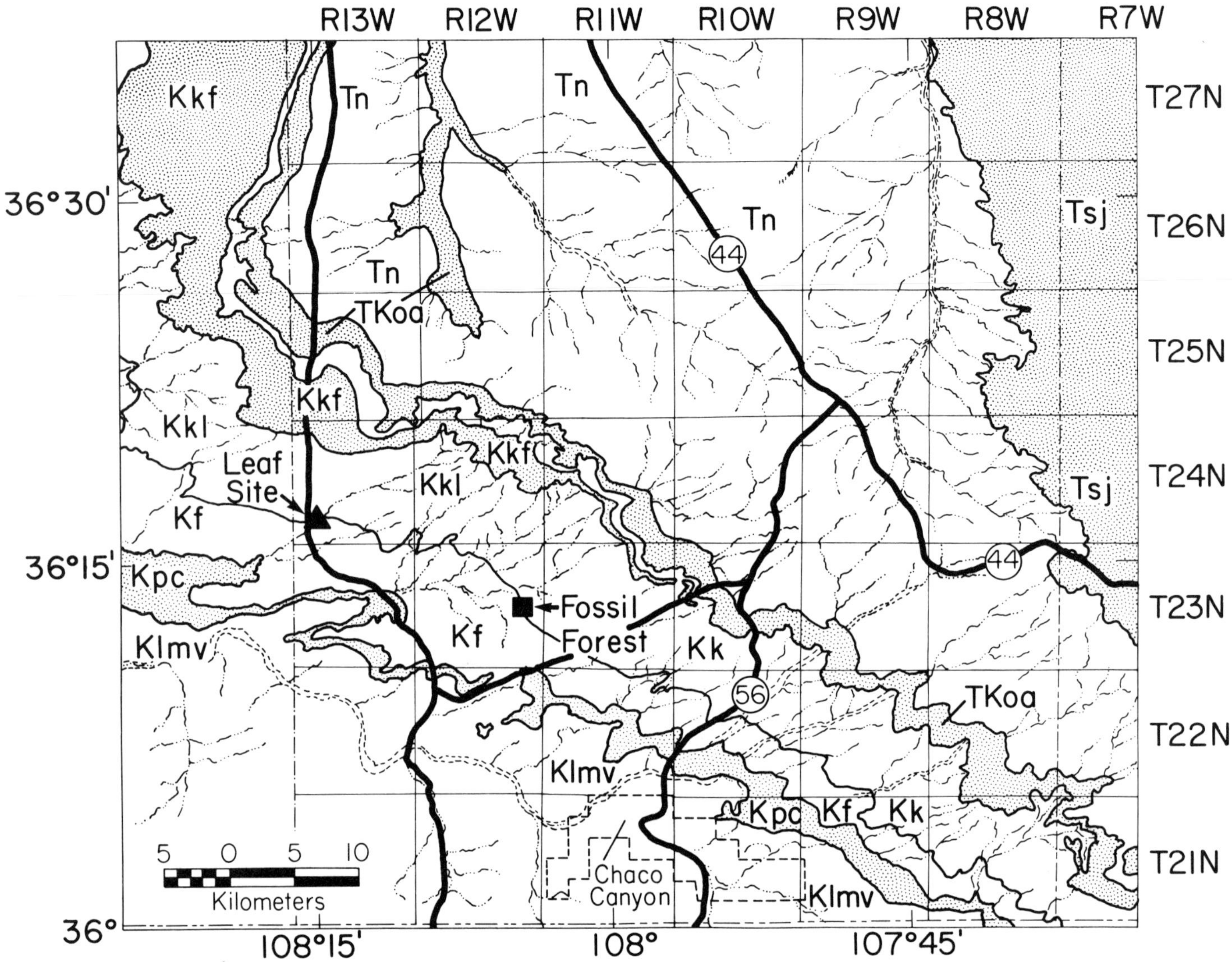

Figure 1. Index map showing the location of the Fossil Forest study area and its proximity to exposures of the Fruitland and Kirtland formations. The fossil leaf locality of Robison and others (1982), located in the Bisti area, is also shown. Abbreviations include: Tsj, San Jose Formation; Tn, Nacimiento Formation; TKoa, Ojo Alamo Sandstone; Kkf, Farmington Sandstone Member of Kirtland Shale; Kkl, Lower Shale Member of Kirtland Shale; Kk, undifferentiated Kirtland Shale; Kf, Fruitland Formation; Kpc, Pictured Cliffs Sandstone; Klmv, undifferentiated Lewis Shale and Mesaverde Group.

event destroyed most of the trees because they are rooted in a single stratigraphic horizon.

A number of mammal-producing localities have been discovered in the Fossil Forest study area. This report deals with part of the mammalian fauna recovered to date from our Quarry 1. As determined by Hunt, Quarry 1 lies 0.7 m (2.3 ft) above the main stump field horizon and occurs at the base of a 1.7 m (6.23 ft) thick channel sandstone. The microvertebrate-rich horizon is a clay-pebble-rich, poorly sorted, medium-grained sandstone. Exact locality data are on file at the Albuquerque District Office, Bureau of Land Management, Albuquerque, New Mexico, and at the New Mexico Bureau of Mines and Mineral Resources, Socorro, New Mexico. In addition to isolated mammal teeth, Quarry 1 has yielded mammalian mandibular fragments with and without teeth, the first mammalian postcranial material reported from the Cretaceous of New Mexico, gar scales, fish vertebrae and teeth, selachian teeth, amphibian dental and postcranial material, crocodile teeth, lizard teeth and postcranial material, dinosaur teeth and bone fragments, abundant gastropod operculae but a small amount of mollusk shell material, and some carbonized leaf and wood material. Gastropod operculae are dominantly calcitic, whereas shell material is dominantly aragonitic, and the scarcity of shell material is viewed as a diagenetic effect.

In 1979, a joint study was initiated involving the U.S. Bureau of Land Management, Albuquerque District, and the New Mexico Bureau of Mines and Mineral Resources to evaluate the geology and paleontology of the area known as the Fossil Forest, near Split Lip Flats, south of Farmington, New Mexico. The Fossil Forest study area overlies major coal resources but is of paleontologic and stratigraphic significance as well. The paleontologic potential of the area had been noted in a BLM survey conducted in 1977 (Kues and others).

Early Collections from the Fossil Forest

It became apparent to us during our initial studies that the Fossil Forest area had been the scene of intensive and active collecting in the past. Several major quarrying sites have been discovered, and judging from the extent of excavations and remains of burlap and plaster at these sites and the magnitude of dinosaur material still remaining in the ground, significant collections must have been made in the area. In one instance, parts of a horse-drawn wagon and an apparent camping site were discovered. However, no clearly identified collections from the Fossil Forest have been documented in the holdings of any institution. Hunt (1983) reviewed all known data to determine who might possibly have amassed collections from the area and persuasively argued that C. H. Sternberg may well have colllected in the Fossil Forest in the early 1900s.

In 1984, Wolberg and Hunt discovered an additional quarry site that yielded a corroded Marsh-pick head and remnants of soldered cans, in addition to decaying burlap and fragmentary dinosaur bone. An additional camp site was discovered that included soldered cans and a broken, hand-beaten awl point. These artifacts would seem to indicate a date of no later than the immediately post–World War I period and support the identification of Sternberg as collector; he is the only collector known to us to have been active on a large scale in the San Juan Basin at that time.

It is also possible, if not probable, that material removed by later collectors and currently housed in institutions is either unrecognized as coming from the Fossil Forest, or still jacketed, or incorporated into collections with vague locality data.

Deposition and Composition

Fruitland and older Upper Cretaceous formations in the Fossil Forest reflect coastal lowland deposition that fluctuated with several marine transgressions and regressions. During the Late Cretaceous, sediments in the Western Interior of North America were deposited in or near an epeiric sea extending from arctic latitudes to the Gulf of Mexico. Approximately 1660 km (1000 mi) in width and 5000 km (3000 mi) in length, this seaway divided North America into two large landmasses. The San Juan Basin of northwestern New Mexico and southwestern Colorado was situated on the western margin of the seaway and was repeatedly affected by transgressions and regressions of waters of the sea (Fassett and Hinds, 1971).

The final regression of the seaway led to the deposition of the three rock units of most concern to this paper: the Pictured Cliffs Sandstone, the Fruitland Formation, and the Kirtland Shale. The Pictured Cliffs Sandstone consists of delta-front and barrier beach deposits (Erpenbeck, 1979; Cumella, 1981). Organic-rich sediments, including coal deposits of variable lateral extent and thickness, accumulated in delta-plain and back-barrier environments (Erpenbeck, 1979). A paucity of organic-rich sediments characterizes the upper part of the Fruitland Formation and overlying Kirtland Shale, suggesting an increased distance from the Pictured Cliff shoreline as the sea continued to withdraw from the region. Hartman (1981), reporting on a preliminary analysis of Fruitland and Kirtland brackish and freshwater molluscan faunas, suggests a progressive upward trend from brackish to freshwater environments.

Max Bauer (1917) named the Fruitland Formation and Kirtland Shale for exposures along the San Juan River in northwest San Juan County, New Mexico. The Fruitland Formation is mainly composed of interbedded coals, shales, and sandstones. The Kirtland has been divided into three generally recognized members: the Lower Shale Member, the Farmington Sandstone Member, and the Upper Shale Member (Bauer, 1917; Reeside, 1924). The contact between the Fruitland Formation and the Lower Shale Member of the Kirtland Shale is gradational, lacking major unconformities or pronounced changes in lithology. Choice of a boundary between the two formations is definitely arbitrary within the study area. Several possible criteria have been suggested to define a boundary (Fassett and Hinds, 1971).

Stratigraphy

The stratigraphy of the Fossil Forest study area has been investigated in detail by Hunt (1983), who has determined that the only detailed studies that previously considered the area (Bauer and Reeside, 1921; Reeside, 1924) indicate that the Fruitland-Kirtland boundary passes through the Fossil Forest study area. However, for a number of reasons, developed by Hunt, it is apparent that Bauer and Reeside did not actually enter the main part of the study area. Recent stratigraphic and mapping studies by Hunt and members of the U.S. Geological Survey place the study area within the Lower Shale Member of the Kirtland Shale (Hunt, 1983), but we do not consider the matter totally resolved.

Hunt (1983) has convincingly correlated his Fossil Forest stratigraphic section, which includes Quarry 1, to a section in the vicinity of Hunter Wash (Figs. 1 and 2). Robison and others (1982) described a compression flora from the basal part of the Kirtland Shale (leaf locality of Fig. 1). This stratigraphic assessment agrees only in part with criteria for recognition of the Kirtland Shale as developed by O'Sullivan and others (1972), Scott and others (1979), and Mytton (1979). In the past, some stratigraphers have attempted to separate the Fruitland Formation and Kirtland Shale at the top of the uppermost "major" coal horizon.

Rocks above it belong to the Lower Shale Member of the Kirtland Shale; rocks below belong to the Fruitland Formation. This definition creates some problems for surficial geologic maps and has caused us some concern in identifying the formation from which the Quarry 1 fossils came. The major coals involved are lenticular, somewhat discontinuous, and extremely variable in thickness and in quality. Mytton (1979) and other authors also use the presence of sideritic nodules and the higher proportion of sandstones as characteristics of the Fruitland Formation. Both of these characteristics persist higher up in the section above the highest "major" coal seam. There is no general agreement regarding a recognizable formational boundary between the Fruitland and Kirtland formations in this part of the San Juan Basin. This fact has been recognized by Lindsay and others (1981), who identified a 16-m (50-ft) transition zone between the two formations.

Clemens (1973a) identified a series of fossiliferous horizons that produced his microvertebrate fauna. The lowest stratigraphical occurrence was 13 m (40 ft) below the "thin but widespread coal chosen as the top of the [Fruitland] formation." He continues,

> Our largest collections of small vertebrates were obtained from the sand and siltstones of a channel filling. On one side of this channel the uppermost coal of the Fruitland conformably feathers out into the sands of the channel deposit. On the other side, the wall of the channel is cut through the coal which is thereby separated from the sediments of the channel filling by an erosional unconformity. The highest stratigraphical concentration of microvertebrate fossils collected in the Bisti area was obtained from a lenticular silty sandstone approximately 55 ft (17.5 m) above the base of the Lower Shale Member of the Kirtland Shale (Clemens, 1973a, p. 162).

Clemens was obviously not using a "major" coal, which would tend to be irregular and somewhat discontinuous, as the formational contact but a "thin, widespread coal." Our field experience on the south flank of the San Juan Basin has shown that these thin (less than 20 cm—7.9 in—thick) coal seams are the most continuous of all lithologies, with perhaps the exception of volcanic air-fall tuffs (Brookins and Rigby, this volume). We believe that this thin coal, described by Clemens, is the same coal that is located just above the channel-filling sandstone which preserved the fossil material of Quarry 1 in its base. We therefore believe, formational boundary difficulties aside, that the Quarry 1 fossil mammals are very nearly time-equivalent to those of the major site of Clemens (1973a) located near Hunter Wash. If Clemens's definition of the Fruitland-Kirtland boundary is used, then the fossils of Quarry 1 of the Fossil Forest came from the uppermost Fruitland Formation. If the definition of other authors is used, so that the boundary is defined at the top of the highest "major coal," then the fossils of this study came from the lower 16–24 m (50–75 ft) of the Kirtland Shale.

Matrix for this study was collected from Quarry 1 from 1980 to 1982 and transported to Albuquerque and Socorro (no source of water is readily available in the area for screen wash-

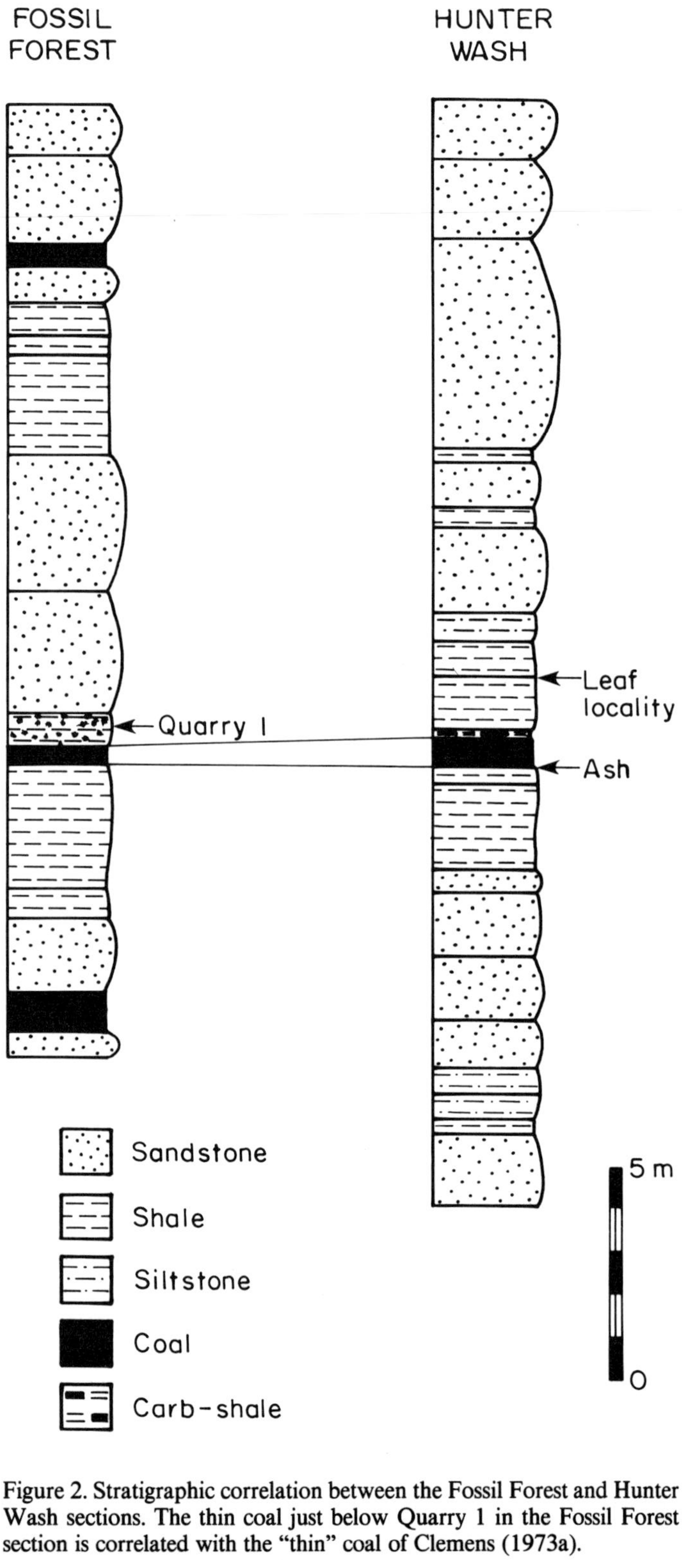

Figure 2. Stratigraphic correlation between the Fossil Forest and Hunter Wash sections. The thin coal just below Quarry 1 in the Fossil Forest section is correlated with the "thin" coal of Clemens (1973a).

TABLE 1. THERIAN MAMMALS FROM QUARRY 1, FOSSIL FOREST, SAN JUAN BASIN, NEW MEXICO

- Class Mammalia
  - Subclass Theria
    - Infraclass Metatheria
      - Superorder Marsupialia
        - Family Didelphidae
          - Subfamily Didelphinae
            - Alphadon halleyi
            - Alphadon parapraesagus, n. sp.
            - Alphadon cf. wilsoni
          - Subfamily Ectocentrocristinae, n. subfam.
            - Ectocentrocristus foxi, n. gen. & sp.
        - Family Pediomyidae
            - Pediomys fassetti, n. sp.
          - ?Pediomyidae incertae sedis
            - Aquiladelphis paraminor, n. sp.
        - Family Stagodontidae
            - cf. Eodelphis
    - Infraclass Eutheria
      - Order Insectivora
        - Superfamily Leptictoidae
            - Gypsonictops clemensi, n. sp.
            - Gypsonictops cf. lewisi
        - Superfamily Palaeoryctoidea
            - Cimolestes lucasi, n. sp.
        - Superfamily Erinaceoidea
            - Paranyctoides cf P. sternbergi
    - Eutheria, incertae sedis
            - Eutherian of uncertain affinities

TABLE 2. ALPHADON HALLEYI MEASUREMENTS

		Length	Anterior Width	Posterior Width
UNM-B1742	M^1	1.75	1.65	1.9
UNM-B1739	$M^{?3}$	----	----	2.35
UNM-B1725	M_1	1.8	1.0	1.0
UNM-B1721	$M_{?3}$	1.8	1.03	0.95
UNM-B1726	M_L	1.8	(0.9)	(1.05)

ing). The matrix disintegrates with water soaking, and we discovered that the addition of a detergent (Alconox) reduced the time required for soaking from several days to four to eight hours. Fossil materials were collected by using techniques described by McKenna (1965). The known therian mammalian fauna from Quarry 1 is listed in Table 1.

SYSTEMATIC PALEONTOLOGY

Symbols and Abbreviations

Specimens indicated by UNM-B catalogue numbers are from U.S. Bureau of Land Management collections housed at the University of New Mexico, Albuquerque, New Mexico. AMNH and UA catalogued specimens are respectively from collections at the American Museum of Natural History, New York, and at the University of Alberta, Edmonton. Unless otherwise noted, all measurements are in millimeters. Superscript and subscript notation indicates upper and lower dental elements in conjunction with standard dental nomenclature. The following are also used in tables: L, length; AW, anterior width (trigonid width); PW, posterior width (talonid width); W, width. Measurements in parentheses indicate approximate measurements where some interpretation was necessary, usually due to breakage.

Subclass *THERIA* Parker and Haswell 1898
Infraclass *METATHERIA* Huxley 1880
Superorder *MARSUPIALIA* (Illiger) 1811
Order *MARSUPICARNIVORA* Ride 1964
Superfamily *DIDELPHOIDEA* (Gray, 1821)
Family *DIDELPHIDAE* Gray 1921
Subfamily *DIDELPHINAE* (Gray, 1921)

***Alphadon halleyi* Sahni, 1972**
(Plate 1, Figs. G–I; Plate 2, Figs. A–C; Plate 3, Figs. A–F; Table 2)

Holotype. AMNH 77367, a lower molar.
Horizon and type locality. Clambank Hollow, Judith River Formation, Chouteau County, Montana.
Known distribution. Judith River Formation, Montana; Oldman Formation, Alberta; lowermost Kirtland Shale.
Referred specimens. UNM-B1742, M^1; UNM-B5508, fragmentary M^1; UNM-B1739, fragmentary $M^{?3}$; UNM-B1725, M_1; UNM-B1721, $M_{?3}$; UNM-B1726, M_L.
Revised diagnosis. Small species of *Alphadon* nearly the same size or slightly larger than *A. lulli*; upper molar cusp B large, cusp C small (much smaller than *A. marshi*), cusp D large but smaller than B and low ridge present in cusp E position; cusp A large, separated from B by a significant notch; cusp B close to paracone; paracone and metacone nearly equal in size (paracone is larger than metacone in *A. lulli*) and separated by a wide valley (not narrow as in *A. lulli*); lower molar hypoconulids more separate than in *A. lulli*; lower molars smaller, talonids proportionately smaller, and paraconid more anteriorly projecting than in *A. wilsoni.*
Description. All referred specimens are small (Table 2) and most closely approximate *A. lulli* and *A. halleyi* in size. M^1 has a well-defined stylar shelf, narrower than in *A. lulli,* with stylar cusp A separated from cusp B by a shallow notch. Cusp A, although worn, was probably a separate distinct cusp approximately the size of or slightly larger than D, although placed lower on the crown. Cusp B is the largest of the stylar cusps and is slightly smaller than the paracone. An anterior stylar crest from cusp B terminates at the notch between cusp B and cusp A. There is no noticeable crest descending the posterior slope of cusp B. A low, anteriorly arcuate crest connects the paracone with cusp B. Cusp C is small and variable in its appearance, based upon the limited sample described here. It can either be a distinct cusp, smallest of the stylar row, or may be present as a small cuspule

PLATE 1.

Alphadon cf. *wilsoni,* UNM-B1743, M^2, Figs. A–C: A, occlusal view, stereo pair; B, anterior view; C, posterior view. UNM-B1740, partial M^3, Figs. D–F: D, oclusal view, stereo pair; E, anterior view; F, posterior view. Scale bar represents 0.5 mm.

Alphadon halleyi, UNM-B1742, M^1, Figs. G–I: G, occlusal view, stereo pair; H, anterior view; I, posterior view. Scale bar represents 0.5 mm.

Pediomys fassetti n. sp., UNM-B1734, M^3, holotype, Figs. J–L: J, occlusal view, stereo pair; K, anterior view; L, posterior view. UNM-B1737, $M^{?2}$, Figs. M–O: M, occlusal view, stereo pair; N, anterior view; O, posterior view. UNM-B1738, M^1, Figs. P–R: P, occlusal view, stereo pair; Q, anterior view; R, posterior view. Scale bar represents 0.5 mm.

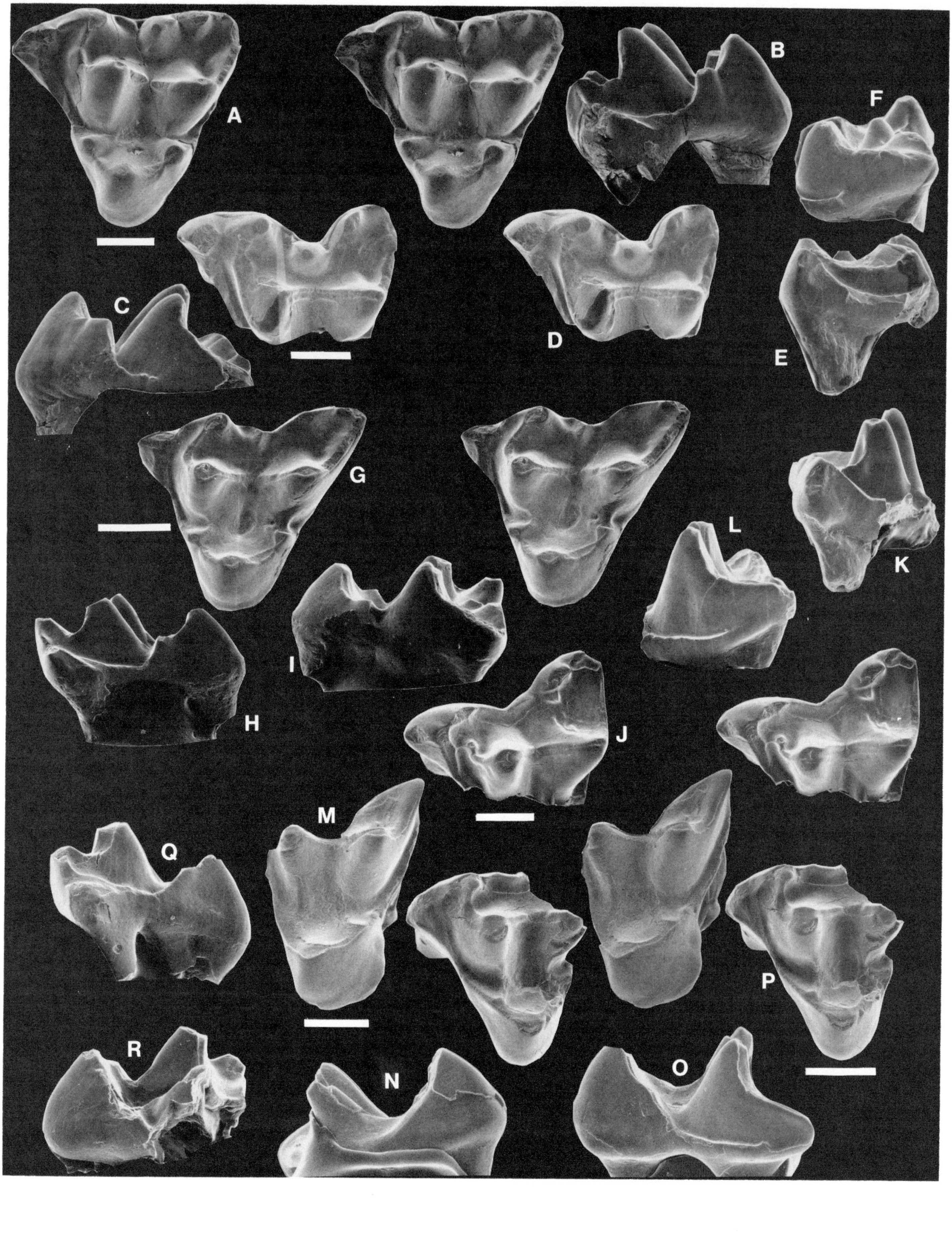
A
B
F
C
D
E
G
L
K
I
H
J
M
Q
P
R
N
O

PLATE 2.

Alphadon halleyi, UNM-B1739, $M^{?3}$, Figs. A–C: A, occlusal view, stereo pair; B, anterior view; C, posterior view. Scale bar represents 0.5 mmm.

Alphadon parapraesagus n. sp., UNM-B1741, M^{U}, Figs. D–F: D, occlusal view, stereo pair; E, anterior view; F, posterior view. Scale bar represents 0.5 mm.

Paranyctoides cf. *Paranyctoides sternbergi,* UNM-B1735, P^4, Figs. G–I: G, occlusal view, stereo pair; H, labial view; I, anterior view. UNM-B1730, $M_{?3}$, Figs. J–L: J, occlusal view, stereo pair; K, labial view; L, lingual view. Scale bar represents 0.5 mm.

Gypsonictops clemensi, n. sp., UNM-B1719, M_3, holotype, Figs. M–O: M, occlusal view, stereo pair; N, labial view; O, lingual view. UNM-B1736, P^3, Figs. P–R: P, occlusal view, stereo pair; Q, anterior view; R, posterior view. Scale bar represents 0.5 mm.

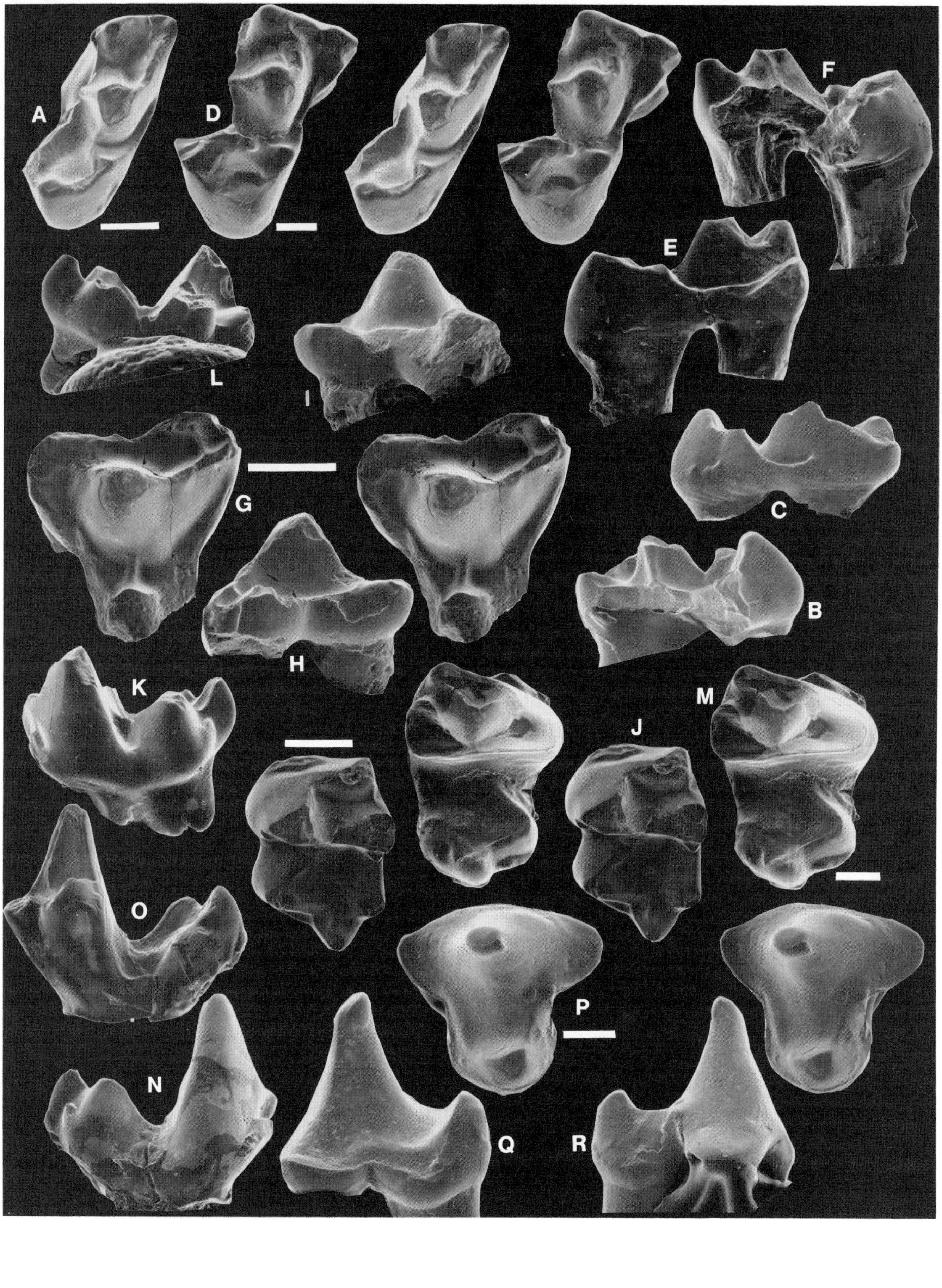
A
D
F
E
L
I
G
C
H
B
K
J
M
O
P
N
Q
R

PLATE 3.

Alphadon halleyi, UNM-B1726, M_L, Figs. A–C: A, occlusal view, stereo pair; B, lingual view; C, labial view. UNM-B1721, $M_{?3}$, Figs. D–F: D, occlusal view, stereo pair; E, labial view; F, lingual view. Scale bar represents 0.5 mm

Aquiladelphis paraminor n. sp., UNM-B1718, M_2, Figs. G–I: G, occlusal view, stereo pair; H, labial view; I, lingual view. UNM-B1731, M_3, Figs. M–O: M, occlusal view, stereo pair; N, lingual view; O, labial view. Scale bar represents 0.5 mm.

Alphadon cf. *wilsoni,* UNM-B1717, M_3, Figs. J–L: J, occlusal view, stereo pair; K, lingual view: L, labial view. Scale bar represents 0.5 mm.

Pediomys fassetti n. sp., UNM-B1725, M_1, Figs. P–R: P, occlusal view, stereo pair; Q, labial view; R, lingual view. UNM-B1724, $M_{?3}$, Figs. S–U: S, occlusal view, stereo pair; T, labial view; U, lingual view. Scale bar represents 0.5 mm.

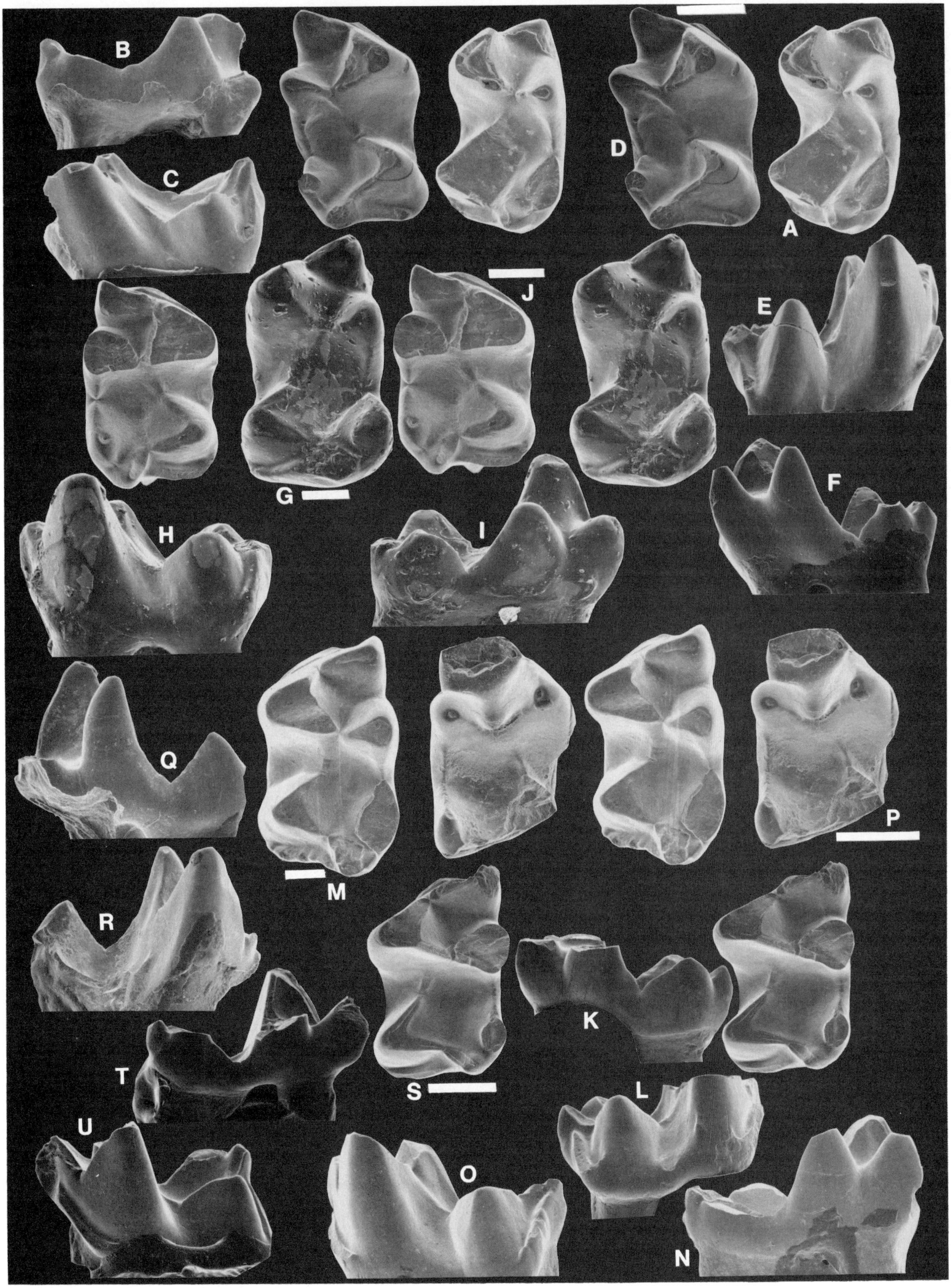
B
C
D
A
J
E
G
H
I
F
Q
M
P
R
K
T
S
L
U
O
N

situated slightly anterior to the shallow ectoflexus. Cusp D is large and bladelike but about two-thirds the size of cusp B. A small cuspule may be present in the position of cusp E. The paracone and metacone are conical, vertical, of subequal size, and not inclined. The metastylar wing is prominent and slightly notched at the posterior margin of the metacone. The trigon basin is very deep. The paraconule and metaconule are large, separate distinct cusps and are partially inflated. The metacone and metaconule are partially fused at their bases while still preserving the anterior metaconule wing connecting the apex of the metaconule and the internal base of the metacone. The paraconule, however, is separate and not fused with the paracone. The external conule wings form the respective anterior and posterior cingulae. The posterior cingulum is continuous labially to the base of the metacone. The anterior cingulum is continuous to stylar cusp A, although strongly deflected toward the roots but not to the depth of the trigon basin. The protocone is lower but broader than either paracone or metacone.

A large part of an assumed M^3 is also present in the collection. It is larger than M^1 but does not differ significantly from it in previously described features. However, a small cusp E may be present, developed as a swelling at the labial termination of the metacrista, and the metacrista is slightly more notched at the base of the metacone.

Lower molars referred to *A. halleyi* from the Fossil Forest are essentially the size of *A. lulli* and *A. halleyi*. Remeasurement of AMNH 77367 (holotype of *A. halleyi*) has necessitated, in part, the revised diagnosis. The tooth is 1.8 mm (0.07 in) long and has a 1.1 mm (0.04 in) width through the talonid. Sahni (1972) reported a length of 2.1 mm (0.08 in) and a talonid width of 1.2 mm (0.05 in). Were it not for the differences in the upper dentition, *A lulli* and *A. halleyi* would be difficult to separate on the basis of the lower dentition only, as noted by Fox (1979a). The M_2s referred by Fox (1979a) to *A. halleyi* from the Oldman Formation are larger than those of *A. lulli* (Clemens, 1966), yet the M_1 is about the same size in both species. The referred lower molars from the Fossil Forest are smaller than those of the Oldman Formation and resemble those of *A. lulli* in that regard. However, despite small sample size, morphological differences are apparent: For example, the talonid of M_1 is broader than the trigonid in *A. lulli*. M_{2-4} of *A. halleyi* have the hypoconulid more medially placed than in *A. lulli* and have less well developed anterior and posterior cingulids.

Discussion. Three species of *Alphadon* are present in Quarry 1 of the Fossil Forest, this being the smallest. Clemens (1973) did not note the presence of a small species of *Alphadon* in the Hunter Wash fauna. In the Fossil Forest sample, the broadness of the protocone serves to separate *A. halleyi* from *A. creber* and to some extent from *A. lulli*. The paraconules and primary molar cusps are equidistantly separated. This differs from the condition seen in most species of *Alphadon* in which the metaconule and metacone are more closely positioned when compared to the placement of the paracone and paraconule. The closer positioning of the metacone and metaconule in M^{1-3} is seen in *A. lulli*.

TABLE 3. ALPHADON PARAPRAESAGUS MEASUREMENTS

		Length	Anterior Width	Posterior Width
UNM-B5338	M_3	3.20	1.90	1.70
UNM-B5338	M_4	3.00	1.70	1.40
				Width
UNM-B1741	M^4			3.6

In UNM-B1721, a right $M_{?3}$, the trigonid is high and inclined anteriorly and has a massive protoconid; the metaconid is slightly larger than the paraconid; the paraconid is anteriorly compressed and medial in position between proto- and metaconids. The trigonid basin is deep and only open lingually between the posterior base of the paraconid and the anterior base of the metaconid. The hypoconid is very large and tall. The cristid obliqua meets the postvallid medially and low. The entoconid is smaller than the hypoconulid and is joined to it by a crest or ridge. The hypoconulid is strongly lingually positioned, not posteriorly projecting, and is joined to the hypoconid by a pronounced ridge. The posterior margin of the tooth is arcuate for the positioning of the next posterior molar and preserves a prominent, posterior shelf that slopes labially and ventrally. The tooth is smaller than *A. halleyi* (Sahni, 1972) but very comparable to *A. halleyi* of Fox (1979a, p. 100).

Fox (1979a) reported the occurrence of *A. halleyi* in the Oldman Formation and separated specimens referred to *A. halleyi* from other related species of *Alphadon* present in his fauna on the basis of size. Reference of the lower molars to his taxon must be questioned. Also, small upper molar sample size requires comparisons of a typological nature with other known taxa.

Alphadon parapraesagus n. sp.
(Plate 2, Figs. D–F; Plate 4, Figs. A–C; Table 3)

Etymology. Named for its assumed close relationship with *A. praesagus*.
Holotype. UNM-B5338, a lower dentary fragment with M_{3-4}.
Horizon and type locality. Quarry 1, Fossil Forest, lowermost Kirtland Shale, New Mexico.
Referred specimens. UNM-B1741, an upper molar fragment.
Diagnosis. Similar to *A. praesagus* but with cristid obliqua meeting the postvallid low and labial to the notch between the protoconid and metaconid (meets on the protoconid); metaconid is not smaller than the paraconid in occlusal view; the hypoconulid is more twinned with the entoconid than posteriorly projecting; the trigonid is slightly shorter anteroposteriorly than in *A. praesagus*.
Description. $M^?$ (UNM-B1741) is a fragmentary right upper molar with protocone, paracone, and stylar cusps A and B preserved, with some irregular enamel swelling on the posterior slope of the protocone that may be some indication of a hypo-

TABLE 4. ALPHADON CF. WILSONI

		Length	Width	
UNM-B1743	M^2	2.20	2.2	
UNM-B1740	M^3	2.10	---	
		Length	Anterior Width	Posterior Width
UNM-B1732	M_2	2.10	1.40	1.30
UNM-B1723	M_2	2.20	1.30	1.20
UNM-B1717	M_3	2.15	1.30	1.30

cone. The paracone is large, is positioned posterior to the protocone, is situated between stylar cusps A and B, and maintains a prominent postparacrista. The protocone preserves a thin external cingulum just discernible in occlusal view at high magnification. A prominent crest is present from stylar cusp B to the base of the protocone as is an anterior cingulum from stylar cusp A to the base of the protocone.

The lower jaw preserves M_{3-4} and is virtually identical to similar material figured by Fox (1979a) from the Oldman Formation of Alberta. The New Mexico specimen is smaller when compared to average measurements published by Fox. Both *A. praesagus* and *A. parapraesagus* share a strongly projecting hypoconulid, erect trigonids, trenchant trigonid cusps, smaller size, and weaker cingulids when compared to *A. rhaister. A. parapraesagus,* however, has a cristid obliqua meeting the postvallid at the base of the protoconid and not medially at the notch, as in *A. praesagus.*

Discussion. UNM-B1741, the referred partial upper molar, is somewhat similar in observable morphology to M^1 of *A. rhaister,* but the tooth most closely resembles *A. praesagus* from the Judith River and Oldman formations. The Quarry 1 tooth has, however, a proportionately larger paracone when compared to the size of the paraconule, which suggests that the New Mexico tooth is not in fact M^1 but M^2 or M^3. If so, it is of substantially smaller size than M^2 or M^3 of *A. rhaister.* It differs from *A. rhaister* in some other features as well: (1) the paracone and stylar cusp B are partially fused at their common base, with the connecting crest anteriorly deflected (most noticeable in the stylar region); (2) all cusps are more inflated; and (3) the protocone base shows only traces of cingulae.

Clemens (1973) noted the occurrence of a species of *Alphadon* comparable to *A. rhaister* in the Hunter Wash fauna. He included in that designation specimens published by Lillegraven (1969) from the upper part of the Edmonton Formation and those identified as *A. rhaister* by Sahni (1972) from the Judith River Formation. The reader is also referred to Fox (1979a) for additional discussion. Clemens (1973a) thought that the two Yale specimens (YPM 14911 and YPM 14912) referred by him (Clemens, 1966) to *A.* cf. *rhaister* are of a different species, previously unnamed. They are of substantially smaller size when compared to other specimens referred by Clemens to *A. rhaister.* It is very possible that these two teeth represent upper dental elements of a form very close to if not conspecific with *A. parapraesagus* n. sp.

The material from the Fossil Forest is definitely smaller than *A. rhaister,* based upon measurements of teeth in the lower jaw. The morphology of the referred upper molar agrees quite closely with that of M^2 or M^3 of *A. praesagus.* It also is most similar to the Yale specimens, *A.* cf. *rhaister* described by Clemens (1966). Clemens thought that the Yale specimens probably came from a locality stratigraphically below larger faunas studied by him in the upper part of the Lance Formation. If true, the presence of a form identified here as *A. parapraesagus* from the Fossil Forest may suggest an older age of the fauna (perhaps Edmontonian–early Lancian) and may closely approximate the age of the Judith River fauna described by Sahni (1972) and of the Oldman Formation described by Fox (1979a).

Alphadon cf. *wilsoni*
(Plate 1, Figs. A–F; Plate 3, Figs. J–L; Table 4)

Horizon and locality. Quarry 1, Fossil Forest, lowermost Kirtland Shale, New Mexico.
Referred specimens. UNM-B1743, M^2; UNM-B1740, partial M^3; UNM-B1723, M_2; UNM-B1732, M_2; UNM-B1717, M_3.
Description. UNM-B1743 is provisionally identified as an M^2 because of its similarity to M^2s described by Lillegraven (1969). The tooth is anteroposteriorly longer than M^3 and is only slightly wider than long. A full complement of stylar cusps is present with cusp B being the highest and largest. Cusp A is separated from B by a notch in the crests connecting the two cusps. Cusp A is positioned lower on the crown than B but is at about the same relative elevation when compared to the remaining stylar cusps. Cusp C is positioned posterior to the shallow ectoflexus and is the only stylar cusp that appears to be inflated. Cusp C is separated from cusp D by a distinct notch and is virtually unconnected by crests to cusp B. Cusp D is only slightly inflated but is somewhat anteroposteriorly elongated. An elongate swelling on the labial extension of the postmetacrista can be considered as cusp E although it is positioned on the postmetacrista and not on the edge of the stylar shelf directly posterior to cusp D. The trigon is only slightly deeper than the level of the stylar shelf when compared to the much deeper basin in *A. marshi.* The conules are distinct cusps with both external conule wings noticeable and confluent with the respective anterior and posterior cingulae. The internal conule wings are not present, and only the bases of the cones and conules fuse to form a slight ridge above the trigon basin. The protocone is not as high as either metacone or paracone and does not have any cingulae around its lingual base.

M^3 (UNM-B1740) is similar to the M^2 in most respects but differs in having a deeper ectoflexus, cusp C positioned slightly anterior to the deepest penetration of the ectoflexus, cusp E positioned directly posterior to cusp D on the labial margin, and cusp A placed higher on the crown, approximating the height of cusp

PLATE 4.

Alphadon parapraesagus n. sp., UNM-B5338, dentary fragment with M_{3-4}, holotype, Figs. A–C: A, occlusal view, stereo pair; B, labial view; C, lingual view. Scale bar represents 2 mm.

Gypsonictops cf. *lewisi,* UNM-B5545, dentary fragment with M_{1-2}, Figs. D–G: D, occlusal view, stereo pair; E, same, lingual view; F, same, labial view; G, labial view showing the alveoli and roots of the anterior dentition from the canine to the trigonid of M_1. Scale bar represents 2 mm.

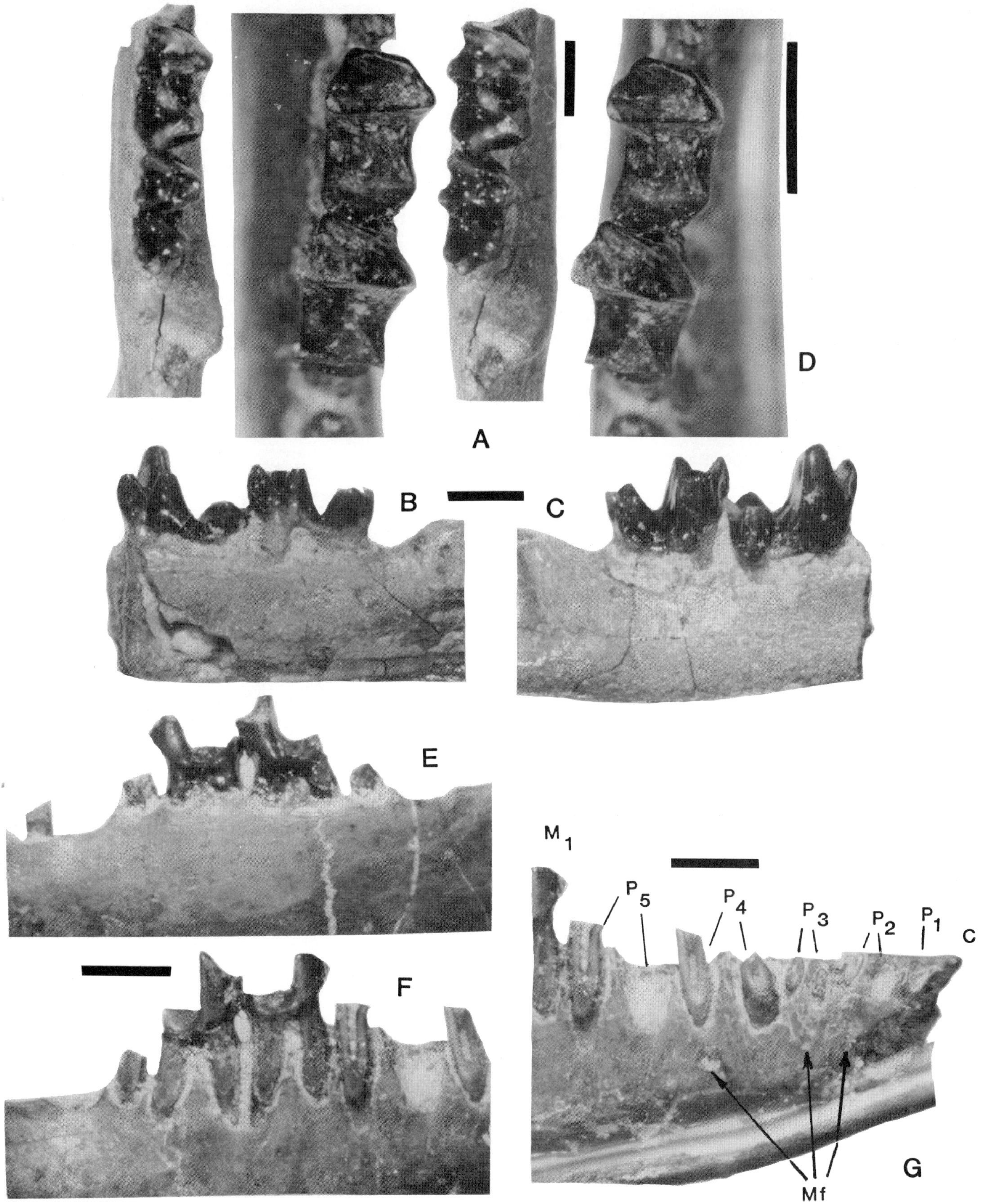
D
A
B
C
E
M1
P5
P4
P3
P2
P1
C
F
Mf
G

B. Cusp D is elongate, well separated from cusp C. The conules and protocone have been broken away.

The referred lower molars are virtually identical to those described by Lillegraven (1969) and are referred to *A.* cf. *wilsoni.*

Subfamily *ECTOCENTROCRISTINAE,* n. subfam.

Type genus. *Ectocentrocristus,* n. gen.
Diagnosis. Relatively large fossil North American didelphid marsupials; stylar shelf broad with trenchant cusps; stylar cusps A–D prominent, large and conical or bladelike; posterior stylar cusps located at the labial edge of upper molars; upper molar conules appear as swellings high on the anterior and posterior protocristae; lower molars with narrow talonids; cristid obliqua meets trigonid near the base of the metaconid.

***Ectocentrocristus,* n. gen.**
(Text Fig. 3)

Etymology. Named for the prominent, labially directed centrocrista of upper molars.
Type. *Ectocentrocristus foxi* n. sp.
Diagnosis. The same as that of the type and only species.

***Ectocentrocristus foxi* n. sp.**

Etymology. Named for Richard C. Fox who was the first to recognize the unique nature of the type specimen.
Type. AMNH 77372, a left upper molar from the Judith River Formation.
Horizon and type locality. Clambank Hollow locality, upper part of the Judith River Formation, Chouteau County, Montana (Sahni, 1972).
Known distribution. Upper Judith River Formation, Montana (Sahni, 1972); Fossil Forest Quarry 1, lowermost Kirtland Shale, New Mexico.
Referred specimens. AMNH 77371, left M_L; UNM-B5485, upper molar fragment.
Diagnosis. Stylar shelf wide and sloped internally and toward the deep trigon basin; broad stylar shelf; stylar cusps A–D large and prominent with stylar cusp C largest; cusp E absent or very small; cusp A distinct, situated more toward the base and connected to cusp B by an uninterrupted crest; medial junction of centrocrista deflected to the stylar margin between cusps C and D; ectoflexus between stylar cusps C–D; metaconule larger than paraconule; both appear as swellings on the pre- and postprotocrista; conules possess only external conule wings, which are placed near the protocone apex; protocone and conules form an anteroposteriorly directed linear lingual crest; metacone larger than paracone and both are higher than the more massive protocone; lower molars have open trigonids and narrow talonids with the cristid obliqua meeting the trigonid at the base of the metaconid; hypoconulid distinct and largest of hypoconulid/entoconid pair.

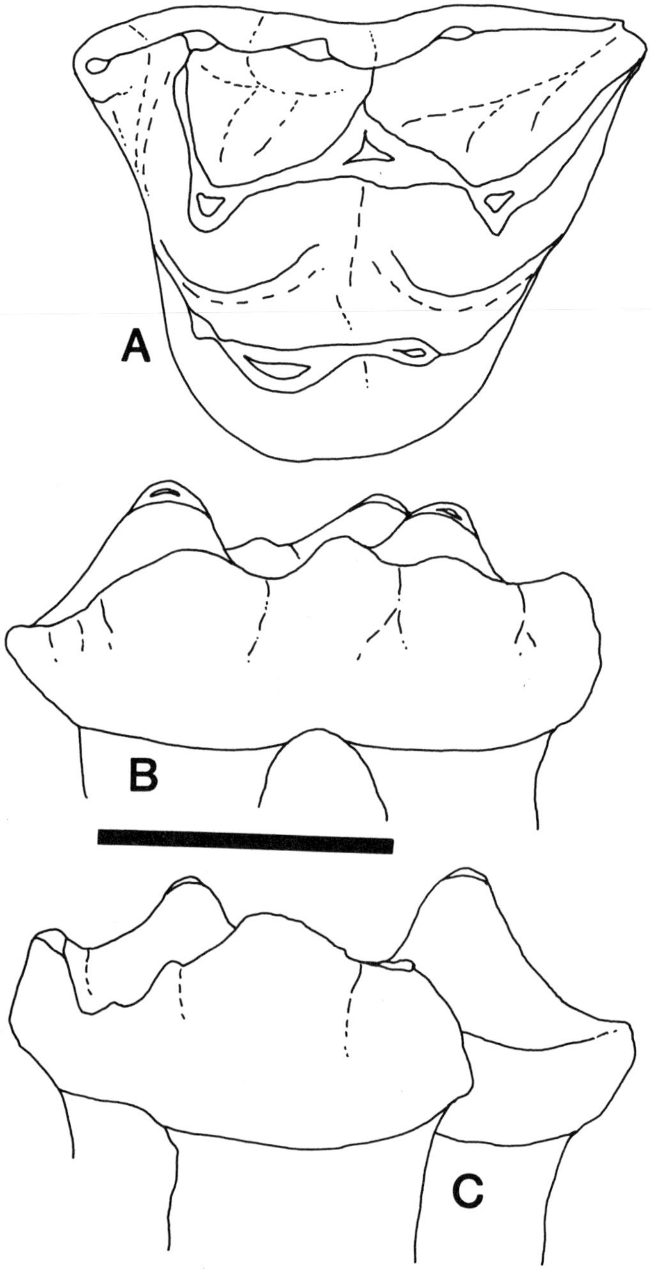

Figure 3. *Ectocentrocristus foxi* n. sp., AMNH 77372, an upper molar, holotype: A, occlusal view; B, labial view; lingual view. Scale bar represents 2 mm.

Description. Fox (1979b, p. 96) described AMNH 77372 and 77371. A single upper tooth fragment has been recovered from screening concentrates from the Quarry 1 in the Fossil Forest. The tooth fragment (UNM-B5485) is virtually identical to morphology observable in the type specimen from the Judith River Formation. It preserves the protocone, all of the paraconule, part of the metaconule, most of the paracone and associated crests, the trigon basin, and the base of the metacone. The preserved fragment is 2.5 × 2.0 mm (0.1 × 0.08 in) and is estimated to be from a

TABLE 5. PEDIOMYS FASSETTI n. sp. MEASUREMENTS

		Length	Width	
UNM-B1738	M^1	(1.50)	1.55	
UNM-B1737	$M^{?2}$	(1.60)	2.20	
UNM-B1734	M^3	1.95	(2.4–2.6)	
		Length	Anterior Width	Posterior Width
UNM-B1725	M_1	(1.50)	0.90	(0.09)
UNM-B1724	$M_{?3}$	1.75	1.10	1.05

tooth 3.5–4.0 mm (0.14–0.16 in) in both length and width. The conules are very small, inflated, possess no internal conule wings, and are located as swellings on the protocristae, immediately anterior and posterior to the protocone apex. The trigon basin is deep and terminates well below the level of the anterior and posterior cingulae. Labial terminations of the cingulae are not preserved. The crests associated with the paracone document that the centrocrista is strongly labially flexed; the termination is not preserved. The preserved part of the labial slope of the paracone documents the presence of a prominent crest directed toward the stylar shelf and presumably toward cusp B. The paracone exterior slope descends toward the stylar shelf and strongly posterior, as in the type specimen. The metacone of UNM-B5485, other than the border of the lingual base, is not preserved. Stylar cusps and shelf are also not preserved.

Discussion. Because the known material consists totally of isolated teeth, considerable thought was given to the possibility that they might be deciduous teeth of some poorly known taxon. Lillegraven (1969) and Clemens (1966) have illustrated teeth thought to be deciduous teeth of didelphid marsupials. These described teeth all show a narrowed protocone-conule region of the molars, even though the stylar region shows larger and more prominent cusps. *Ectocentrocristus foxi* has a broad protocone and conule region in which the conules are placed nearly anterior and posterior to the protocone apex, and the upper teeth do not show the pronounced triangular occlusal view of described DP^3s. The referred lower molar certainly cannot be a deciduous tooth. The tooth does not possess the anteriorly projecting paraconid so typical of all other known didelphid DP_3s.

We agree with Fox (1979a) and formally recognize the uniqueness of these specimens. The taxonomic position of this marsupial is debatable because it compares only remotely to known didelphid marsupials and even less so with other eutherian dental patterns. It differs in the pattern of stylar cusps, placement of the conules relative to the protocone, and nature of the strong labial position of the centrocrista. Fox (1979a, p. 96) did not find any marsupial remains in the Oldman Formation collections that resembled the morphology observed in AMNH 77372 or 77371. The collection from the Fossil Forest Quarry 1 has produced a single fragment (UNM-B5485) that is virtually identical in observable morphology to the type specimen. It preserves enough of the conule/protocone relationship and enough of the centrocrista morphology to confidently refer the specimen to *Ectocentrocristus foxi.*

Family PEDIOMYIDAE (Simpson, 1927a)

Pediomys fassetti n. sp.
(Plate 1, Figs. J–R; Plate 3, Figs. P–U; Table 5.)

Etymology. Named for James E. Fassett for his contributions to the study of San Juan Basin Cretaceous deposits.
Holotype. UNM-B1734, an M^3 fragment.
Horizon and type locality. Quarry 1, Fossil Forest, lowermost Kirtland Shale, New Mexico.
Referred specimens. UNM-B1738, an M^1; UNM-B1737, an $M^{?2}$; UNM-B1724, an $M_{?3}$; UNM-B1725, a broken M_1.
Diagnosis. Small pediomyids with cusp A large, high, smaller than—yet poorly separated from—large cusp B; cusp C absent or minute; cusp D present, isolated and anteroposteriorly elongated; cusp D prominent yet much smaller than cusp B; cusp E present but small.
Description. M^1 (UNM-B1738) is known from a single specimen in which the posterior stylar shelf from the point of the very weak ectoflexus has been broken away; length is therefore estimated (Table 5). The tooth preserves a prominent cusp B, which is about twice the size of cusp B as seen in *P. exiguus.* Cusp B is about one-half the size of the paracone and is connected to adjacent stylar cusps by distinct anteroposteriorly directed crests. A crest connects cusp B and the paracone. It is reduced when compared to species of *Alphadon.* It is not fully terminated at the base of cusp B or deflected toward cusp A as in other pediomyids. Cusp A is assumed to have been large (cusp A apex has been broken away), placed high on the crown, and moderately separated from cusp B, as judged from the position of the anterior cingulum. A prominent labial stylar crest descends the posterior margin of cusp B to the position of the very weak or absent ectoflexus. This crest is assumed to be continuous to the base of cusp D. Cusp C is either not present or is at best minute. The paracone (larger than the metacone) and metacone are large, conical, flattened somewhat labially, and slightly divergent. Both conules and associated wings are present but weakly developed. The paraconule is distinctly taller than the metaconule, although both are similarly placed labially on the slope of the protocone. The anterior paraconule wing is continuous with the precingulum, whereas the posterior conule wing terminates on the basinal side of the paracone. The posterior metaconule wing is continuous with the postcingulum, whereas the anterior metaconule wing terminates on the basinal edge of the metacone. Both internal conule wings continue well onto the paracone and metacone bases. The protocone, though broader, is not as high as either paracone or metacone and has no cingular development around its lingual base.

$M^{?2}$ (UNM-B1737) has been abraded and spalled so that the morphology of the stylar shelf is not preserved. The preserved parts of the stylar shelf are not inconsistent with morphology observed on M^1 or M^3. It is referred with some confidence because it shares morphologic similarity to M^1 and M^2: (1) high, conical, slightly divergent paracone and metacone with the paracone smaller than the metacone; (2) placement of the conules and conule wings; and (3) implied reduction of the stylar shelf.

M^3, the holotype (UNM-B1734), is the labial half of an M^3. The stylar cusps, shelf, paracone, and metacone are well preserved. Stylar cusp A is large, robust, partially fused or twinned with cusp B, and located at the labial terminus of the prominent large anterior cingulum. Cusp B is large, somewhat elongate, and located anterior to the primitive position opposite the paracone. Cusp B is continuous with a well-defined labial crest, on the margin of the stylar shelf, to a position near the base of cusp D. Only the most minute enlargement of the crest just posterior to the reasonably well-defined ectoflexus could be considered as a cusp C. Cusp D is smaller than Cusps "A-B" but larger than other stylar cusps. It is structurally isolated from the anterior crest and from a small yet distinct cusp E located on the extreme labial posterior margin. Cusp D is anteroposteriorly elongated yet is not trenchant. A small cuspule is located at the anterior lingual basal margin of cusp D. The metacingulum is large, prominent, and well defined and meets the metacone midway on the posterior slope. The metacone is slightly larger and taller than the paracone. The paracone and metacone are high, conical, slightly inflated labially, and they are partially fused well above the trigon basin at their common, slightly divergent base. Each preserves the labial remains of the internal conule wings well up on their slopes, the posterior paraconule wing being highest. The preparacrista is prominent and sigmoidal and leads labially to the base of Cusps "A-B." Precingulum and postcingulum are well developed.

There are no lower premolars in the collection that can be confidently referred to *P. fassetti.* Two lower molars are unquestionably pediomyid and are of the appropriate size (UNM-B1725, M_1; UNM-B1724, $M_{?3}$). Both teeth possess a prominent cristid obliqua that terminates anteriorly at the base of the protoconid, which is the typical pediomyid condition. It is, however, not placed as far labially as in most species of *Pediomys.* Both teeth have the metaconid placed posterior to the protoconid.

Most of the paraconid on M_1 is broken away, but it is medially anterior and low on the trigonid and is lower than the smaller metaconid. Anterior and posterior trigonid carnassial notches are present. The talonid notch is lower than the low point on the portion of cristid obliqua preserved. The entoconid is well developed. The remainder of the talonid has been broken away. In M_3, the paraconid is lingual in position anterior to the metaconid. The entoconid is larger than the hypoconulid, which is not posteriorly projecting. The hypoconid is the largest talonid cusp. Pre- and postcristids are well developed and slope labioventrally. Both are confluent with the well-developed ectocingulid. The $M_{?3}$ differs from the M_1 in its larger size, more anteroposteriorly compressed trigonid, elevated paraconid (metaconid and paraconid are at the same relative basal height), paraconid and metaconid fused at their lingual margins to produce a small but true trigonid basin, and talonid notch at about the same height as the low point on the cristid obliqua.

Discussion. There can be little doubt that these specimens represent a species of *Pediomys* that is more primitive in all generally accepted characteristics currently used to group species within the Pediomyidae. It is more primitive than any other known species with the possible exception of *P. exiguus.* It possesses a partially reduced stylar shelf, most noticeable opposite the paracone when compared with *Alphadon.* The stylocone (cusp B) is large (approaching one-half the size of the paracone) yet is more anteriorly placed and more closely associated with cusp A, which is nearly equal in height. When compared to *Alphadon,* which retains cusp A low on the crown, such a character is considered derived and serves to define *Pediomys.* The lower molars have the cristid obliqua labially displaced and terminating at the base of the protoconid, although not nearly to the extent seen in Lancian species of *Pediomys.* The crest connecting the paracone with cusp B is reduced or fully terminated at the base of cusp B on M^3. *Alphadon* retains what is probably the primitive condition by maintaining this crest as a major shearing surface. Such features substantiate the conclusions of Fox (1971, 1979b) that *Pediomys* developed a crushing dentition and relied less on shear function than did *Alphadon.*

Cusp C has been virtually lost; if it is present in some teeth, it is minute. The use of stylar cusp C as a systematic character by authors is problematic in that it seems to appear and disappear frequently within accepted lineages. We suspect that it is either not useful as a taxonomic character or that currently accepted phylogenetic relationships of *Pediomys* species are not clearly understood. For example, Fox (1979b) described a partial M^3 (US-14813) that he referred to *Pediomys* cf. *P. elegans.* UA-14813 does in fact resemble *P. fassetti,* but it possesses a rather prominent cusp C lingual to the stylar shelf marginal crest and would be excluded from conspecific relationship with *P. fassetti* using standard taxonomic practice. However, other characters like the larger, by comparison, cusp B of *P. fassetti* and the substantially wider stylar shelf are generally accepted as primitive characteristics. Can this be taken as evidence that cusp C had a second derivation within at least part of the Pediomyidae? Perhaps, but the question may not be answerable on the basis of current evidence.

P. fassetti most closely resembles *P. exiguus* from the early Campanian Milk River Formation (Fox, 1971). Both taxa are of similar size and preserve relatively large stylocones (a character that separates them from all other known species of *Pediomys*), a weak ectoflexus on the anterior upper molars, and "no" cusp C. *P. fassetti* differs, however, at least in the larger size of cusp B and the discrete, non-bladelike cusp D. It should be noted (Clemens, pers. written communication, 1985) that these features are of questionable significance (same size, weak ectoflexus), are shared primitive characters (presence of B and D), or are characters of debatable polarity (absence of C). No lower molars of *P. exiguus*

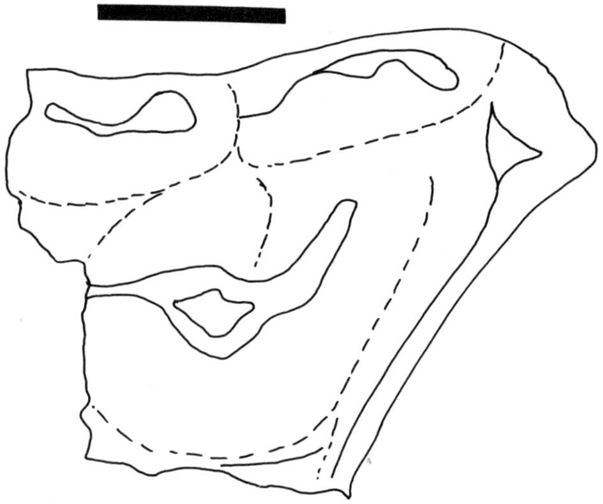

Figure 4. *Aquiladelphis paraminor* n. sp., UNM-B5542, an upper molar fragment, holotype, occlusal view. Scale bar represents 1 mm.

were referred by Fox and therefore comparison is not possible. Furthermore, the type specimen of *P. fassetti,* an M^3, is an additional dental element that Fox did not have at his disposal when he named *P. exiguus.*

?Pediomyidae *incertae sedis*
Aquiladelphis Fox, 1971

Holotype. *A. incus* Fox 1971
Known distribution. Upper Cretaceous: Milk River Formation, Alberta; Oldman Formation, Alberta; Fruitland Formation, New Mexico.
Included species. *Aquiladelphis incus* Fox 1971: *A minor* Fox, 1971; *A. paraminor,* n. sp.
Revised diagnosis. Medium-sized to large North American pediomyids; upper molars with two stylar cusps B; cusp C as large as or larger than stylar cusps B; cusp D larger than B cusps and smaller than or equal to C; paracone and metacone labiolingually compressed, low, and massive and may be only slightly divergent; conules strong; trigon basin broad; lower molars with low heavy trigonid cusps; talonid basins open to broad and shallow; hypoconulid separate and not positioned close to entoconid.

Aquiladelphis paraminor, n. sp.
(Plate 3, Figs. G–I, M–O; Text Fig. 4; Table 6.)

Pediomys cf. *P. hatcheri,* Fox 1979b.

Etymology. Named for its similarity to *A. minor* Fox, 1971.
Holotype. UNM-B5542, an upper molar fragment.
Horizon and type locality. Quarry 1, Fossil Forest, lowermost Kirtland Shale, New Mexico.

TABLE 6. AQUILADELPHIS PARAMINOR, n. sp. MEASUREMENTS

		Length	Anterior Width	Posterior Width
UNM-B5559	M_1	2.9	1.60	1.5
UNM-B1718	M_2	3.05	1.65	1.75
UNM-B1731	M_3	3.55	2.00	2.15

Known distribution. Lowermost Kirtland Shale, New Mexico; upper part of the Oldman Formation, Alberta.
Referred specimens. UNM-B5559, M_1; UNM-B1718, M_2; UNM-B1731, M_3; UNM-B5543, M_3, talonid.
Diagnosis. A pediomyid species of medium to large size, similar to *A. minor* Fox (1971); upper molars with two stylar cusps B, the anterior being smaller and closely approximating the position of cusp A; the primary cusp B is very large and may be the largest stylar cusp (cusps D and E are not preserved); cusp C is large but neither so large nor so high as primary cusp B; cusp C bilobed, the anterior lobe being larger; cusp A is large and is only weakly separated from cusp B; the crest between the paracone and stylar cusp B is fully terminated at the base of cusp B.
Description. The holotype (UNM-B5542) is fragmentary, yet preserves the necessary morphology to distinguish it from other species of *Aquiladelphis.* It preserves stylar cusps A–C, the paracone, the base of part of the metacone, and part of the anterior cingulum. Other cusp designations were considered, including the possibility that the fragment was the posterolabial part of the tooth with the major cusp being the metacone. This alternative was not accepted because all known Cretaceous therians and metatherians do not have a very strongly posteriorly projecting, isolated metastyle as this specimen would have shown.

The fragment has a length of 2.45 mm (0.1 in) and a width of 2.4 mm (0.09 in). Based upon comparisons with *A. minor* and *A incus,* the holotype is a fragment of a tooth of dimensions approximating *A. minor* and is certainly smaller than those of *A. incus* (Fox, 1971). Stylar cusp A is large and poorly separated from the anterior and smaller of cusps B. Cusp A is positioned high (away from the roots) on the stylar crest and forms a prominent termination of the stylar crest and of the large well-defined continuous anterior cingulum. Stylar cusps B are unequal in size. The more anterior of the two is smaller and appears as a distinct separate cusp on the anterior of the larger B cusps. The cusp A separation is only slightly stronger than the separation between the two B cusps. Although broken labially, the primary B cusp is very large but somewhat smaller than the paracone. Cusp C is labially placed and smaller than primary cusp B. Cusp C is bilobed with the very weak ectoflexus, if present, situated labial to the mid-region of cusp C. The posterior lobe of cusp C is smaller and more trenchant than the more conical anterior part. The crest between the paracone and the primary cusp B is fully terminated at the base of the paracone and does not continue onto the base of the primary cusp B. A weak crest descends the slope of the primary cusp B but does not approach the base of the cusp and is decidedly smaller and less well defined when compared to the labial crest of the paracone. The paracone is large and conical and preserves the anterior half of the centrocrista, which is directed toward the metacone. If the structural continuity continues around the base of the metacone, as observed morphology tends to indicate, then the posterior border of the stylar shelf is narrow and would closely approximate the condition seen in *A. minor.*

Three lower molars are referred to *A. paraminor.* These are some of the largest teeth in the collection from the Fossil Forest (Table 6). M_1 (UNM-B5559) has an anteriorly expanded trigonid with the paraconid placed forward when compared to the other lower molars. It is most similar to UA-5534 illustrated by Fox (1971, Pl. 6, Fig. B) and questionably identified by him as an M_2. The paraconid on UNM-B5559 is smaller but broader than the metaconid, is more labially placed than in *A. minor,* and has a prominent carnassial notch that separates the crests of the paraconid and the protoconid. An equally well-developed notch separates the protoconid and metaconid at the same relative position at the base of the protocone. The protoconid is very large compared to other trigonid cusps and similar in size to that of *A. minor* but is less inflated. The metaconid is very similar to UA5534 yet less inflated. The most significant differences between *A. minor* and *A. paraminor,* seen in M_1, are observed in the talonid: It is about as wide as the trigonid in *A. paraminor* but distinctly wider than the trigonid in *A. minor.* The trigonid labial length is approximately two-thirds the length of the tooth in *A. paraminor,* whereas in *A. minor* it forms only about one-half the length of the tooth. The M_1 talonid of *A. paraminor* is not nearly as broad and open as that of *A. minor.* The hypoconulid of *A. paraminor* is medially placed and only shares part of a common base with the entoconid (not the "twinned" condition typically seen in other Cretaceous marsupials). The cristid obliqua is well defined and terminates anteriorly on the medial posterior slope of the protoconid. Anterior and posterior cingulids are prominent and are well defined.

The referred M_2 (UNM-B1718) shows a remarkable similarity to M_1 but differs in that the trigonid is less anteroposteriorly expanded and is slightly larger. This tooth fills a morphologic gap between the two lower teeth of *A. minor* illustrated by Fox (1979b) and provides some additional justification for our reidentification of UA5534 as M_1. M_2 closely resembles M_1 and M_3 (description follows) in several features: (1) two similarly placed trigonid carnassial notches are present, (2) talonids are not noticeably wider than the trigonids, (3) the hypoconulid is somewhat isolated from the entoconid and is medial in position, (4) the cristid obliqua terminates anteriorly on the medial posterior slope of the protoconid, and (5) anterior and posterior cingulids are prominent and large.

The referred M_3 (UNM-B1731) has a wear facet on the posterior margin of the tooth and cannot be M_4. Its trigonid is more closed anteroposteriorly than either M_1 or M_2. We therefore assume it to be M_3. A prominent cristid obliqua terminates on the medial side of the protoconid. The hypoconulid is distinct, not "twinned" but more closely positioned near the entoconid than in more anterior molars. The trigonid is somewhat worn, and most of the protocone is not preserved. The paraconid is smaller than the metaconid, which in turn is much smaller than the protoconid. The paraconid is trenchant and somewhat anteriorly reflected at its lingual margin to accommodate the next anterior tooth. Carnassial notches are present between the protoconid and paraconid and between the protoconid and metaconid. The notches occur at equal elevations on the base of the protoconid.

Several talonids and trigonid fragments could be referred questionably to this taxon because the termination of the cristid obliqua on the trigonid is fully observable. They would not be confused with those of *Eodelphis,* however, because the distance between the cristid obliqua and anterior crest of the entoconid at the base of the trigonid is significantly greater than in specimens of *Eodelphis.* Also, the hypoconulid is more posteriorly projecting than in *Eodelphis* molars. Their large size, however, could be indicative of either taxon.

Discussion. This species closely resembles *A. minor* (Fox, 1971) but it differs from it in the placement, relative size, and differentiation of cusps B, the relative size of cusp C (C being smaller than B), the assumed reduction of the stylar shelf posterior to cusp C, and the narrower, less open talonids. There seems little question that *A. paraminor* is most closely related to *A. minor* among known Cretaceous pediomyids. It resembles Lancian pediomyid species in having primitive characteristics that serve only to unite species within the Pediomyidae.

The lower molars are here referred with some confidence because of similarities to those of *A. incus* and *A. minor* described by Fox. There is little else in the Fossil Forest fauna with which to confuse them. They have the metaconid larger than the paraconid, unlike *Eodelphis.* The position of the cristid obliqua clearly indicates pediomyid affinities, and the large size of the lower molars eliminates the need for comparison with lower molars of a much smaller species, *Pediomys fassetti* n. sp., also present in the fauna. However, the pronounced lack of "twinning" of the hypoconulid and entoconid may suggest nonpediomyid affinities. This would support Clemens's assertion (Clemens, pers. written communication, 1985) that *Aquiladelphis* is not a pediomyid at all but may belong to an unnamed higher taxonomic category.

Fox (1979b) identified UA-14816, an M_2 or perhaps M_3, as *Pediomys* cf. *P. hatcheri.* The tooth, obviously pediomyid as Fox suggests, closely resembles UNM-B1718, an M_2, as described above. The teeth are virtually identical in observable morphol-

TABLE 7. EUDELPHIS sp. MEASUREMENTS

		Length	Anterior Width	Posterior Width
UNM-B5487	P_3	5.1	3.4	---
UNM-B5486	$M?_2$	---	(2.3)	2.1
UNM-B5540	$M?_3$	---	2.6	---

ogy. UA-14816 cannot belong to *A. incus* or *A. minor* because of its narrower talonid and proportionately much larger trigonid and heavier cusps. We therefore refer UA-14816 to *A. paraminor.*

Family STAGODONTIDAE Marsh, 1889b

cf. *Eodelphis* Matthew, 1916
(Table 7.)

Referred specimens. UNM-B5544, an M^U fragment; UNM-B5488, *?P^U; UNM-B5487, P_3; UNM-B5486, $M_{?2}$ fragment; UNM-B5540, $M_{?3}$; UNM-B5539, M_L fragment.*
Horizon and locality. Quarry 1, Fossil Forest, lowermost Kirtland Shale, New Mexico.
Description. The upper molar fragment (UNM—B5544) is questionably referred. It is grouped here because of its large size and because it most closely resembles the anterior stylar shelf region of UA-7039, an M^1 of *Eodelphis* sp. (Fox, 1981) from the Oldman Formation. The worn single-rooted P^U is also questionably referred.

The elements of the lower dentition have been referred to *Eodelphis* rather than *Didelphodon* because they share the following characteristics with *Eodelphis:*

1. Size much smaller than *Didelphodon* and only slightly smaller than *E. cutleri* and *E. browni* (Fox, 1981); closely resembling *Eodelphis* sp. of Fox (1971).
2. Higher trigonids with pronounced higher protoconids.
3. Presence of a pronounced anterior cingulum that is much smaller than that of *Didelphodon* and slightly smaller and less complete lingually than in any known species of *Eodelphis.*
4. Pronounced trenchant trigonid cusps when unworn.
5. Trigonids not anteroposteriorly compressed.
6. Relative height of lower molar trigonid cusps most closely approaches that of *Eodelphis,* although the assumed height of the paraconid more closely approaches that of the metaconid than in other known species of *Eodelphis.*
7. Position of the anterior termination of the cristid obliqua at the trigonid notch, which more closely approaches the condition seen in *Eodelphis* rather than *Didelphodon.*
8. Posterior cingulid, where preserved, prominent yet much narrower than in *Didelphodon* and even substantially narrower than in species of *Eodelphis.*
9. Hypoconulid, although twinned with the entoconid, a separate and distinct cusp (a character that allows immediate separation from *Ectocentrocristus* new genus, this paper).
10. Maintains more or less complete rugose cingulids, which more closely approximates the condition seen in *Eodelphis.*

Discussion. The referred materials, because of the above shared characteristics, are most closely related to the Stagodontidae. Yet they possess several characteristics that may be more primitive, based upon group comparisons with *Alphadon* and *Pediomys,* than in other known representatives of the family. These features include the more lingual termination of the cristid obliqua at the trigonid notch (not at the base of the protoconid as in known stagontid species), high trigonids, metaconid more closely approaching the assumed size of the paraconid, trenchant trigonid cusps, and trigonid equal to or wider than the talonid.

Of forms now known, the Fossil Forest stagodontid most closely approaches the condition seen in a specimen from the Milk River Formation described by Fox (1971). However, the New Mexico form differs from it in several important aspects: The posterior cingulid is not so well-developed, the trigonids are not so high as the Milk River form, and the metaconid is larger when compared to the protoconid in the Fossil Forest form. This is probably a primitive *Eodelphis*-like species: a new species based upon current knowledge.

Subclass THERIA Parker and Haswell 1897
Infraclass EUTHERIA Gill, 1872
Order INSECTIVORA Illiger 1811
Suborder PROTEUTHERIA (Romer, 1966)
Family LEPTICTIDAE Gill 1872
Subfamily GYPSONICTOPINAE Van Valen 1967
***Gypsonictops clemensi,* n. sp.**
(Plate 2, Figs. M–R; Text Fig. 5.)

***Gypsonictops* ?n. sp., Clemens 1973.**

Etymology. Named for William A. Clemens, Jr., the discoverer of the species and a well-known student of Mesozoic mammals.
Holotype. UNM-B1719, isolated M_3.
Horizon and type locality. Quarry 1, Fossil Forest, lowermost Kirtland Shale, New Mexico.
Referred specimens. UNM-B1736, abraded P^3; UNM-B5556, ?P^U; upper molar fragments UNM-B5499, 5551–5554; UNM-B5548, P_5; UNM-B5555, M_L talonid.
Diagnosis. A large species of *Gypsonictops* similar in size to *G. illuminatus*; P^3 without metacone; upper molars anteroposteriorly compressed; protocone not anteriorly displaced as in *G. illuminatus* but like *G. lewisi* in which molars appear more erect rather than anteriorly inclined; well-defined hypocone present as a swelling on the prominent posterior cingulum; anterior cingular prominent; paracone much larger and higher than metacone; both are conical, high, and fused basally throughout much of the height of the metacone; paraconule and metaconule large, of equal size, and placed equidistantly from the labial margin; paraconule placed much higher than the metaconule on the preprotocrista; stylar shelf prominent with a small stylar cusp E known to be present and separate from the labial termination of the metacrista; paraconid relatively large for *Gypsonictops* yet substantially smaller than metaconid; trigonids with paraconid and metaconid basally fused and high to form a true trigonid basin; paraconid labially placed and farther from the metaconid than in all species of *Gypsonictops;* trigonid and talonid form about equal parts of lower molars; significant constriction present immediately posterior to the triconid; talonid cusps, including a

large entoconulid, are present, high and not as flared as other species. Measurements of the holotype: length, 2.80 mm (0.11 in); anterior width, 1.95 mm (0.08 in); posterior width, 1.50 mm (0.06 in).

Description. P^3 (UNM-B1736) has been so abraded that the surface detail is not observable. Anterior and posterior cingulae appear to have been present. The protocone is large and distinct. The stylar shelf appears to have been complete around the base of the paracone. A small parastyle could have been present but certainly not one as large as that of *G. illuminatus.* Although worn, no structure is seen that would indicate the presence of a metacone on the posterior edge of the large, very high conical paracone. A small metastyle may have been present but evidence to support this is not observable.

A fragment of an upper premolar is questionably referred to this species (UNM-B5556). The fragment approximates the size of *G. clemensi* n. sp., is assumed to be eutherian, and resembles, in observable morphology, the posterior part of P^3 of *G. illuminatus* (assuming four upper premolars; P^1 being first). A small metastyle is present: the stylar cingula is complete around the base of the assumed paracone. A posterior cingula is visible, terminating at the base of the metastyle.

Several fragments of upper molars are present in the collection. The relative width of the base of the protocone and the lingual placement of the hypocone gives some evidence to the position in the tooth row of each of the larger lingual fragments. Stylar fragment placement in the tooth row has been more difficult, but in general, the width of the stylar shelf at the base of the metacone has been used. The wider stylar shelf indicates a more posterior position in the tooth row (except M^3). Using the stylar shelf criterion, a single fragment of P^4 is present in the collection (UNM-B5499). No significant differences were observed between the New Mexico form and *G. illuminatus* from Alberta.

M^3 is not preserved in the collection. M^1 and M^2 are represented by several fragments. Each shows narrower lingual protocones, hypocones, and anterior cingulae than can be seen in *G. illuminatus.* A most striking feature is present in upper teeth fragments: the total erect and upright appearance of the teeth when compared to other species of *Gypsonictops.* The protocone is only slightly anteriorly inclined in M^2 and is medially lingually inclined in M^1 in contrast to the inclination seen in this part of other *Gypsonictops* molars. The conules are large in both teeth, are not inflated, and are preserved as the elevated confluences of the respective conule wings. The medial conule wings meet their respective labial-cusps on the medial basinal side. The paraconule is significantly higher, away from the roots, when compared with the metaconule. This feature corresponds to a pronounced "notch" in lower molars so that the anterior termination of the cristid obliqua occurs much lower on the trigonid base than in other species of *Gypsonictops.* It is suggestive of the condition seen in some species of *Cimolestes.* A relatively large hypocone is present as a lingual terminal swelling of the prominent posterior cingulum. The cingulum terminates labially at the anterior base of the metaconule. The anterior cingulum is narrower than the posterior cingulum and terminates anterolabial to the base of the paraconule where it is deflected slightly toward the continuation of the anterior paraconule wing. As judged by preserved fragments, all indications are that the paracone and metacone were large, erect, high cusps that were at least partially fused at their bases.

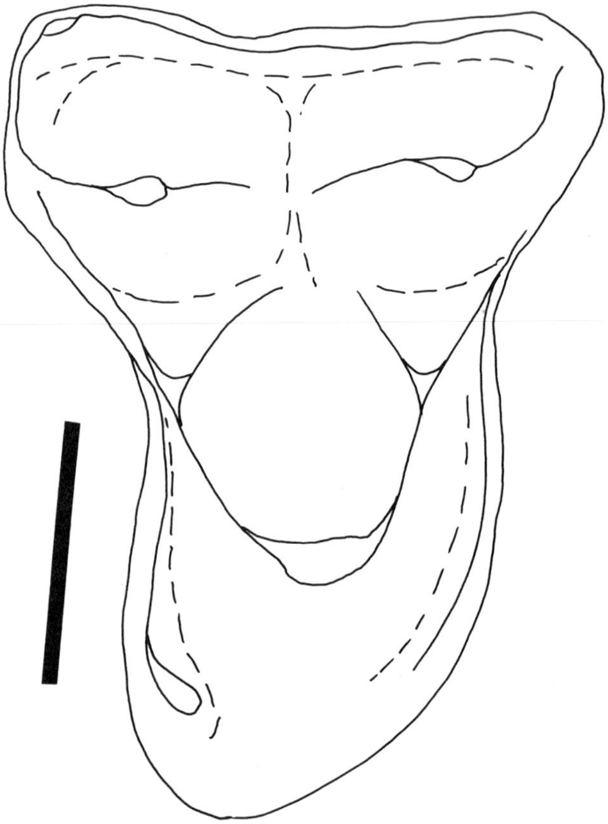

Figure 5. *Gypsonictops clemensi* n. sp., composite reconstruction of an M^1. Scale bar represents 1 mm.

A partial P_5 is preserved in the collection. It preserves the paraconid, protoconid, metaconid, and part of the cristid obliqua. The posterior part has been broken away. It is large and hence cannot be confused with *G. lewisi,* also present in the collection. The protoconid and metaconid are typically worn, and the posterior slope of the protoconid is worn to the base of the trigonid. The posterior slope of the metaconid is less worn. It appears as though the protoconid was the larger of the two cusps. The paraconid is a low lingual swelling of the anterior cingulid that is complete to the labial base of the protoconid. Most of the talonid has been broken away.

The lower molars are substantially larger than, but similar in construction to, *G. lewisi* in the positioning of the paraconid and relative size of the trigonid cusps. *G. clemensi* does not possess the more pronounced medial placement and reduced size of the paraconid as in *G. illuminatus.* The paraconid in *G. clemensi* is substantially smaller than the metaconid although not as small as in Lancian species of *Gypsonictops.* The paraconid and metaconid

TABLE 8. GYPSONICTOPS cf. LEWISI MEASUREMENTS

		Length	Anterior Width	Posterior Width
UNM-B5549	P_5	(2.3)	1.35	---
UNM-B5545	M_1	2.2	1.4	1.2
	M_2	(2.1)	1.6	(1.3)
UNM-B5557	M_L	---	---	1.6
UNM-B5558	M_3	---	---	1.35

are somewhat as in Lancian species of *Gypsonictops*. The paraconid and metaconid are somewhat appressed and are partially fused at their bases so that the lingual margin is elevated to produce a true trigonid basin. The metaconid is large yet noticeably smaller than the protoconid. The available lower molars preserve a pronounced constriction of the tooth immediately posterior to the junction of the trigonid and talonid. They also have a large entoconulid at the labial margin of the talonid notch. The remaining primary talonid cusps are large, separate, and distinct but not so broadly flared as in other species of *Gypsonictops*. The talonids of lower molars preserve the broad talonid wear facet typical of *Gypsonictops* and not that of *Cimolestes*. Strong anterior and posterior cingulids are present and continue to the labial margin. The hypoconulid on M_3, large and well developed, is not strongly posteriorly projecting and resembles *G. illuminatus* in this regard. The hypoconulid position differs from *G. illuminatus*, however, in that it is more closely positioned toward the entoconulid although not approaching the hypoconulid/entoconulid "twinning" seen in marsupials.

Discussion. Clemens (1973) mentioned the presence of two species of *Gypsonictops* within the Hunter Wash fauna. Without elaboration, he identified one as being the size of *G. illuminatus* but differing from it in several morphological features. He suggested that the form so identified was probably a new species. We here formally recognize that distinction.

G. clemensi does in fact resemble *G. illuminatus* in size as Clemens (1973) suggested, yet it retains some features that are generally regarded as primitive with respect to other eutherians: (1) the partially appressed metaconid and paraconid; (2) the relatively large size and labial placement of the paraconid; (3) the upright and narrow molar protocones; (4) the reasonably complete and broad, compared to other species of *Gypsonictops*, stylar cingulum, and (5) tall, upright paracone and metacone. *G.* cf. *lewisi* and *G. clemensi* are both taken to be primitive forms of the genus and may in fact document a convergence of some characteristics (i.e., the paraconid-metaconid appression) within two separate lineages of species currently contained within the genus *Gypsonictops*. Fox (1979c) suggested that *G. lewisi* and *G. hypoconus* are closely related and may in fact be two elements of a single lineage that should be tested with additional fossil discoveries. It is quite possible that *G. clemensi* and *G. illuminatus* may show a similar close relationship, although morphological differences are much greater between *G. clemensi* and *G. illuminatus* than between *G. hypoconus* and *G. lewisi*.

Gypsonictops cf. *lewisi*
(Plate 4, Figs. D–G; Table 8.)

Referred specimens. UNM-B5545, dentary fragment with M_{1-2} and alveoli for C-M_3; UNM-B5549, P_5; UNM-B5557, M_L talonid; UNM-B5522, M_L trigonid; UNM-B5558, M_3 talonid.

Horizon and locality. Quarry 1, Fossil Forest, lowermost Kirtland Shale, New Mexico.

Description. A dentary fragment (UNM-B5545), the most complete in the collection, is referred to *Gypsonictops*. It preserves the posterior alveolar wall of the canine and a circular alveolus for the single rooted P_1, two larger alveoli for a double rooted P_2, a pair of much smaller alveoli for P_3, two large alveoli for P_4 and two large alveoli for P_5. M_1 and M_2 are present and are substantially worn. Two alveoli for M_3 are also present in the dentary. Judged from the spacing and size of the alveoli, the P_1 was single rooted and much smaller than P_2. P_2 was much larger than either P_1 or P_3. P_4 and P_5 were large, long teeth that must have been as large as or larger than M_1 and M_2. A referred P_5, UNM-B5549, preserves a reasonably large, distinct, low paraconid and a massive and more or less equally proportioned, transversely opposed metaconid and protoconid. The tooth has been labially spalled and posteriorly broken, so exact dimensions are not known. A large wear facet appears on the posterior slope of the protoconid and to a much lesser extent on the posterior slope of the metaconid. The talonid is worn but appears to have possessed a large hypoconid and at least one other talonid cusp, the hypoconulid.

The largest mental foramen is present on the dentary beneath the anterior part of the posterior root of P_4. In *G. lewisi*, *G. hypoconus*, and *G. illuminatus* it appears beneath P_5. Two more foramina are present beneath P_{2-3} on UNM-B5545.

Two reasonably complete lower molars are preserved in the dentary (UNM-B5545). Both are worn, and M_2 has been damaged posteriorly. Both teeth are within the size range of *G. hypoconus* (Clemens, 1973) yet are on the large end of that range. Unlike *G. hypoconus*, they possess the more anteriorposteriorly expanded trigonids of *G. lewisi* as described by Fox (1979a) and Sahni (1972). The paraconids of M_1 and M_2, although worn, are anteriorly projecting (a condition not present in *G. hypoconus*) and are more labially placed than in *G. hypoconus*. Anterior and posterior cingulids are not equally developed. The anterior cingulid is preserved only as a small lingual shelf that served as a positioner for the next anterior tooth. The posterior cingulid is small and not well defined on M_1 and M_2 and is not present on M_3. The trigonid part of lower molar teeth is proportionally much smaller in the Fossil Forest form than in *G. hypoconus* from the type Lance Formation (Clemens, 1973) and in *G. lewisi* from the Oldman Formation. Relative sizes of the trigonid cusps are difficult to determine because of wear. However, the talonid cusps are flared, but to a lesser extent than in *G. lewisi*, and the entoconid is placed slightly posterior to the hypoconid.

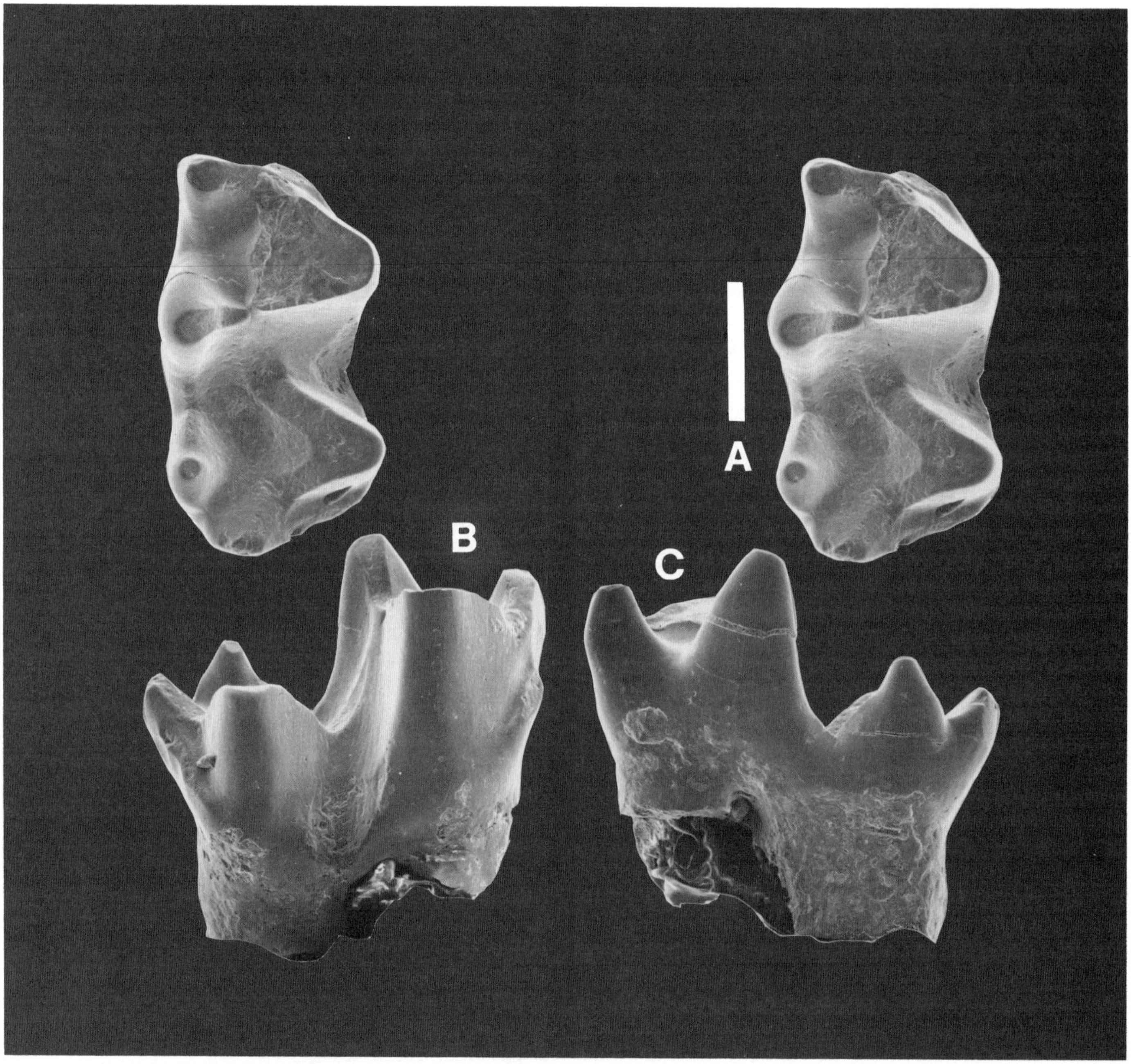

Figure 6. *Cimolestes lucasi* n. sp., UNM-B1728. M_1, holotype, A–C: A, occlusal view, stereo pair; B, labial view; C, lingual view. Scale bar represents 0.5 mm.

Discussion. Clemens (1973a) reported the occurrence of two species of *Gypsonictops* in the Hunter Wash fauna. He identified one as *G. hypoconus,* the other "was the size of *G. illuminatus* but differed from it in several morphologic features and is probably a new species" (see *Gypsonictops clemensi* n. sp.). Sahni (1972) described and named *G. lewisi* from the Judith River Formation of Montana. The smaller *Gypsonictops* species from the Fossil Forest, presumably the same identified as *G. hypoconus* by Clemens (1973a), most closely resembles *G. lewisi.* It differs from *G. lewisi,* however, in some aspects. (1) The anterior depth of the mandibular ramus is much greater (attributed to individual variation by Clemens, 1973b). (2) P_1 is only single rooted, but the character is known to be variable. (3) P_1–P_5 are somewhat longer (as judged by alveolar distances). (4) Molar talonids are longer when compared to trigonids. (5) The largest mental foramen is under P_4 instead of P_5 (the character is also variable).

These teeth may in fact represent a new species of *Gypsonictops,* but it cannot be justified on the basis of current material.

TABLE 9. CIMOLESTES LUCASI n. sp. MEASUREMENTS

		Length	Anterior Width	Posterior Width
UNM-B5546	M^3	(1.40)	(2.0)	(1.6)
UNM-B1728	M_1	1.65	0.85	0.85
UNM-B1722	M_2	(1.6)	1.15	0.95

TABLE 10. PARANYCTOIDES cf. P. STERNBERGI MEASUREMENTS

		Length	Anterior Width	Posterior Width
UNM-B1735	P^4	1.30	1.35	1.05
UNM-B1730	M_3	1.70	1.10	1.00
UNM-B5547	M_L	----	----	(1.20)

Superfamily PALAEORYCTOIDEA (Winge, 1917)
Family PALAEORYCTIDEA (Winge, 1917) Simpson, 1931

Cimolestes lucasi n. sp
Text Fig. 6A–C; Table 9.

Etymology. Named for Spencer G. Lucas, student of Cretaceous and Tertiary deposits of the San Juan Basin.
Holotype. UNM-B1728, an M_1.
Horizon and Type locality. Quarry 1, Fossil Forest, lowermost Kirtland Shale, New Mexico.
Referred specimens. UNM-B5546, an M^3; UNM-B1722, a partial M_2.
Diagnosis. Smallest known species of *Cimolestes,* being substantially smaller than *C. propalaeoryctes*; upper molars lingually broad for species of *Cimolestes*; stylar shelf not posteriorly complete on M^3; conules relatively large and lingually placed; paracone much larger than metacone; lower molars with open trigonids; cristid obliqua terminates anteriorly on the posterior labial slope of the metaconid; talonids of molars are nearly as wide as trigonids except on M_2 where the talonid is noticeably narrower (as in *C. propalaeoryctes*); entoconid and hypoconid more medially positioned and closer to each other than in other species.
Description. UNM-B5546 is identified as M^3 and resembles other species of *Cimolestes* where M^3 is known. The protocone and lingual portion of the tooth is much broader than in other species of *Cimolestes.* The paracone is substantially larger than the metacone. The stylar shelf is distinct to a position labial to the central part of the centocrista; it is not present around the base of the metacone. The metaconule on M^3 is more labially positioned than the paraconule. It is just lingual to the metacone, whereas the paraconule is just labial to the protocone. No posterior cingulum is present, nor are lingual cingulae around the base of the protocone. The anterior labial margin has been broken away so that the parastyle and remaining stylar shelf are not preserved.

There are two lower teeth belonging to *C. lucasi* in the collection. UNM-B1728, the holotype, is interpreted as being an M_1 because of its similarity to M_1 of *C. propalaeoryctes.* As in *C. propalaeoryctes,* the trigonid is anteriorly expanded, and the metaconid is slightly posterior to the protoconid. This is not seen in UNM-B1722, the other referred lower molar. The M_1 paraconoid is large but smaller than the metaconid and anteriorly extended. Two trigonid carnassial notches—anterior and posterior—are present. The trigonid forms the largest part of the tooth and is not as high as that of the M_2 (UNM-B1722); an additional condition shared with *C. propalaeoryctes.* The three major talonid cusps are present on M_{1-2} with the small but distinct entoconid and large hypoconid placed closer to each other (narrower talonid basin) than in other species of *Cimolestes.* The paraconid, metaconid, and entoconid are nearly aligned, which is suggestive of marsupial affinities. However, the hypoconulid is relatively large, posteriorly projecting, but smaller than the hypoconid. The hypoconid is nearer to the entoconid than the hypoconid but definitely not "twinned" in the marsupial fashion. The entoconid of UNM-B1722 is larger than that of UNM-B1728. Well-defined anterior and posterior cingulids occur on both M_1 and M_2, and extend to the basal labial margin. The M_2 has a proportionately narrower talonid when compared with M_1, but the trigonid is much higher by comparison. The distinctive feature of both teeth is the cristid obliqua that terminates on the posterior base of the metaconid, but not nearly to the extent seen in Ectocentrocristinae (new subfamily, this report) or in most species of *Pediomys.*
Discussion. Clemens (1973) listed a eutherian species, smaller than *C. incisus,* as cf. *Cimolestes* sp. in his faunal list for the Hunter Wash fauna and probably has specimens of *C. lucasi* in that collection. This species is interesting from a number of standpoints. The relationship between *C. lucasi* and *C. propalaeoryctes* may be sufficiently documented to construct a new genus to include those two species and to further segregate other species currently contained within *Cimolestes* into new genera. Also, the narrowness of the protocone in most species of *Cimolestes* appears to be derived with respect to other eutherian forms. In this respect, the Fossil Forest cimolestid is primitive in that it retains a relatively broad protocone.

Suborder LIPOTYPHLA Haeckel 1866
Superfamily ERINACEOIDEA (Fischer Waldheim) 1817
cf. Family NYCTITHERIIDAE Simpson 1928
Paranyctoides cf. *P. sternbergi* Fox, 1979c
(Plate 2, Figs G–L; Table 10.)

Holotype. UA-14822, incomplete right dentary with P_4M_1.
Horizon and type locality. Oldman Formation, near Irvine, Alberta.
Known distribution. Oldman Formation, Alberta; lowermost Kirtland Shale, New Mexico.

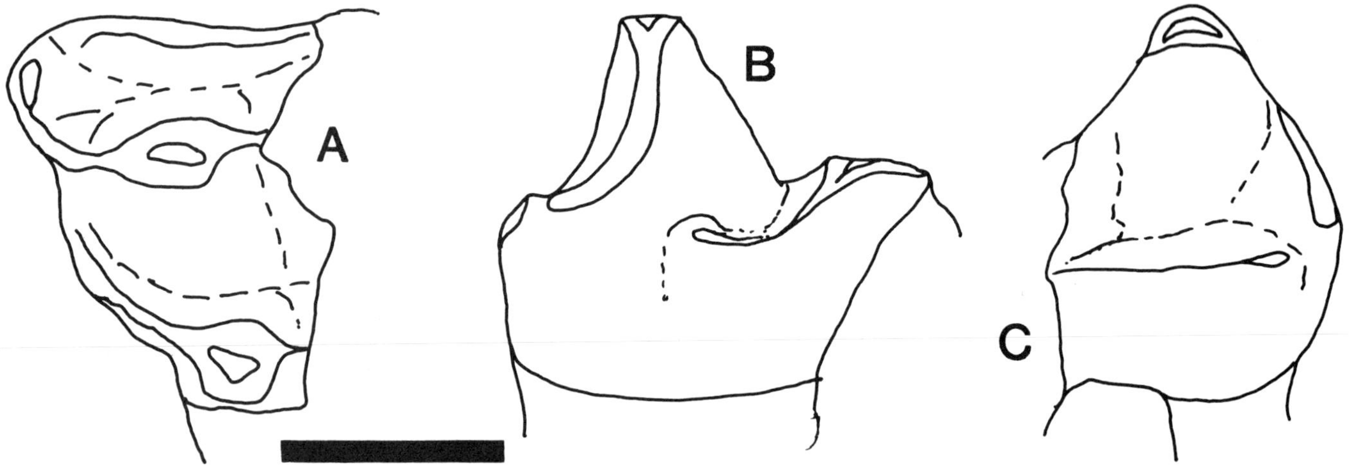

Figure 7. Eutherian of uncertain affinities. UNM-B5550, an upper molar fragment: A. occlusal view; B, anterior view; C, labial view. Scale bar represents 1 mm.

Referred specimens. UNM-B1735, a P^4; UNM-B1730, an $M_{?3}$; UNM-B5547, a lower molar talonid.

Description. An isolated P^4 (UNM-B1735) is questionably referred to *P.* cf. *P. sternbergi.* It is eutherian, not metatherian, because of its reduced stylar shelf and prominent protocone. The stylar shelf is complete, although reduced, and possesses a stylar cusp in the D position and strong anterior and posterior cingulae. The main cusp, the paracone, is conical although slightly flattened labially. No metacone is observed; crests simply descend the anterolabial and posterolabial slopes of the paracone. A small notch is seen on the posterior crest approximately one-third the way down the slope.

UNM-B1730 is a left lower molar (M_3); the trigonid is slightly wider than the talonid and is only moderately higher. The trigonid cusps are closely spaced with a low paraconid, not lingual to the midline, and a well-developed metaconid, the tip of which is broken. The protoconid is the tallest trigonid cusp. The talonid is broadly basined with a well-developed hypoconid, entoconid, and hypoconulid. The hypoconulid is bulbous and projects posteriorly; it is placed nearer the entoconid than the hypoconid. The hypoflexid is prominent; the cristid obliqua joins the protoconid and continues posteriad to the base of the hypoconulid as a postcristid. The precingulid is well developed and sharply sloping. The postcingulid is short but prominent and forms a sloping shelf behind the hypoconulid.

Discussion. UNM-B1730, an M_3, possesses all of the characteristics used by Fox to define *P. sternbergi.* UNM-B1730 is somewhat larger than the published measurements of M_3 (Fox, 1979c), and the hypoflexid is not quite so deep (inferred from illustrations of M_2). It does, however, possess a distinct lingual lean of the crown, low trigonid cusps, a reasonably deep hypoflexed notch, high entoconid, and strong posteriorly projecting hypoconulid (not twinned with the entoconid). UNM-B5547, a questionably referred talonid, retains an assumed deep hypoflexid and size distinctly smaller than that of lower molars referrable to other eutherian taxa in the fauna.

Fox (1970) described, but did not name, a eutherian upper molar from the Milk River Formation that was distinct in a number of features from other Cretaceous eutherians. Clemens (1973) identified an upper molar from the Hunter Wash fauna that he thought was a representative of the lineage discovered by Fox. There are additional teeth in the collection from the Fossil Forest that do, in fact, resemble the tooth described by Fox, but they are much larger.

The above P^4 is much too small to be related to the unnamed eutherian form from the Fruitland Formation, yet it is about the size of the form from the Milk River Formation. The protocone is smaller than in all species of *Gypsonictops* and *Cimolestes* and is placed not nearly so lingually as in those genera. The P^4 is substantially reduced in height when compared with *Gypsonictops* and *Cimolestes* and as such agrees with the description of Fox (1979c) for *P. sternbergi,* even though the species was defined only on lower dental elements. At best, the reference of the P^4 to *P.* cf. *sternbergi* is questionable.

Eutherian of uncertain ordinal affinities
(Text Fig. 7.)

Referred specimens. UNM-B5550, an M^U fragment; UNM-B5498; an M^U fragment.

Description and discussion. Two M^U fragments from Quarry 1 within the Fossil Forest are closely related to an unnamed eutherian described by Fox (1970) from the Milk River Formation of Alberta. Clemens (1973a) noted the presence in the Hunter Wash fauna of a member of the lineage described by Fox. We suspect that the same taxon mentioned by Clemens is also present in Quarry 1 and probably represents a new species closely related to the unnamed form described by Fox.

Both referred fragments preserve the metacone and related stylar region. UNM-B5550 preserves the metaconule, associated crests, and part of the paracone. Although fragmentary, these teeth show the enlarged conules described by Fox (1970), substantially narrowed stylar region typical of eutherians, and prominent posterior metaconule wing forming the prominent posterior cingulum that is not confluent with crests associated with the protocone. A small cusp D is located on the labial stylar shelf crest. It is confluent with that crest and is simply an elongate swelling labially opposite the metacone. No other stylar cusps are preserved.

The paracone and metacone are relatively high but not so high as in some other eutherians. They are partially fused at their common base, although the paracone is not totally preserved, and it appears that the paracone was substantially larger than the metacone. The stylar shelf narrows and may actually be discontinuous around the base of the paracone as in *Pediomys.* The stylar shelf is much more reduced than that of *Pediomys,* the conules are much larger in the unknown eutherian form, and it is about 40 percent larger than *Pediomys fassetti* n. sp. described in this paper.

Although very similar to the form described by Fox, the Fossil Forest form is distinctly larger. It is not clear whether this observed larger size is a result of comparing different dental elements or if the Fossil Forest form is measurably larger. It is also possible, but unlikely, that these teeth are elements of the upper dentition of *Paranyctoides* cf. *P. sternbergi.*

AGE OF THE FRUITLAND AND KIRTLAND FORMATIONS

North American Cretaceous mammalian faunas are placed within loosely defined "land mammal ages," each of which is typified by a characteristic fauna. Of particular interest to this discussion will be those land mammal ages surrounding the Campanian-Maastrichtian and Cretaceous-Tertiary boundaries. In order of increasing age, these units are Bugcreekian (see Sloan, this volume), Lancian, Edmontonian (*sensu* Clemens and others, 1979), Judithian, and Aquilian. The Edmontonian age is thought to straddle the Campanian-Maastrichtian boundary. Reasonably large fossil mammal faunas are known from the Bugcreekian, Lancian, Judithian, and Aquilian intervals. The only known Edmontonian mammalian fauna, which is meager at best, is from the St. Mary River Formation (Sloan and Russell, 1974).

It has only been in the last few years that enough information has been available to make Cretaceous, mammal-based, biostratigraphic correlations. Clemens (1973a) briefly described the mammalian fauna he collected along Hunter Wash and concluded that the Hunter Wash fauna was older than faunas studied by him from the type Lance Formation, Wyoming (Clemens, 1964, 1966, and 1973b). Other faunas used in comparison with the Fossil Forest therian mammals include several faunas from the Hell Creek Formation in the area of the Fort Peck Reservoir (Sloan and Van Valen, 1965; Van Valen, 1978; and Archibald, 1982); the fauna from the Edmonton Formation (now the Scollard Formation) (Lillegraven, 1969); the fauna from the Judith River Formation, Lethbridge Coal Member, described by Sahni (1972); the Judithian equivalent fauna from the Lethbridge Coal Member of the Oldman Formation in Canada (now also called the Judith River Formation) described by Fox (several publications between 1970 and 1981); and the Aquilian mammalian fauna from the Milk River Formation of Canada (Fox, 1970 and 1971). It is possible to compare the fauna collected by us in the Fossil Forest with other known mammalian faunas and be able at least to determine a relative age of the Quarry 1 mammals. All of these faunas, including to some extent the mammals from the Fossil Forest (see Fassett, this volume) lie within continental deposits that interfinger with marine rocks for which ammonite zones are known.

Mammalian faunal comparisons provide the following conclusions:

1. The Fossil Forest fauna is not closely comparable to Bugcreekian faunas because of the lack of diversity of ungulate taxa and notable differences between species of *Alphadon, Pediomys,* and other insectivore species.

2. Most of the taxa here described retain characters that are thought to be primitive (i.e., pediomyids with the cristid obliqua not fully labially directed). These are characters that Clemens (1966 and 1973b) used to define higher taxonomic categories. The Fossil Forest forms, however, do not show the derived features that Clemens used to define species within those higher categories and as such can be easily separated from taxa described by Clemens (1966 and 1973b) from the type Lance Formation. The reader is referred to detailed discussions regarding primitive and advanced character states in each of the taxonomic descriptions of the present work. Clemens (1973a) thought the small Hunter Wash fauna was more primitive than materials he had studied from the Lance Formation. *Alphadon parapraesagus* described here may also occur in a Yale University locality in the Lance Formation but from a locality stratigraphically lower than those from which Clemens (1964, 1966, and 1973b) obtained his larger collections (see Clemens, 1966, *Alphadon* cf. *rhaister*). *Alphadon* cf. *wilsoni* described here resembles *A. wilsoni* (Lillegraven, 1969) from the Scotland formation; however, on the basis of a very limited sample size from the Fossil Forest, some differences seem to persist between the two samples. Therefore, the Quarry 1 fauna from the Fossil Forest may have two taxa that may be shared with Lancian faunas in Wyoming and Alberta.

3. The most favorable comparison of the Fossil Forest therian mammals is with the Judith River fauna described by Sahni (1972) and with the Oldman fauna described by Fox (1977, 1979a, 1979b, 1979c, and 1981). The species level taxonomy, particularly of the Fossil Forest forms, is notably different from the other two faunas. At this time it is not known if these differences are time significant or are caused by geographic variation. The overall faunal similarity, however, is quite striking. The presence of several taxa (*Aquiladelphis* n. sp. and the eutherian of uncertain affinities) may suggest an older age for the Quarry 1

mammals because of their similarity to forms described by Fox (1970 and 1971) from the Early Campanian Milk River Formation. However, the bulk of the fauna appears to be most closely related to known Judithian species.

The formational contact (see Brookins and Rigby, this volume, for a discussion of local stratigraphic problems) of the Fruitland and Kirtland formations on the southwest flank of the San Juan Basin, as judged on the basis of mammalian biostratigraphy, is approximately equivalent to the age of the Lethbridge Coal Member of the Judith River Formation. This requires that the Fossil Forest Quarry 1 fauna be Middle to Upper Campanian in age. This determination is consistent with a number of other lines of evidence discussed in Brookins and Rigby; Fassett; Fassett, Lucas, and O'Neill; and Newman, elsewhere in this volume. This conclusion necessitates a revision in the numbering of magnetochrons by Lindsay and others (1981) as revised by Butler (1984).

If the Fossil Forest Quarry 1 therian mammals are Late Campanian in age as we suspect, then at least two alternative magnetostratigraphic correlations must now be addressed (Fig. 8). Each alternative correlation considers the two long normal zones and their medial small reversed zone as a single normal zone; either magnetochron 33 (alternative 1) or 32 (alternative 2). The therian mammalian fauna, here described, is more consistent with stratigraphic alternative 1 and much less with alternative 2.

Data from Brookins and Rigby (this volume) indicate that the age of the Quarry 1 mammals must be older than average K-Ar radiometric dates (73.5 ± 2.7 Ma, new decay constants) obtained from air-fall tuffs at a higher stratigraphic level than Quarry 1. These mammals similarly must be younger than 74.5 ± 2.7 Ma, the age (corrected to new decay constants) reported by Obradovich and Cobban (1975) for the underlying *Baculites scotti* ammonite zone, which contains the Huerfanito Bentonite Bed in the Lewis Shale (Cobban, 1973). Quarry 1 mammals, therefore, must be temporally located very near the Campanian-Maastrichian boundary.

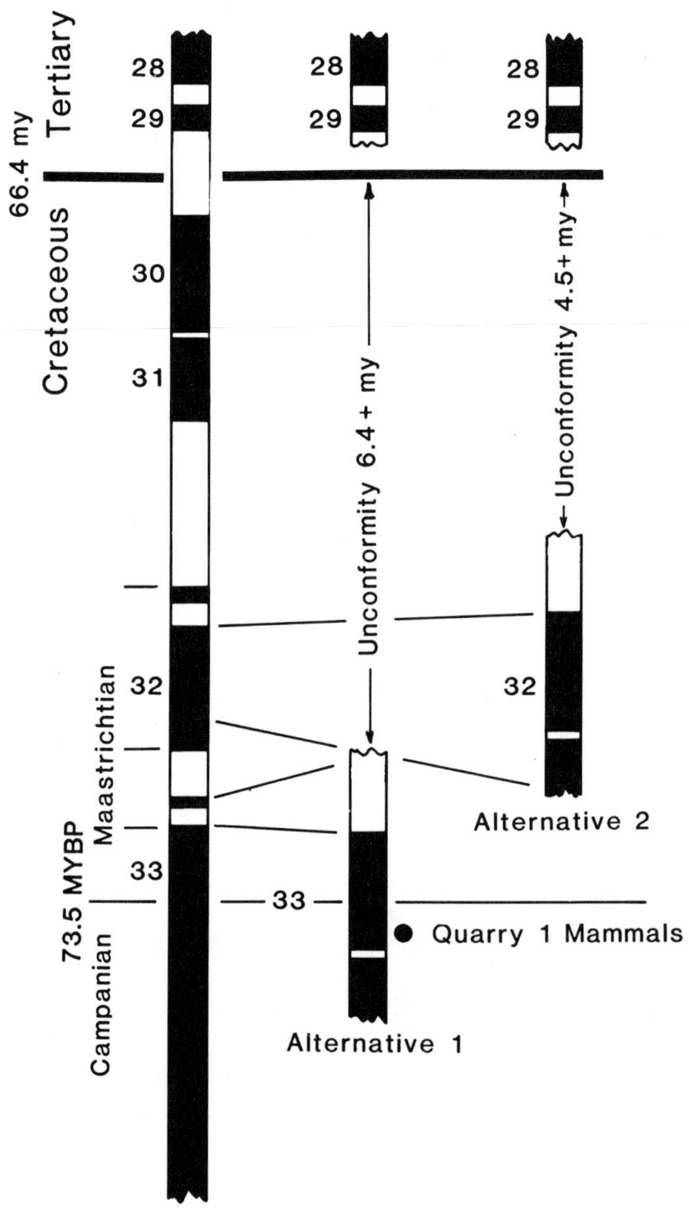

Figure 8. Alternative correlations of magnetozones with the standard reference section modified from LaBrecque and others (1977). The "Judithian" mammals from Quarry 1 are plotted in their approximate stratigraphic position in Alternative 1. Both correlations require that an unconformity be present between Quarry 1 and "Puercan" Tertiary mammal localities higher in the section (Rigby and Lucas, 1977).

ACKNOWLEDGMENTS

We express our thanks to our joint researchers and field assistants in the Fossil Forest projects, without whose help none of the work would have been possible. These include Adrian Hunt, Coleman Robison, Mary Beth Lucas, Niall Mateer, Julie Menack, Joseph Hartman, Tom Lehman, Richard Precek, Barry Kues, Spencer Lucas, Joy McKinney, Elizabeth Oswald, George Engleman, Mike O'Neill, David Frailey, and numerous volunteers from the Albuquerque area rock and mineral clubs. This work was supported by the Albuquerque District, U.S. Bureau of Land Management, Paul Applegate, District Manager; by the New Mexico Bureau of Mines and Mineral Resources, Frank Kottlowski, Director; and by a donation of the Hill Foundation to the Dean Clarence Manion Fund of the University of Notre Dame. Scanning electron microscope illustrations were prepared by Carl Allen of the Institute of Meteoritics, University of New Mexico, Klaus Keil, Director. We thank Robert E. Sloan, William A. Clemens, Malcolm C. McKenna, and Judy A. Schiebout for review of the manuscript.

REFERENCES CITED

Archibald, J. D., 1982, A study of Mammalia and geology across the Cretaceous-Tertiary boundary in Garfield County, Montana: University of California Publications in Geological Sciences, v. 122, 286 p.

Bauer, C. M., 1917, Contributions to the geology and paleontology of San Juan Basin, New Mexico; (1) Stratigraphy of a part of the Chaco River Valley: U.S. Geological Survey Professional Paper 980, p. 271–278.

Bauer, C. M., and Reeside, J. B., 1921, Coal in the middle and eastern parts of San Juan County, New Mexico: U.S. Geological Survey Bulletin 716, p. 155–248.

Butler, R. F., 1984, Paleomagnetic and rock-magnetism of continental deposits, San Juan Basin, New Mexico: Geological Society of America Abstracts with Programs, v. 16, p. 217.

Clemens, W. A., Jr., 1964, Fossil mammals of the type Lance Formation, Wyoming; pt. I, introduction and Multituberculata: University of California Publications in Geological Sciences, v. 48, p. 1–105.

—— , 1966, Fossil mammals of the type Lance Formation, Wyoming; pt. II, Marsupialia: University of California Publications in Geological Sciences, v. 66, p. 1–122.

—— , 1973a, The roles of fossil invertebrates in interpretation of Late Cretaceous stratigraphy of the San Juan Basin, New Mexico, *in* Fassett, J. E., ed., Cretaceous and Tertiary rocks of the southern Colorado Plateau: Durango, Colorado, Four Corners Geological Society Memoir, p. 154–167.

—— 1973b, Fossil mammals of the type Lance Formation, Wyoming; pt. III, Eutheria and summary: University of California Publications in Geological Sciences, v. 94, 102 p.

Clemens, W. A., Jr., Lillegraven, J. A., Lindsay, E. H., and Simpson, G. G., 1979, Where, when, and what; A survey of known Mesozoic mammal distribution, *in* Lillegraven, J. A., Kielan-Jaworowska, Z., and Clemens, W. A., Jr., eds., Mesozoic mammals; the first two-thirds of mammalian history: University of California Press, p. 7–59.

Cobban, W. A., 1973, Significant ammonite finds in uppermost Mancos shale and overlying formations between Barker Dome, New Mexico, and Grand Junction, Colorado, *in* Fassett, J. E., ed., Cretaceous and Tertiary rocks of the southern Colorado Plateau: Four Corners Geological Society Memoir, p. 148–153.

Cumella, S. P., 1981, Sedimentary history and diagenesis of the Pictured Cliffs Sandstone, San Juan Basin, New Mexico and Colorado: Texas Petroleum Research Committee, Report no. UT 81-1, 219 p.

Erpenbeck, M. F., 1979, Stratigraphic relationships and depositional environments of the Upper Cretaceous Pictured Cliffs Sandstone and Fruitland Formation, southwestern San Juan Basin, New Mexico [M.S. thesis]: Lubbock, Texas Tech University, 78 p.

Fassett, J. E., and Hinds, J. S., 1971, Geology and fuel resources of the Fruitland Formation and Kirtland Shale of the San Juan Basin, New Mexico: U.S. Geological Survey Professional Paper 676, 76 p.

Fox, R. C., 1970, Eutherian mammals from the early Campanian (Late Cretaceous) of Alberta, Canada: Nature, v. 227, p. 630–631.

—— , 1971, Marsupial mammals from the early Campanian Milk River Formation, Alberta, Canada: Zoological Journal of the Linnean Society, v. 50, Supplement 1, p. 145–164.

—— , 1977, Notes on the dentition and relationships of the Late Cretaceous insectivore *Gypsonictops* Simpson: Canadian Journal of Earth Sciences, v. 14, no. 8, p. 1823–1831.

—— , 1979a, Mammals from the Upper Cretaceous Oldman Formation, Alberta; I. *Alphadon* Simpson (Marsupialia): Canadian Journal of Earth Sciences, v. 16, no. 1, p. 91–102.

—— , 1979b, Mammals from the Upper Cretaceous Oldman Formation, Alberta; II. *Pediomys* Marsh (Marsupialia): Canadian Journal of Earth Sciences, v. 16, no. 1, p. 103–113.

—— , 1979c, Mammals from the Upper Cretaceous Oldman Formation, Alberta; III. Eutheria: Canadian Journal of Earth Sciences, v. 16, no. 1, p. 114–125.

—— , 1981, Mammals from the Upper Cretaceous Oldman Formation, Alberta; V. *Eodelphis* Matthew, and the evolution of the Stagodontidae (Marsupialia): Canadian Journal of Earth Sciences, v. 18, no. 2, p. 350–365.

Hartman, J. H., 1981, Mollusca from Upper Cretaceous Fruitland and Kirtland Formations, western San Juan Basin, New Mexico [abs.]: American Association of Petroleum Geologists Bulletin, v. 65, p. 560.

Hawley, J. W., and Love, D. W., 1981, Overview of geology as related to environmental concerns in New Mexico, *in* Wells, S. G., and Lambert, W., eds., Environmental geology and hydrology in New Mexico: New Mexico Geological Society Special Publication No. 10, p. 1–10.

Hunt, A., 1983, Stratigraphy, sedimentology, taphonomy, and magnetostratigraphy of the Fossil Forest area, San Juan Basin, New Mexico [M.S. thesis]: Socorro, New Mexico Institute of Mining and Technology, 380 p.

Kues, B. S., Froehlich, J. W., Schiebout, J. A., and Lucas, S. G., 1977, Paleontological survey, resource assessment, and mitigation plan for the Bisti–Star Lake area, northwestern New Mexico: Unpublished report to the U.S. Bureau of Land Management, Albuquerque District Office, 525 p.

LaBrecque, J. L., Kent, D. V., and Cande, S. C., 1977, Revised magnetic polarity time scale for Late Cretaceous and Cenozoic time: Geology, v. 5, p. 330–335.

Lillegraven, J. A., 1969, Latest Cretaceous mammals of upper part of Edmonton Formation of Alberta, Canada, and review of Marsupial-placental dichotomy in mammalian evolution: Paleontological Contributions, Lawrence, University of Kansas, v. 50, p. 1–122.

Lindsay, E. H., Butler, R. F., and Johnson, N. M., 1981, Magnetic polarity zonation and biostratigraphy of Late Cretaceous and Paleocene deposits, San Juan Basin, New Mexico: American Journal of Science, v. 281, p. 390–435.

McKenna, M. C., 1965, Collecting microvertebrate fossils by washing and screening, *in* Kummel, B., and Raup, D., eds., Handbook of paleontological techniques: San Francisco, W. H. Freeman & Co., p. 193–203.

Mytton, J. W., 1979, Preliminary geologic map of Chaco Canyon 1°X1/2° Quadrangle, showing coal zones of Fruitland Formation, San Juan, Rio Arriba, and Sandoval Counties, New Mexico: U.S. Geological Survey Miscellaneous Field Studies Map MF-1104, scale 1:100,000.

Obradovich, J. C., and Cobban, W. A., 1975, A time scale for the Late Cretaceous of the Western Interior of North America, *in* Caldwell, W.G.E., ed., The Cretaceous system in the Western Interior of North America: Geological Association of Canada Special Paper 13, p. 31–54.

O'Sullivan, R. B., Repenning, C. A., Beaumont, B. C., and Page, H. G., 1972, Stratigraphy of the Cretaceous rocks and the Tertiary Ojo Alamo Sandstone, Navajo and Hopi Indian reservations, Arizona, New Mexico, and Utah: U.S. Geological Survey Professional Paper 521, 65 p.

Reeside, J. B., Jr., 1924, Upper Cretaceous and Tertiary formations of the western part of the San Juan Basin of Colorado and New Mexico: U.S. Geological Survey Professional Paper 134, p. 1–70.

Rigby, J. K., Jr., and Lucas, S. G., 1977, Fossil mammals from the Ojo Alamo Sandstone: New Mexico Geological Society Guidebook, 28th Field Conference, San Juan Basin, Supplement, p. 55–56.

Robison, C., Hunt, A., Wolberg, D., 1982, New Late Cretaceous leaf locality from lower Kirtland Shale Member, Bisti area, San Juan Basin, New Mexico: New Mexico Geology, v. 4, no. 3, p. 42–45, 47.

Sahni, A., 1972, The vertebrate fauna of the Judith River Formation: American Museum of Natural History Bulletin, v. 147, no. 6, p. 323–412.

Scott, G. R., O'Sullivan, R., Mytton, J. W., 1979, Reconnaissance geologic map of the Alamo Mesa West quadrangle, San Juan County, New Mexico: U.S. Geological Survey Miscellaneous Field Study Map ME-1074, 1 sheet, scale 1:24,000.

Sloan, R. E., and Russell, L. S., 1974, Mammals from the St. Mary River Formation (Cretaceous) of southwestern Alberta: Life Sciences Contributions, Royal Ontario Museum, v. 95, p. 1–20.

Sloan, R. E., and Van Valen, L., 1965, Cretaceous mammals from Montana: Science, v. 148, p. 220–227.

Van Valen, L., 1978, The beginning of the age of mammals: Evolutionary Theory, v. 4, no. 2, p. 45–80.

MANUSCRIPT ACCEPTED BY THE SOCIETY MARCH 24, 1986

Printed in U.S.A.

Geological Society of America
Special Paper 209
1987

Lithofacies relationships and depositional environment of the Tertiary Ojo Alamo Sandstone and related strata, San Juan Basin, New Mexico and Colorado

Pamela G. L. Sikkink
U.S. Geological Survey
MS 913, Box 25046, Denver Federal Center
Denver, Colorado 80225

ABSTRACT

Lithofacies analysis of the Tertiary Ojo Alamo Sandstone and related strata in the San Juan Basin indicates that Laramide (Late Cretaceous–early Tertiary) volcanism and uplift north of the present-day San Juan Basin controlled sedimentation patterns of Upper Cretaceous and lower Tertiary rocks. Eight major lithofacies reflect changes in sedimentation that occurred during this time.

The Ojo Alamo Sandstone is characterized in most areas of the San Juan Basin by a pebbly, trough-crossbedded lithofacies. A related channel-form sandstone and shale facies makes up the Ojo Alamo at Mesa Portales. Both lithofacies include both sediment derived from north of the present-day San Juan Basin and sediment eroded and reworked from (1) a carbonaceous shale and channel-form sandstone facies, (2) a shale and volcaniclastic sandstone facies, and (3) a volcaniclastic conglomerate and sandstone facies.

The pebbly, trough-crossbedded lithofacies, which was deposited by streams on alluvial plains, differs in grain size, pebble composition, and transport direction on the east and west sides of the present-day basin. At least two distinct source areas for the streams are suggested by these differences. One source is in the area of the present-day Needle Mountains and western San Juan Mountains. A second source is located in the area of the central to eastern San Juan Mountains of southwest Colorado. Sediments deposited by alluvial streams in the western San Juan basin include sand- and pebble-size material. Initially, Ojo Alamo streams carried up to 25 percent volcanic pebbles reworked from the Animas Formation or from Upper Cretaceous andesitic flows in the source area. Later streams, however, carried an increasing percentage of quartz pebbles over volcanic pebbles.

Lithofacies of the Ojo Alamo in the eastern San Juan Basin include channel sandstone and conglomerates and a channel-form sandstone and shale facies. Compared to sediments of the western alluvial complex, the eastern sediments (mapped as Ojo Alamo Sandstone, upper part of the Animas Formation, and Nacimiento Formation) are finer grained, contain few pebbles, contain less than 1 percent volcanic pebbles, and show different transport directions. Mudstone interbeds are thicker and more abundant, especially at Mesa Portales where an accompanying down-dip change in the alluvial system contributes to formation of the channel-form sandstone and shale lithofacies.

INTRODUCTION

During the past 70 years, one of the most persistent debates concerning San Juan Basin geology has centered around the age, nomenclature, and lithologic correlation of the Ojo Alamo Sandstone. One reason for disagreement is that few scientists have studied the overall facies relationships, physical processes, and sedimentary characteristics of the unit throughout its depositional extent in the basin (Fig. 1). In the past decade, interest in the Ojo Alamo has increased. For one thing, the Ojo Alamo has become important because of its stratigraphic position at or near the Cretaceous-Tertiary boundary. The "Cretaceous-Tertiary boundary problem," as it is frequently referred to, renewed interest in dating the Ojo Alamo Sandstone and sparked new controversy when various dating methods gave conflicting results. Second, the Ojo Alamo Sandstone became economically important in the late 1970s when a low-grade, disseminated deposit of uranium was discovered in the formation near Cuba, New Mexico.

This report describes the lithology, sedimentary structures, transport directions, and petrography of the Ojo Alamo Sandstone and proposes a genetic model for the origin of the formation. In addition, the study examines the lithofacies relationships of the Ojo Alamo to adjacent Cretaceous and Tertiary formations, such as the Fruitland Formation, Kirtland Shale, Animas Formation, and Nacimiento Formation, and proposes a depositional model for Late Cretaceous–early Tertiary sedimentation in the San Juan Basin.

METHODS OF STUDY

Twenty-five stratigraphic sections were measured in the San Juan Basin. Section sites were selected on the basis of the amount of exposure, completeness of stratigraphic section, accessibility, and even geographic distribution around the basin (Fig. 1, Table 1). Descriptions of these sections provide detailed data on lithology, mineralogy, texture, sedimentary structures, biological constituents, and sediment transport directions in the Ojo Alamo Sandstone; these data are incorporated into this report. The measured sections also include units mapped as Fruitland Formation, Kirtland Shale, Animas Formation, and the lower portion of the Nacimiento Formation. Several lithologic descriptions from stratigraphic sections previously measured in the eastern and northern basin by Anderholm (1980), Baltz (1967), Baltz and others (1966), and Wood and others (1948) have also been incorporated into this report. These descriptions supply supplemental information in areas where detailed stratigraphic sections were not measured by the author and provide comparisons with stratigraphic sections described for this report. Finally, 11 mineralized core sections from Mesa Portales and Eagle Mesa were described, then sampled to determine their petrography and geochemistry (Fig. 1, loc. 21–27). Rock descriptions and samples from all measured sections and core holes were grouped into eight distinct lithofacies and then into depositional facies.

Because the study area is large, five 1 × 2 degree geologic maps were required to prepare the base map for this report (Fig. 1). These maps provided a convenient scale for plotting locations of measured sections, transport-direction data, and lithofacies. Where more detailed geologic maps and stratigraphic studies existed in the basin, they were used in field studies. These included Baltz (1967), Barnes (1953), Barnes and others (1954), Fassett (1966), Hinds (1966), Larson and Cross (1956), O'Sullivan and others (1972), and Wood and others (1948).

Trough crossbed axes were measured on as many complete or nearly complete trough crossbeds as possible at 43 locations throughout the outcrop area (Fig. 1). In addition to measurements on Ojo Alamo rocks, crossbed measurements were made on sandstone lenses within the Animas conglomeratic sandstones near Durango (Fig. 1, loc. 44) and on sandstones of the shale and volcaniclastic sandstone facies (mid to upper Animas Formation) in the northern basin (Fig. 1, loc. 38). These measurements were made to determine if changes in transport direction occurred between the time of lower Animas deposition and Ojo Alamo deposition.

At each locality, an average mean vector direction was determined for that location using methods described by Davis (1973). Where more than one conglomeratic sandstone bed occurs (locs. 22, 25, 38, and 44), trough axes were measured in each bed to determine if changes in transport direction occurred within the formation through time. These beds are designated as upper, middle, and lower on the summary diagrams. A composite rose diagram and vector mean for the entire Ojo Alamo–type depositional system were constructed after individual-location analyses were completed.

Pebble composition and pebble imbrication studies were conducted at various locations throughout the basin. These included rock type, size, shape, and orientation of individual pebbles within gravel beds. One hundred pebbles were counted and measured at each locality for representative size and percentage composition.

In addition to the field data, 162 petrographic samples (89 from outcrop and 73 from core), 114 geochemical analyses, 10 carbon- and oxygen-isotope analyses, five Scanning Electron Microscope (SEM) samples, and a number of geophysical logs and X-ray analyses from both Cretaceous and Tertiary rocks form the data base.

PREVIOUS WORK

The history of the nomenclature and definition of the Ojo Alamo Sandstone has been long and controversial because most studies on it have been areally or topically restricted. Barnum Brown (1910), of the American Museum of Natural History, first applied the name Ojo Alamo to an interval of shale containing dinosaur bones with an overlying conglomerate near Ojo Alamo Trading Post. Brown did not define a basal contact on his "Ojo Alamo beds"; he named them only to identify stratigraphically a vertebrate fossil–bearing layer in the Cretaceous rocks of the San Juan Basin. Bauer (1916) first defined upper and lower contacts

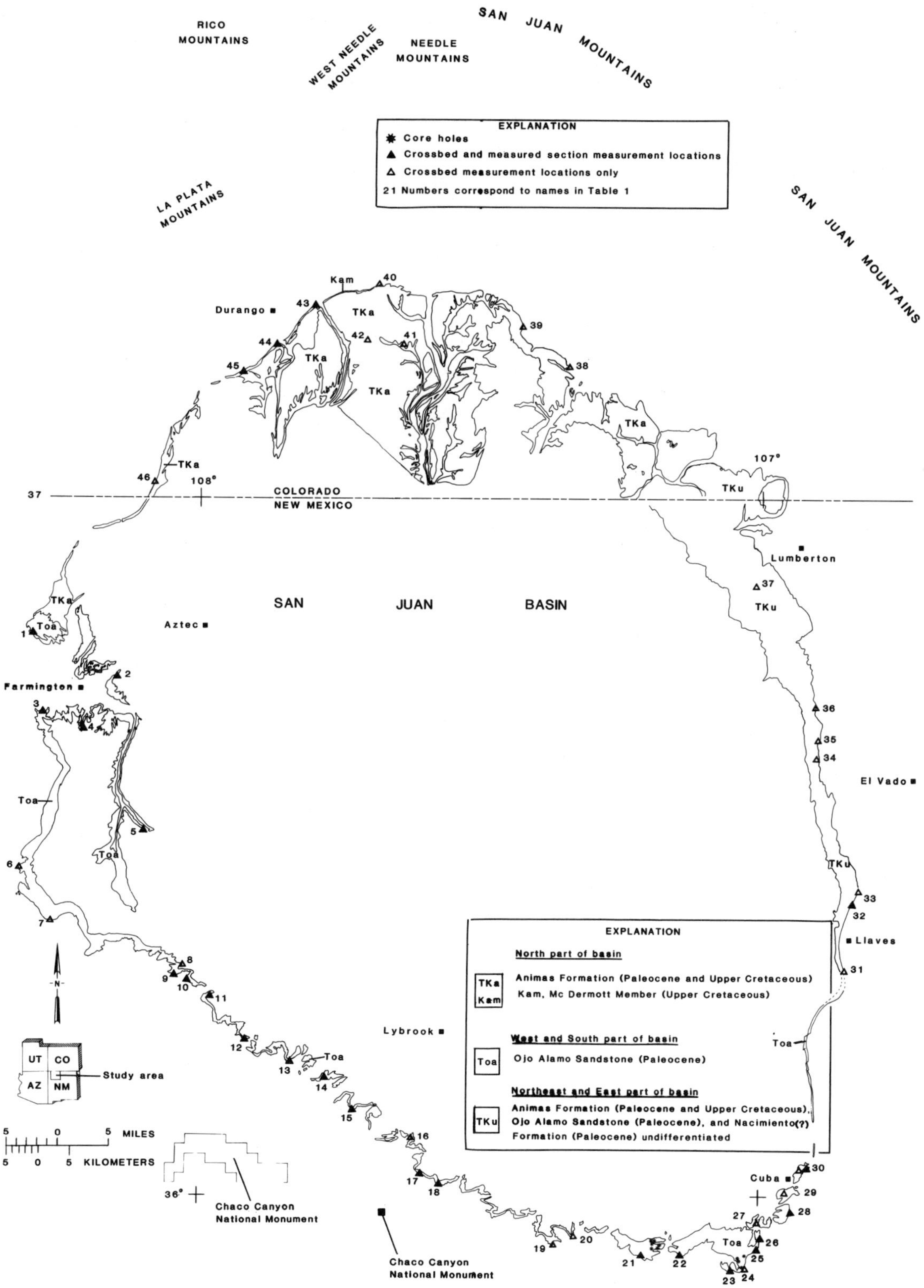

Figure 1. Index map showing locations of study area, core holes, crossbed measurements, and stratigraphic sections. Base map and geology from Wyant and Olson (1978), Manley and others (1978), Haynes and others (1972), Steven and others (1974), and O'Sullivan and Beikman (1963). Numbers correspond to location names in Table 1.

TABLE 1. LOCATION NAMES FOR FIGURE 1

No.	Location	No.	Location
1	Pinon Mesa West	24	Arroyo Jarido Spring
2	Animas River	25	Mesa Portales East
3	Kirtland, N.M.	26	Arroyo Piedre Lumbre
4	Head Canyon	27	Arroyo Chijuilla
5	Gallegos Canyon	28	Arroyo Hondo
6	Pinabete Arroyo	29	J.R. Davidson
7	Brimhall Wash	30	Vallecitos Rio Puerco
8	Ojo Alamo Trading Post	31	Arroyo Blanco
9	Ojo Alamo Wash S. Mesa	32	Chupadera Spring
10	De Na Zhin	33	Chupadera Spring-82
11	Huerfano S.W.	34	Stinking Lake
12	Tsun Je Zhin	35	Ross Spring
13	Kimbeto	36	Boulder Lake
14	Betonnie Tsosie Wash	37	Dulce Lake
15	Escavada Wash	38	Fossett Gulch
16	Lybrook Southeast	39	Pargin Mountain
17	Ojo Socorro	40	Rules Hill
18	Canada Alamos	41	Gem Village
19	Casi Cejo del Raton	42	Dry Creek
20	Ojo Encino Mesa	43	Terry Palmer Ranch
21	Eagle Mesa	44	Loma Linda
22	Fork Rock Mesa	45	Indian Creek
23	Mesa Portales South	46	McDermott Arroyo

of a unit he called Ojo Alamo Sandstone at Ojo Alamo Wash. He determined the unit to contain an upper and lower conglomerate with a middle shale member. Since 1910, other workers have argued over the definition and age of the stratigraphic section at Ojo Alamo Trading Post and over the original interpretation of Brown in 1910 (Fig. 2). These workers include Sinclair and Granger (1914), Reeside (1924), Dane (1936), Baltz and others (1966), O'Sullivan and others (1972), and Fassett (1973, 1974). Others who have worked on the Ojo Alamo in other parts of the San Juan Basin include Anderson (1960), Baltz (1967), Brimhall (1973), Fassett (1966), Fassett and Hinds (1971), Gilmore (1916), Hinds (1966), Powell (1973), and Rigby and Lucas (1977). Excellent summaries of the early history of the Ojo Alamo nomenclature problem at the Ojo Alamo Trading Post are presented by Baltz and others (1966), Clemens (1973), and Fassett (1973).

U.S. Geological Survey usage of the name "Ojo Alamo Sandstone" is that of Baltz and others (1966). The original type Ojo Alamo Sandstone of Bauer (1916) is restricted to include only the upper conglomeratic, trough-crossbedded, coarse-grained, arkosic sandstone bearing silicified logs and unconformably overlying the Kirtland Shale. This report follows the Ojo Alamo Sandstone restricted usage at the type section (Fig. 1, loc. 9) and throughout the southern outcrop area (Fig. 1, loc. 7–20; Fig. 3). In the southeastern area, near Cuba, New Mexico (Fig. 1, loc. 21–30), it follows the usage of Fassett (1966). Here the name Ojo Alamo Sandstone refers to a unit that is composed of a number of lenticular, medium- to coarse-grained, trough-crossbedded sandstones that contain petrified wood and clay balls and are separated by shales. The sandstones have scour bases and lenticular or irregular shapes due to scouring by similar channel deposits. They unconformably overlie shales assigned to the Kirtland Shale and Fruitland Formation, undifferentiated. In the northwestern outcrop area near Farmington, New Mexico (Fig. 1, loc. 1–7; Fig. 3), a trough-crossbedded, coarse-grained, conglomeratic, sheetlike sandstone lithofacies with abundant

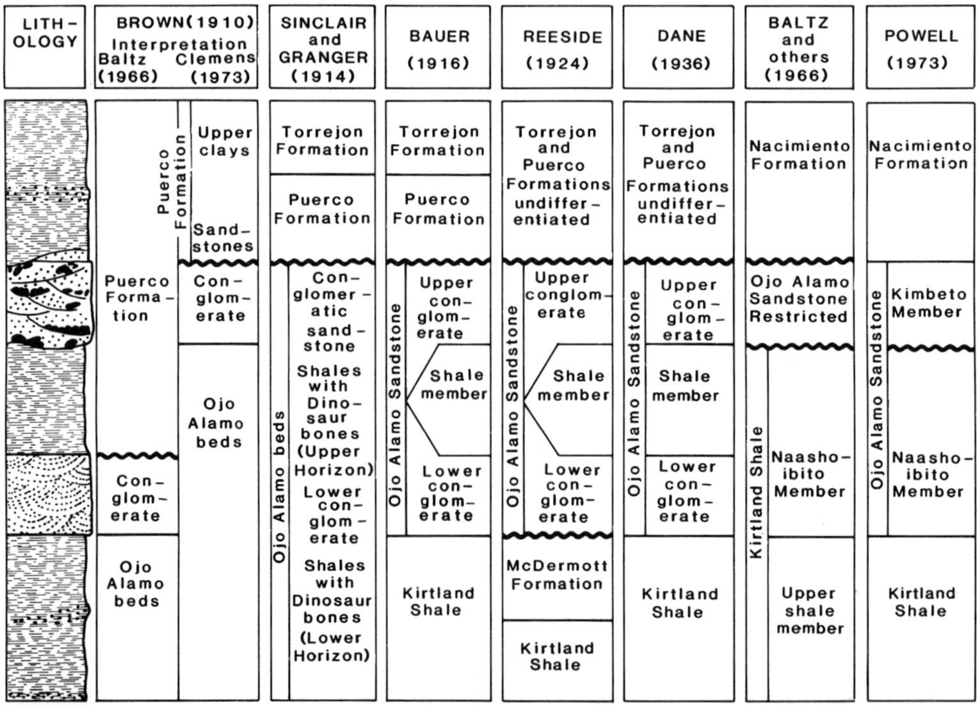

(schematic not to scale)

Figure 2. History of definition and age designation of the Ojo Alamo Sandstone (modified from Clemens, 1973, and Fassett, 1973). Terminology is for Ojo Alamo Arroyo area.

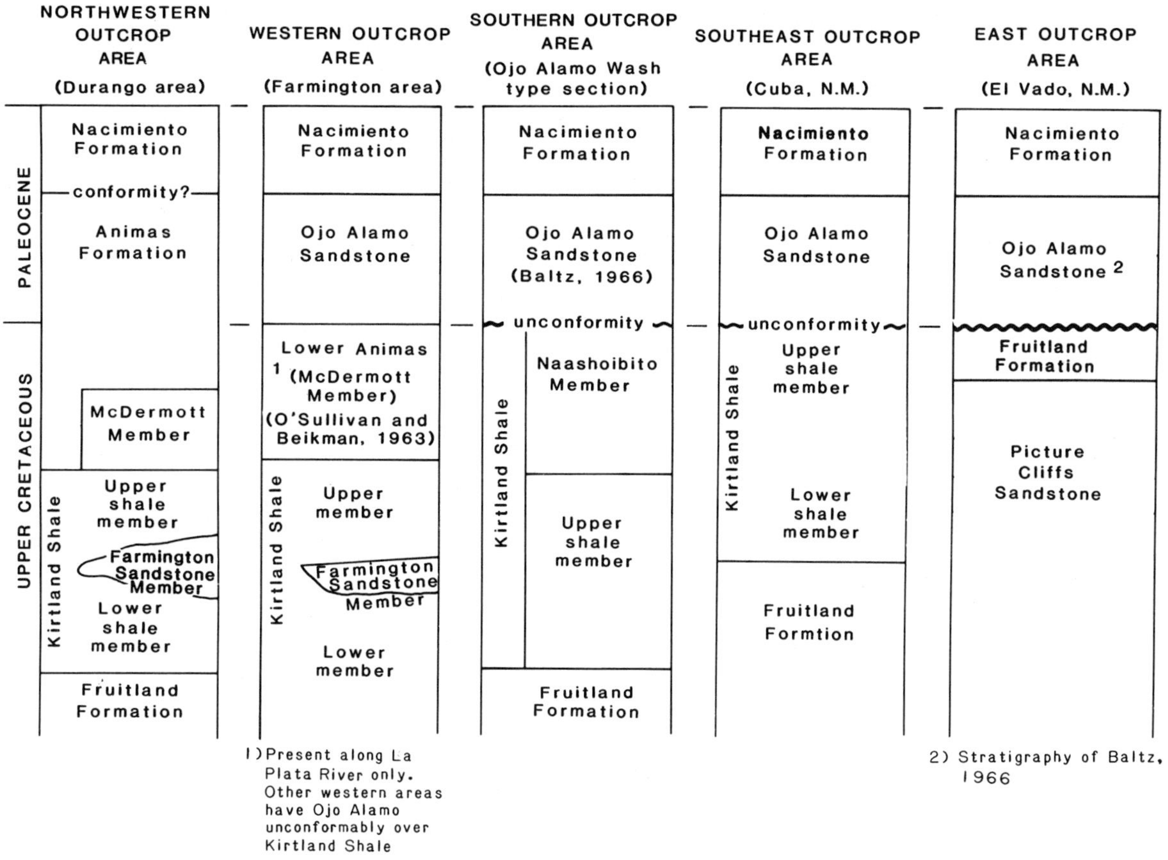

Figure 3. Generalized stratigraphic columns of Cretaceous and Tertiary rocks located throughout the study area, San Juan Basin, New Mexico. Compiled from Steven and others (1974), Barnes and others (1954), O'Sullivan and Beikman (1963), Baltz (1966), Wyant and Olson (1978), Fassett and Hinds (1971), Hinds (1966), Fassett (1966), and Manley and others (1978). Thicknesses not to scale.

quartzose pebbles and petrified wood is herein called Ojo Alamo Sandstone. In the eastern outcrop area (Fig. 1, loc. 31–37; Fig. 3), the name Ojo Alamo is applied to the first medium- to coarse-grained, trough-crossbedded, arkosic sandstone bed above the Kirtland Shale. In this area, several beds of channel sandstone, separated by shales, exist. The upper beds are mapped as Nacimiento Formation (Manley and others, 1978).

Powell (1972) published an interpretation of the origin of the sandstone, conglomerate, and shale beds named "Ojo Alamo." His interpretation was based on data from more than 7,000 crossbeds, one measured section, and results of general size-distribution measurements and heavy-mineral studies of the Ojo Alamo throughout much of the San Juan Basin. Powell was also caught up in the nomenclature problem. He renamed the units in his thesis and later publications (Powell, 1972, 1973). He called the upper conglomerate interval the Kimbeto Member of the Ojo Alamo Sandstone. The lower conglomerate and shales were renamed the Naashoibito Member of the Ojo Alamo Sandstone (Fig. 2). Powell concluded that the Kimbeto Member was deposited on an alluvial plain and that Kimbeto sediments originated in the vicinity of the present-day Needle Mountains; he did not discuss the origin of the "Naashoibito Member."

Other studies of the Ojo Alamo include palynological studies of a core in the northern San Juan Basin (Tschudy, 1973) and outcrop samples from the southern basin area (Anderson, 1960), paleomagnetic studies of the Cretaceous-Tertiary boundary (Butler and others, 1976; Lindsay and others, 1977, 1978, and 1981), and geochemical studies for high concentrations of iridium at or near the base of the Ojo Alamo Sandstone (Orth and others, 1981). These authors offered conflicting ages for the Ojo Alamo, from Cretaceous—based on paleomagnetics—to early Tertiary—based on palynology. The iridium anomaly at the Cretaceous-Tertiary boundary in the Raton Basin of New Mexico and Colorado is not present in the San Juan Basin (Orth and others, 1981).

MAJOR FACIES

The Cretaceous and Tertiary rocks examined in this study are divided into eight major lithofacies. Figure 4 (parts 1, 2, and

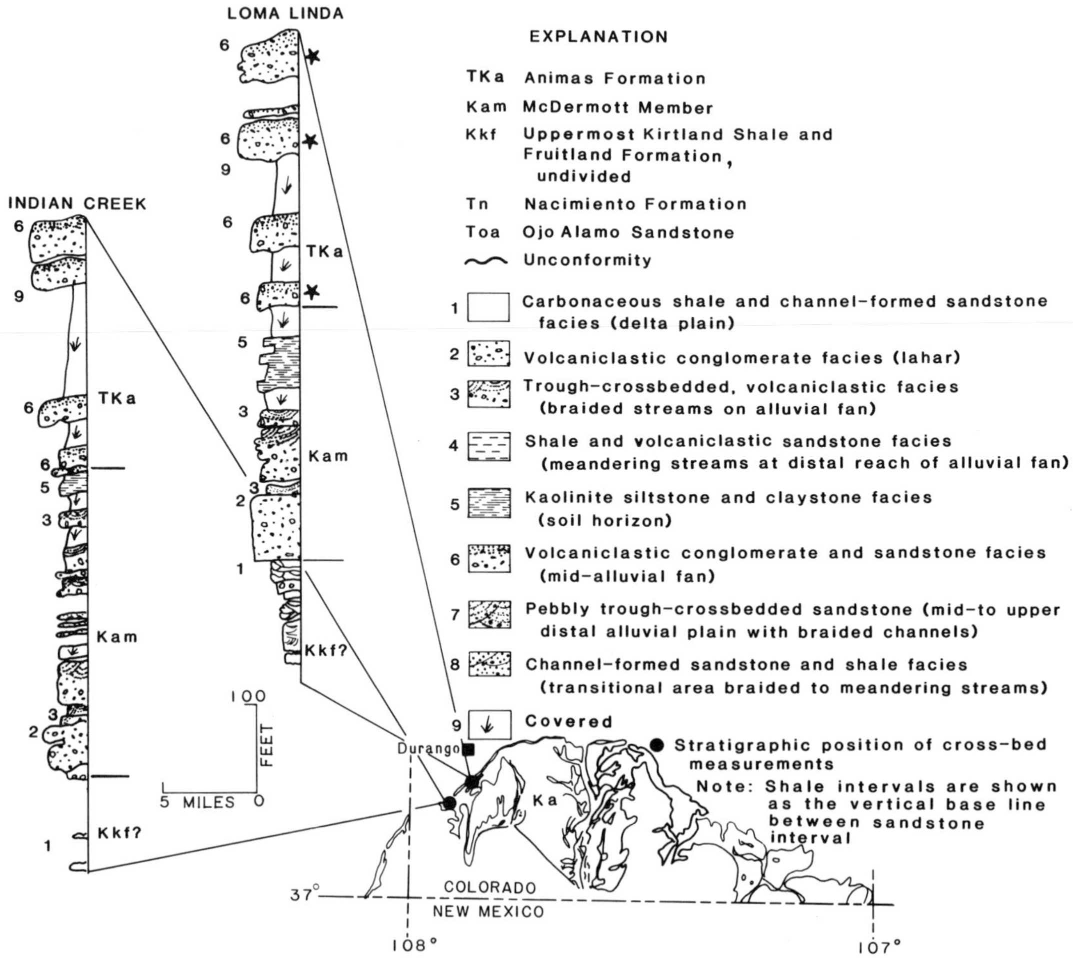

Figure 4 (this and facing page). Generalized stratigraphic sections and major lithofacies at selected sites around the north, west, south, and southeast parts of the San Juan Basin, New Mexico and Colorado. Sections include Kirtland Shale, Fruitland Formation, Ojo Alamo Sandstone, and Animas Formation. Depositional facies are shown.

3) shows these lithofacies within the stratigraphic sections. The eight major lithofacies are: (1) carbonaceous shale and channel-form sandstone facies (lowest in the stratigraphic sections); (2) volcaniclastic, conglomeratic facies; (3) trough-crossbedded, volcaniclastic sandstone facies; (4) kaolinitic siltstone and claystone facies; (5) shale and volcaniclastic sandstone facies; (6) volcaniclastic conglomerate and sandstone facies; (7) pebbly, trough-crossbedded facies; and (8) channel-form sandstones and shale facies. The Ojo Alamo Sandstone in the San Juan Basin is composed of only the pebbly, trough-crossbedded facies (7) and a related facies near Mesa Portales, the channel-form sandstones and shale facies (8). The remaining lithofacies are important, however, to an understanding of the depositional events that preceded Ojo Alamo deposition and affected its composition and deposition.

Carbonaceous Shale and Channel-form Sandstone Facies

This facies composes the uppermost part of the Fruitland Formation-Kirtland Shale, which underlies the Ojo Alamo in most areas of the San Juan Basin. It consists primarily of varicolored shales and siltstones interbedded with thin, laterally extensive, fine-grained sandstones and minor coal (Fig. 4). The sandstones of this lithofacies contain a wide variety of minerals and rock fragments and are finer grained than sandstones of the other lithofacies discussed. An example of the sandstone petrology of this lithofacies is shown in Plate 1, Figure 1. This lithofacies also contains a variety of marine and nonmarine vertebrate and invertebrate fossils, including dinosaur, turtle, crocodile, fish, mollusks, and mammals.

According to Fassett and Hinds (1971), this lithofacies represents deposition on a delta plain. Meandering streams flowed generally northeast on the delta plain to the Pictured Cliffs sea, on a fairly flat surface that was dominated by overbank shales and swampy areas rich in fossils.

Volcaniclastic Conglomerate Facies

The volcaniclastic conglomerate facies crops out near Durango (Fig. 4). It makes up the basal portion of the Cretaceous

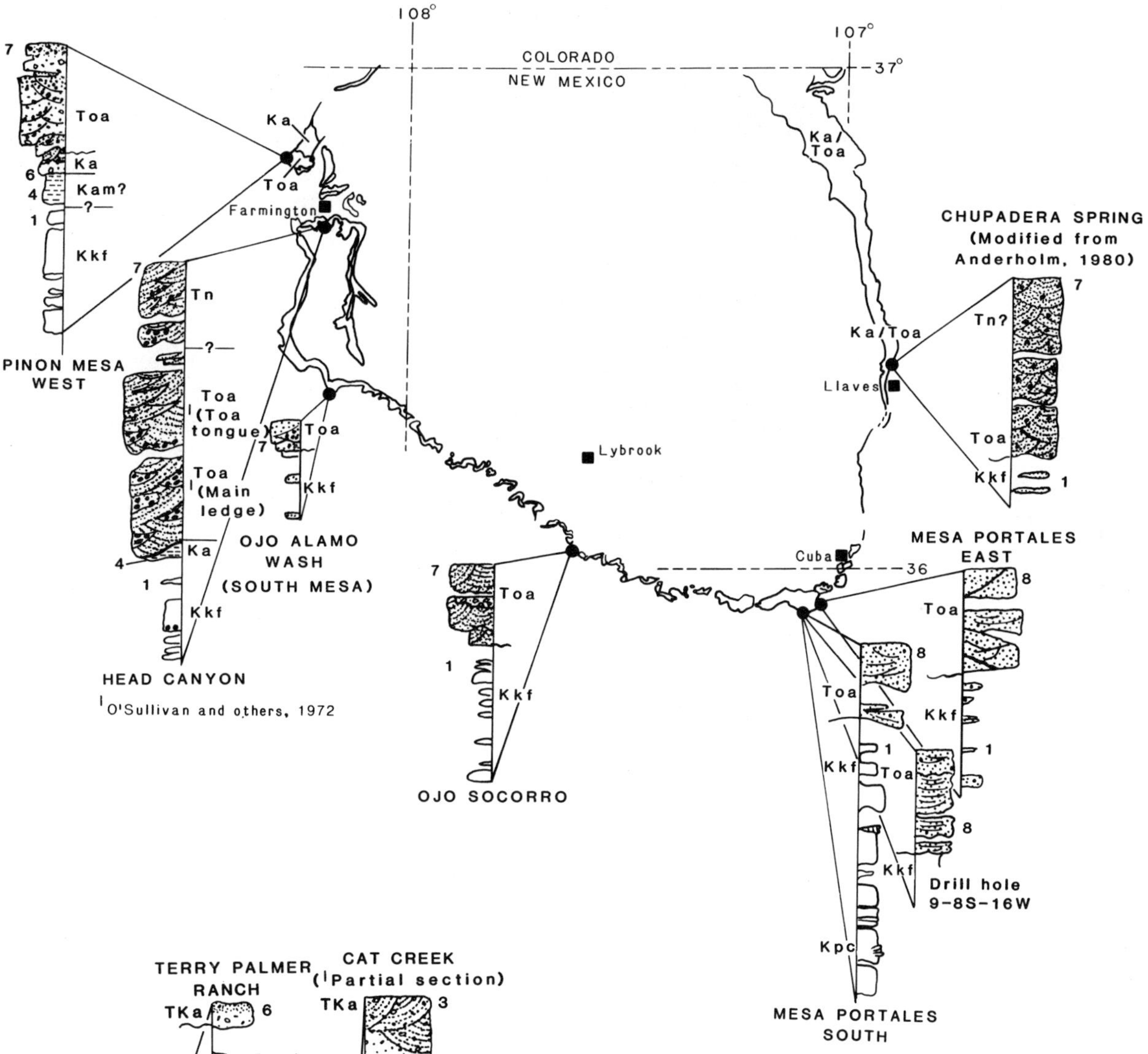
108°
107°
COLORADO
NEW MEXICO
37°
Ka
Toa
Farmington
Ka/Toa
CHUPADERA SPRING
(Modified from Anderholm, 1980)
Tn?
Toa
Kkf
Llaves
Lybrook
Cuba
36
PINON MESA WEST
Kam?
Kkf
Tn
Toa (Toa tongue)
Toa (Main ledge)
OJO ALAMO WASH (SOUTH MESA)
HEAD CANYON
O'Sullivan and others, 1972
OJO SOCORRO
MESA PORTALES EAST
Drill hole 9-8S-16W
Kpc
MESA PORTALES SOUTH

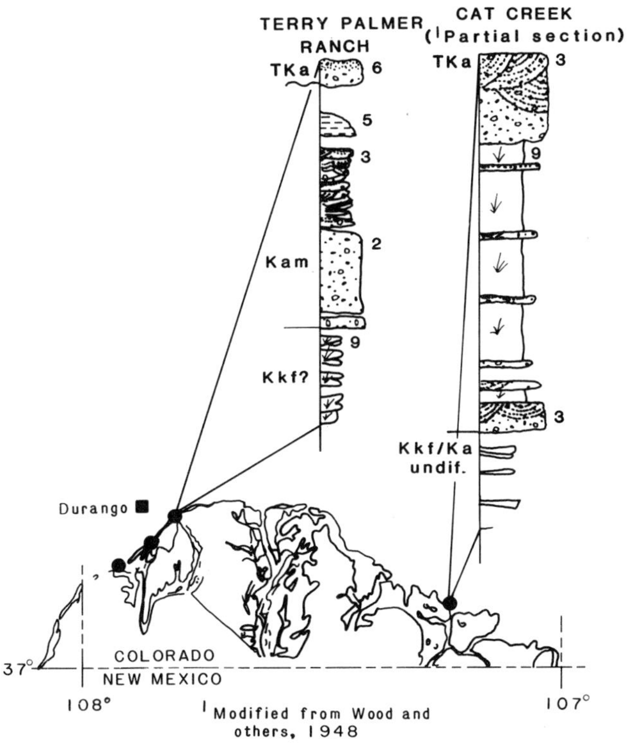
TERRY PALMER RANCH
CAT CREEK (Partial section)
TKa
Kam
Kkf?
Kkf/Ka undif.
Durango
COLORADO
NEW MEXICO
37°
108°
107°
Modified from Wood and others, 1948

Figure 5. Shale and volcaniclastic sandstone facies (TKa) and pebbly, trough-crossbedded facies (Toa), Pinon Mesa West section. Note discontinuous sandstones of Animas Formation (TKa), thick shale deposits, and upper rippled area (below large ledge). Thickness of Animas Formation is approximately 33.3 m (110 ft).

McDermott Member of the Animas Formation in the San Juan Basin. This lithofacies consists of one or two purple, conglomeratic beds. The conglomerate is characterized by angular to subangular boulders up to 60 cm (24 in) in diameter. It is unsorted, matrix supported, and polymodal. Felsic volcanic debris dominates all parts of the lithofacies. It is composed mainly of augite and zoned and unzoned plagioclase. Its mineralogy and stratigraphic characteristics indicate that this facies was deposited as a lahar prior to Ojo Alamo deposition. The source of the volcanic debris flow(s) was probably in the vicinity of the present-day La Plata Mountains (Fig. 1).

Trough-crossbedded, Volcaniclastic Sandstone Facies

This lithofacies includes the middle to upper portion of the McDermott Member of the Animas Formation near Durango. It is composed mainly of purple, trough-crossbedded sandstone with minor mudstone interbeds. The sandstones are fairly wide and lenticular, with abundant cut-and-fill structures. The mudstone interbeds are discontinuous, calcareous, cut by the channel sandstones, and locally burrowed and contain herbaceous carbonaceous debris. Compositionally, this lithofacies is composed primarily of volcanic material, including augite and zoned and unzoned plagioclase.

The trough-crossbedded, volcaniclastic sandstone lithofacies represents deposition by braided streams flowing southward on a fairly steep slope. These streams drained areas of extensive volcanic debris, including the lahar deposits of the volcaniclastic conglomerate facies. Their source was probably also in the La Plata Mountains area.

Kaolinitic Siltstone and Claystone Facies

This thin lithofacies crops out only in the Durango area (Fig. 4). It consists of dark red, green, and brown mostly massive-bedded siltstones and claystone. Outcrops have a speckled appearance because they are composed of abundant kaolinite "worms" and chalcedony in an iron-dominated matrix. At the Terry Palmer Ranch and Indian Creek sections (Fig. 1, loc. 43 and 45), small rootlet structures, minor organics, and small, unidentified burrows were also found in this facies.

This lithofacies represents a paleosol at the top of the McDermott Member of the Animas Formation. It is significant because it indicates that at least the northwestern San Juan Basin underwent a period of nondeposition between volcanic activity near La Plata and later Laramide deposition.

Shale and Volcaniclastic Sandstone Facies

The shale and volcaniclastic sandstone lithofacies crops out at Pinon Mesa near Farmington, New Mexico (Fig. 4). It consists of light gray and purple shale and siltstone and white sandstone (Fig. 5). The sandstones are lenticular, discontinuous, trough crossbedded and rippled, fine to upper-medium grained, and moderately well sorted. At the upper contact, they are ripple bedded and highly scoured by Ojo Alamo Sandstone (Fig. 6).

Like the trough-crossbedded, volcaniclastic sandstone facies, this facies contains abundant, relatively unaltered volcanic debris in both the channel sandstones and the overbank shales. In addition, it also contains quartzite and chert pebbles, which are not present in the volcaniclastic sandstone facies. Minor weathered volcanic (andesite), chert, and quartzite pebbles as much as 10 cm (4 in) in diameter were identified in the upper part of the facies about 1 m (3 ft) below the Ojo Alamo scour in a one-pebble-thick layer (Fig. 5). They are associated with clay ripups and minor coalified material.

The lower portion of this lithofacies represents a more distal depositional facies than the trough-crossbedded, volcaniclastic sandstone facies described above. It was deposited by meandering streams that flowed over a low-gradient slope characterized by extensive floodplains. Farther south in the San Juan Basin this facies is not present, having been removed by scouring during deposition of the Ojo Alamo Sandstone. The upper rippled portion of the lithofacies may, in part, be a more distal facies of the volcaniclastic conglomerate facies with sandstone lenses (described below).

A distal portion of the shale and volcaniclastic facies also crops out east of Durango at Fossett Gulch (Fig. 1, loc. 38). At this location, it consists only of a thin (<2-m tr 6.6-ft) purple shale with abundant carbonaceous plant material. It represents deposition in a low-energy, floodplain environment.

The interval of shale and volcaniclastic sandstone is mapped as Animas Formation at Pinon Mesa (O'Sullivan and Beikman, 1963). The lower portion of the lithofacies is probably laterally equivalent to the McDermott Member of the Animas near

Durango. The upper portion is probably laterally equivalent to beds mapped as upper Animas near Durango. At Fossett Gulch, the interval was mapped as Animas Formation and Kirtland Shale, intertonguing (Wood and others, 1948).

Volcaniclastic Conglomerate and Sandstone Facies

The volcaniclastic conglomerate and sandstone lithofacies crops out near Durango (Fig. 4). It consists of several andesitic and rhyolitic, sheetlike conglomerates that contain white, tan, and bluish-green lenticular sandstones. The conglomerates are separated by sandstones or siltstones, which form mainly covered slopes.

The conglomerates of this facies are poorly sorted, clast to matrix supported, polymodal, unstratified, and ungraded. The clasts are as much as 25 cm (9 in) in diameter. Amounts of sedimentary and metamorphic rock fragments, quartzite, chert, and clay chips increase upward in the stratigraphic section. Lower conglomerates are dominantly matrix-supported volcanic cobbles with fresh volcanic constituents in the matrix (pyroxene, plagioclase, volcanic glass). They also contain minor scattered green quartzite, vein quartz cobbles, and fragments of siltstone that indicate erosion of some sedimentary and metamorphic terrain during their deposition. Middle conglomerates are also composed predominantly of volcanic pebbles. However, the overall percentage and variety of other pebble types (including red, white, and black chert; green and red quartzites; granite; limestone; and vein quartz) is increased relative to the lower conglomerate. Upper conglomeratic beds contain the greatest percentage of quartz-bearing pebbles (as much as 20 percent), as well as fragments of limestone, siltstone, and claystone clasts.

The sandstones within the conglomerate beds of this lithofacies are small, coarse grained, arkosic, and trough crossbedded and contain petrified wood and coalified material. These sandstones also become more abundant in the upper conglomerate beds. Crossbed directions taken from sandstone lenses within each conglomeratic bed indicate a south to southeast flow direction for the sands (Fig. 4, loc. 44).

Petrographically, this lithofacies consists mainly of volcanic rock fragments containing zoned plagioclase and augite, tuff fragments, and occasional ripups from middle and lower McDermott units. Sedimentary rock fragments increase in upper deposits. Grains are angular to subangular, poorly sorted, and predominately less than 4 mm (0.16 in) in diameter (matrix size) (Plate 1, Fig. 2).

The conglomerate beds of the volcaniclastic conglomerate and sandstone lithofacies represent the mid-fan depositional facies of an alluvial fan originating in the vicinity of the present-day Needle Mountains. The increase in the proportion of quartz pebbles upward in the stratigraphic section reflects erosion of new material in the source area during continued uplift. Lower conglomeratic beds represent erosion and reworking of earlier La Plata volcanic flows and debris; middle and upper beds represent the beginning of erosion of a sedimentary and metamorphic core

Figure 6. Trough-crossbedded, volcaniclastic facies, Pinon Mesa West section. At base of trough-crossbedded unit with clay ripups are irregular patches of pebble lag approximately 6 cm (2 in) thick composed of volcanic, chert, and quartzite pebbles (arrow). Upper rippled deposit is scoured by Ojo Alamo Sandstone (Toa). Lower channel sandstone is approximately 1 m (3 ft) thick.

of the uplift. Sources for quartzite pebbles of this lithofacies may include the Vallecito Conglomerate (Precambrian), Irving Formation (Precambrian), and the Uncompahgre Formation (Precambrian). Sources for the chert may include the Ouray Limestone (Devonian) and Leadville Limestone (Mississippian) and the Molas Formation (Pennsylvanian) (Steven and others, 1974; Larson and Cross, 1956). This lithofacies is mapped as Animas Formation in the northwestern San Juan Basin (Steven and others, 1974).

Pebbly, Trough-Crossbedded Facies and Channel-Form Sandstone and Shale Facies

The pebbly, trough-crossbedded facies crops out on the western and southern edge of the San Juan Basin and is synonymous with the name Ojo Alamo Sandstone (U.S. Geological Survey formal usage) in this area. This facies consists of large- and medium-scale trough-crossbedded sandstones that contain both volcanic and quartzose pebbles. The sandstones are mainly medium to coarse grained. The pebbles are less than 10 cm (4 in) in diameter, aligned along trough crossbedding, well to moderately well rounded, and imbricated in some areas.

The pebbly, trough-crossbedded facies also crops out on the eastern edge of the basin (see Chupadera Spring section, Fig. 4). This facies differs from the western Ojo Alamo, however, in pebble size and composition, transport direction, and the size of minor shale interbeds. This facies encompasses both the Ojo Alamo Sandstone and the lower Nacimiento Formation on the eastern side of the San Juan Basin to Llaves (Fig. 4).

The channel-form sandstone and shale facies crops out in

TABLE 2. PERCENT COMPOSITION AND SIZE DISTRIBUTION OF PEBBLES

Location Name	Formation	V	Q	S	M	O	Maximum Clast Size
Pinon Mesa West	Toa (middle)	16	81	2		1	10 cm
Animas River	Toa (basal)	9	87	3			10 cm
Head Canyon	Toa (basal)	25	70	4	1		10 cm
Ojo Alamo Wash	Toa (basal)	27	70	2	1		5 cm
Ojo Alamo Wash	Toa (upper)	10	88			2	5 cm
Tsun Je Zhin	Toa (basal)	6	94				5 cm
Mesa Portales East	Toa (basal)	1	98			1	.5 cm
Arroyo Bonita	TKu (basal)	1	99				.5 cm
Dulce Lake	TKu (basal)	98					.5 cm
Rules Hill	TKa (middle)	77	10	13			10 cm
Loma Linda	TKa (upper)	75	20	5			10 cm

Toa = Alamo Sandstone
TKa = Animas Formation
TKu = Nacimiento, Ojo Alamo, and Animas Formation, undivided

V = volcanics
Q = quartz, quartzite, and chert
S = sedimentary rocks
M = metamorphic rocks
O = other (e.g., petrified wood)

the Mesa Portales area (Fig. 1, loc. 21–27; Fig. 4). It consists of up to five channel-form sandstones interbedded with shales. In many places, upper channel sands scour through shale interbeds into lower channel sands to form wide sandstone beds. This facies is regarded as a more distal facies of the pebbly, trough-crossbedded facies and is also referred to as Ojo Alamo Sandstone in this paper. Fassett (1966) included these rocks in the Ojo Alamo at Mesa Portales. Baltz, however, regards all these channel-form sandstones as Kirtland sandstones because they are not the continuous, sheetlike sandstones that he defined as Ojo Alamo in the Ojo Alamo Sandstone restricted type section (Baltz, 1983, personal communication). This study uses the terminology of Fassett (1966) in this area for three main reasons. First, the sandstone grains of the channel-form sandstones are larger (medium- to coarse-grained) and much more angular-grained than most sandstones of the Kirtland Shale examined during this study. They also contain polycrystalline quartz, large potassium feldspar grains, metamorphic rock fragments, and some fairly large pyralspite garnets, similar to sandstones of the pebbly, trough-crossbedded facies farther north. Second, transport directions obtained during this study indicate southerly transport directions. Northeasterly transport directions would be expected for Kirtland sandstones. Finally, the channel sandstones show sedimentary structures and bedforms that indicate transitional depositional environments that would be expected between the pebbly, trough-crossbedded lithofacies to the north and a distal facies southward out of the basin.

COMPARISON OF FACIES CHARACTERISTICS

Comparison of the composition, sedimentary structures, and depositional features of the eastern and western pebbly, trough-crossbedded facies and the channel-form sandstone and shale facies are discussed in detail under specific headings below. It should be emphasized that only at Chupadera Spring (Fig. 4) can crossbed measurements and pebble-count data of the eastern pebbly, trough-crossbedded facies be located within a measured stratigraphic section. North of Chupadera Spring (Fig. 1, loc. 40) to Rules Hill there are no detailed stratigraphic sections completed to pinpoint the measurements accurately. The author is fairly confident that the data were taken from within the lower portion of the pebbly, trough-crossbedded facies. The eastern data should be used with caution, however, until sections in this area are measured in detail.

Pebble Size and Composition

In most outcrops in the western and southern areas of the San Juan Basin, the Ojo Alamo contains conglomeratic lenses and/or scattered pebbles aligned along trough crossbeds. The pebbles exhibit size and compositional variations areally and vertically. Pebbles on the west side of the San Juan Basin generally decrease in size and in percentage of volcanic constituents southward from Pinon Mesa (Table 2). The percentage of quartz pebbles increases southward from Pinon Mesa as well as vertically in most stratigraphic sections. Compositionally, the pebbles of the Ojo Alamo from Pinon Mesa to the Fork Rock Mesa section (Fig. 1, loc. 22) include fresh and weathered igneous rocks (andesite, rhyolite, dacite?, and granite), metatuffs, varicolored chert and quartzite, vein quartz, gneiss, slate, sandstone, and claystone clasts, and minor fossiliferous marine limestone clasts as much as 10 cm (4 in) in diameter. Pebble suites sampled in the western area contain as much as 27 percent volcanic pebbles (obtained in basal Ojo Alamo at Ojo Alamo Trading Post site, Fig. 1, loc. 8).

The eastern and northeastern exposures from Dulce Lake to Arroyo Blanco (Fig. 1, loc. 31–37) contain less than 1 percent volcanic pebbles (Table 2). In this area quartzite and chert dominate, and all pebbles found are less than 0.5 cm (0.25 in) in diameter. The Mesa Portales sections (Fig. 1, loc. 21–27) contain rare, isolated pebbles of quartz and iron concretions.

Petrography

Ojo Alamo Sandstone throughout the basin consists of fresh and weathered zoned and unzoned plagioclase (oligoclase), perthitic feldspars, chert, and volcanic, sedimentary, and metamorphic rock fragments including abundant quartzite. Distinctive volcanic rock fragments of feldspar laths in fine-grained matrix, similar to those in the volcaniclastic conglomerate facies, are present in the western samples (Plate 1, Fig. 3). Minor percentages of biotite, muscovite, zircon, tourmaline, sphene, detrital hornblende, epidote, monazite, and barite are also present. Grains are angular to subangular, moderately to poorly sorted, and predominantly less than 2 mm (0.08 in) in diameter. Samples from the pebbly, trough-crossbedded facies in the western and southwestern areas also contain larger, more abundant garnets (pyralspites) than samples from the channel-form sandstone and shale facies at Mesa Portales (Plate 1, Figs. 3 and 4).

Mesa Portales samples contain few to no volcanic rock fragments, especially with feldspar laths in a fine-grained matrix. Dominant fragments in these sediments are polycrystalline quartz and single quartz and potassium feldspar grains. Metamorphic rock fragments are still present (Plate 1, Fig. 4).

Sedimentary Structures and Biologic Constituents

The sedimentary structures in the Ojo Alamo Sandstone, shown in Plate 2, indicate a fluviatile origin. The fluvial channels of the Ojo Alamo, which scoured into underlying Cretaceous units, are marked by irregular bases (Plate 2, Fig. 1). Individual channels are as much as 0.5 km (0.3 mi) wide, lenticular, and dominated by trough crossbeds (Plate 2, Fig. 2). Wedge- and tabular-planar crossbeds also occur but are not common. The trough and planar beds are commonly graded (Plate 2, Fig. 3), and climbing ripples and ripple drift occur on some channel margins. Ripples are best preserved in outcrops near Eagle Mesa and Mesa Portales (Plate 2, Fig. 4). Channel bases often contain clay ripups, clay balls, iron concretions, and pebbles aligned along trough cross-bed sets (Plate 2, Fig. 3). Some of the larger clay ripups or balls contain disseminated plant fragments (Plate 2, Fig. 5).

Sandstones of the Ojo Alamo contain contorted bedding throughout the study area, especially in the southeastern area near Mesa Portales. Some of these contortions are subtle; others are pronounced (Plate 2, Fig. 6). These make obtaining reliable crossbed data difficult, especially where the bedding is also masked by iron stains and liesegang banding.

Other characteristic features of Ojo Alamo exposures include large petrified logs, beds of bentonitic volcanic ash, and abundant carbonaceous material. Petrified logs occur throughout the outcrop area; the largest and best preserved are found between Ojo Alamo Wash and Eagle Mesa (Plate 3, Figs. 1 and 2). Flat-bedded volcanic ash beds up to 1 m (3 ft) thick occur at Mesa Portales (Plate 3, Fig. 3). They are best preserved in outcrops to the southeast where later fluvial activity has not scoured them out. Carbonaceous material is abundant in the sandstones as small disseminated fragments and as coaly pockets.

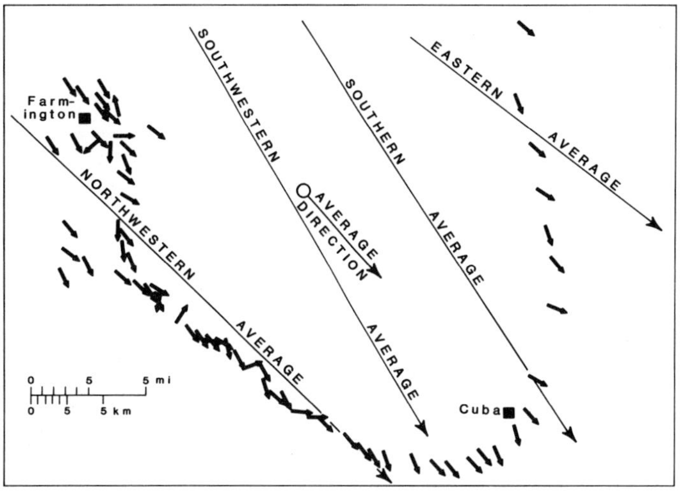

Figure 7. Mean transport directions obtained from the Ojo Alamo Sandstone of Powell (1972, p. 47). Powell suggested a single source northwest of the present-day San Juan Basin.

Fossil bones are rare in the Ojo Alamo, as it is used in this study, although reports of dinosaur bone within the Ojo Alamo have been made by Sinclair and Granger (1914), Powell (1973), and Lucas and Mateer (1983). The few bones discovered during this study are disarticulated and fairly small (Plate 3, Fig. 4). In contrast, shale units beneath the Ojo Alamo bear abundant fossils of dinosaur, turtle, crocodile, and other reptiles. In the past, these fossil-bearing shales have been included within the Ojo Alamo Sandstone (Bauer, 1916). In this report, these shales are considered part of the carbonaceous shale and channel-form sandstone facies of the Kirtland Shale and Fruitland Formation. Depositionally, they are more closely related to this facies than to the pebbly, trough-crossbedded lithofacies or channel-form sandstone and shale facies that is herein called Ojo Alamo Sandstone.

Crossbed Directions

Crossbed directions measured from trough axes in lithologies of the Paleocene Ojo Alamo Sandstone and related Upper Cretaceous rocks throughout the San Juan Basin indicate a northern source area for these Upper Cretaceous–lower Tertiary sediments. These results are different from those of Powell (1972), who proposed a northwestern source area and an average flow direction of approximately S.45E. (Fig. 7). A composite rose diagram constructed from all Ojo Alamo crossbed measurements taken for this study shows an overall mean transport direction of S.4W. for these sediments.

Analysis of the crossbed data collected during this study (Figs. 8, 9; Table 3) also shows two distinct transport-direction trends for sediments of the Ojo Alamo of Powell (Fig. 3). An average of the mean vectors for the first trend indicates a south to

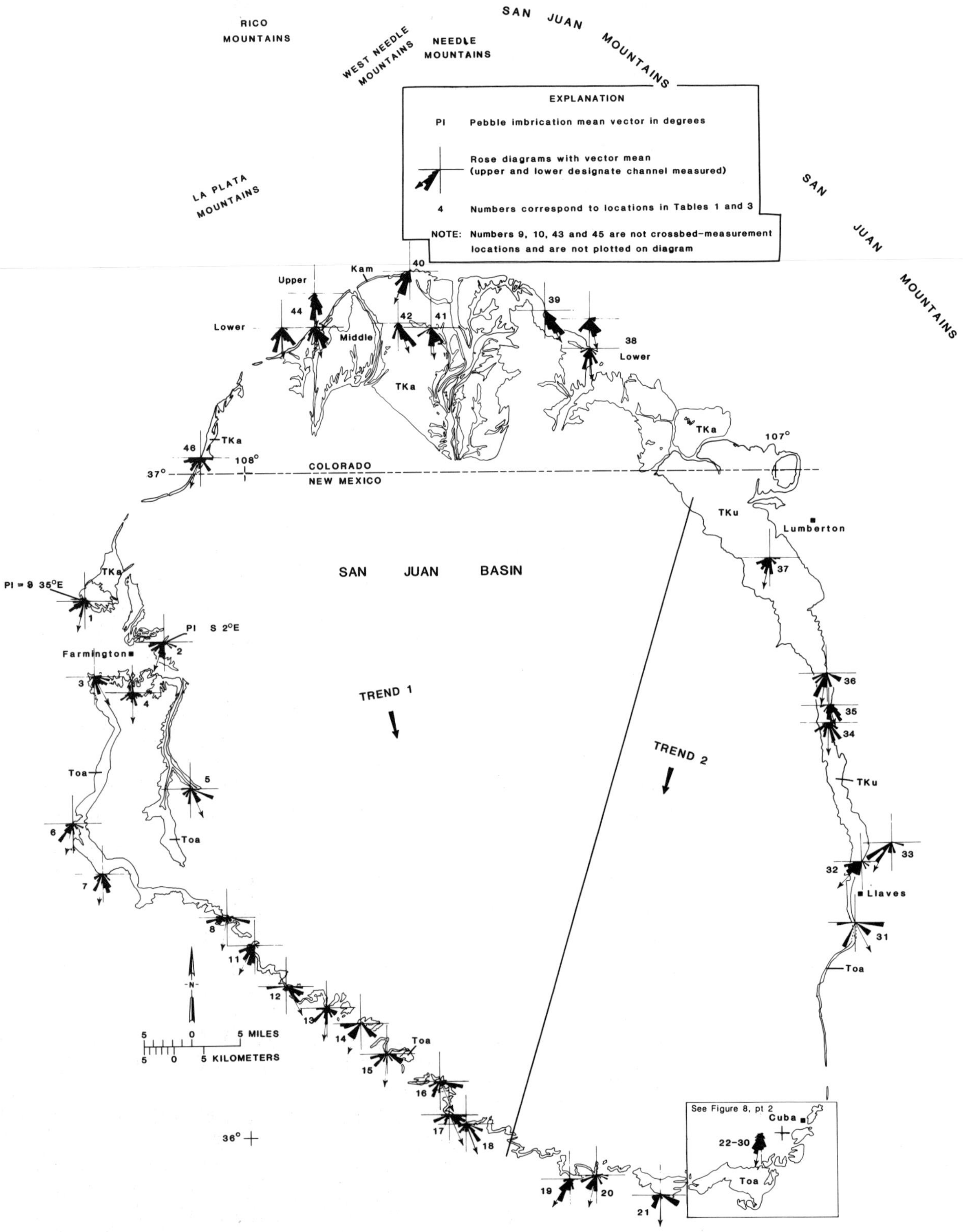
RICO MOUNTAINS
WEST NEEDLE MOUNTAINS
NEEDLE MOUNTAINS
SAN JUAN MOUNTAINS
LA PLATA MOUNTAINS
SAN JUAN MOUNTAINS
EXPLANATION
PI Pebble imbrication mean vector in degrees
Rose diagrams with vector mean (upper and lower designate channel measured)
4 Numbers correspond to locations in Tables 1 and 3
NOTE: Numbers 9, 10, 43 and 45 are not crossbed-measurement locations and are not plotted on diagram
Kam
Upper
Lower
Middle
TKa
Lower
107°
108°
37°
COLORADO
NEW MEXICO
TKu
Lumberton
SAN JUAN BASIN
PI = 9 35°E
PI S 2°E
Farmington
TREND 1
TREND 2
Toa
TKu
Llaves
Toa
5 0 5 MILES
5 0 5 KILOMETERS
36°
See Figure 8, pt 2
Cuba
22-30
Toa

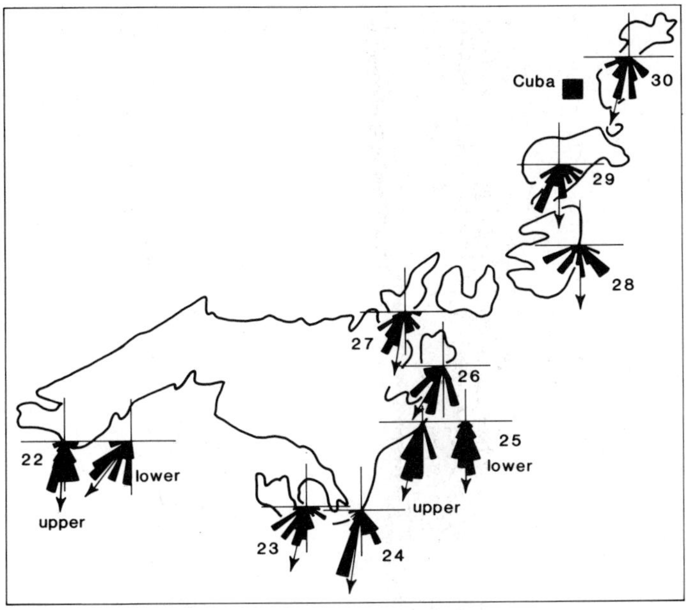

Figure 8 (this and previous page). Summary of crossbed measurement data from the Ojo Alamo Sandstone (Toa), Animas Formation (TKa), and Nacimiento, Ojo Alamo, and Animas Formations undivided (TKu), northern San Juan Basin, New Mexico. Numbers refer to locations listed in Table 3.

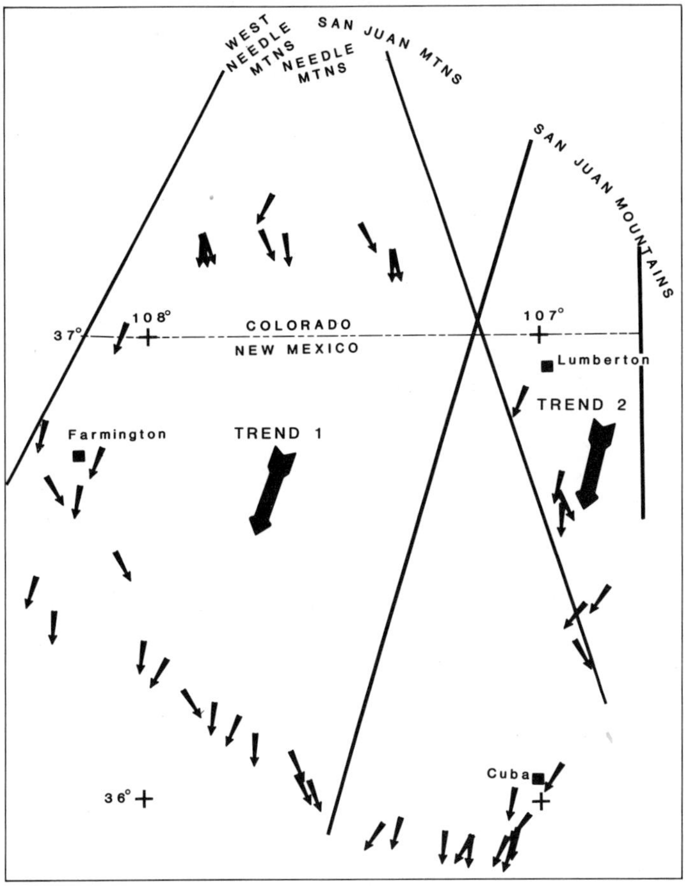

Figure 9. Transport-direction vector means for Ojo Alamo Sandstone and related strata obtained on trough axes. Vectors suggest two source areas for the sediments.

slightly southeast transport direction (S.15E.). This trend corresponds to the pebbly, trough-crossbedded lithofacies in the western outcrop area. It also is similar to crossbed measurements from the volcaniclastic conglomerate and sandstone facies of the Animas Formation near Durango and from the shale and volcaniclastic sandstone facies of the mid to upper Animas Formation in the northern basin (Fig. 1, loc. 38–44). The average of the mean vectors for the second trend indicates a southwest transport direction (S.10W.). This trend includes the channel-form sandstones and shale facies of the Ojo Alamo in the southeastern and eastern outcrop area. Transport directions in the southern outcrop area from thin, pebbly, trough-crossbedded lithofacies are bimodal and vary from east-southeast to west-southwest (Fig. 8, loc. 8, 11–18). Crossbed data for both directional trends are complex and do not indicate a simple point source for the sediments.

Transport directions at Mesa Portales may be included in trend 1 or trend 2. The composite rose diagram for the Mesa Portales area shows an overall S.10W. average flow direction for the channels, with most measurements ranging from S.30W. to S.30E. Most of the individual rose diagrams for the Mesa Portales area (Fig. 8, pt. 2), however, show strong south and southwest components, and only minor southeast components, for stream flow in the Ojo Alamo.

Bedding Morphology

The Ojo Alamo also undergoes a morphological change from north to south. In the northwestern and western portion of the study area it consists of massive ledges over 30 m (100 ft) thick. These ledges consist of individual, stacked, trough-crossbedded, fluvial channels with few or no mudstone interbeds (Fig. 10). To the south, in the Mesa Portales area, the Ojo Alamo consists of as many as five fairly wide (0.8 km—0.5 mi) lenticular sandstone ledges; these are 7 to 9 m (25–30 ft) thick and are separated by 7 to 9 m (25–30 ft) of mudstone (Fig. 11). In places on Mesa Portales, these mudstone interbeds are scoured out by later channels (Fig. 11).

Although no sections were measured in the northeastern part of the study area, examination of Ojo Alamo–type lithologies in this area indicates that the rocks are also composed of medium- to coarse-grained, moderately sorted, feldspathic sandstone ledges with mudstone interbeds. These rocks are lithologically and morphologically more similar to those of the southern area near Mesa Portales than to those of the northern area near Farmington, New Mexico.

Depositional Facies

The pebbly, trough-crossbedded lithofacies represented by the Ojo Alamo Sandstone in the western and southern San Juan

TABLE 3. DATA SUMMARY OF CROSSBED MEASUREMENT DATA AND LOCATIONS IN OJO ALAMO SANDSTONE AND ANIMAS FORMATION IN FIGURE 8.

Location Number	Location Name	n	v	R	R/n
1	Pinon Mesa West	48	223	31	.65
2	Animas River	45	199	30	.66
3	Kirtland, N.M.	46	154	35	.75
4	Head Canyon	54	179	37	.68
5	Gallegos Canyon	16	177	11	.69
6	Pinabete Arroyo	40	192	30	.75
7	Brimhall Wash	30	187	24	.80
8	Ojo Alamo Trading Post	25	189	12	.47
11	Huerfano S.W.	28	206	22	.80
12	Tsun Je Zhin	28	152	18	.64
13	Kimbeto	26	186	18	.67
14	Betonnie Tsosie Wash	25	202	15	.61
15	Escavada Wash	16	185	08	.51
16	Lybrook Southeast (pipeline)	24	157	14	.60
17	Ojo Socorro	36	158	20	.56
18	Canada Alamos	36	161	20	.67
19	Casi Cejo del Raton	41	196	33	.79
20	Ojo Encino Mesa	26	194	16	.63
21	Eagle Mesa	30	182	18	.60
22	Fork Rock Mesa, Upper	27	183	23	.86
	Lower	17	216	15	.91
23	Mesa Portales South	26	194	21	.82
24	Arroyo Jarido Spring	25	188	23	.91
25	Mesa Portales East, Upper	25	193	23	.95
	Lower	33	179	31	.95
26	Arroyo Piedre Lumbre	18	207	15	.85
27	Arroyo Chijuilla	43	191	33	.76
28	Arroyo Hondo	30	181	21	.71
29	J.R. Davidson	27	182	22	.83
30	Vallecitos	25	195	20	.81
(21-30)	Cuba Area Composite	296	190	248	.84
31	Arroyo Blanco	07	154	04	.54
32	Chupadera Spring	12	210	10	.83
33	Chupadera Spring-82	21	221	16	.78
34	Stinking Lake	29	181	22	.77
35	Ross Spring	31	165	22	.71
36	Boulder Lake	10	190	08	.79
37	Dulce Lake	31	192	23	.74
38	Fossett Gulch, Upper	25	168	21	.84
	Lower	15	176	13	.85
39	Pargin Mountain	11	154	11	.98
40	Rules Hill	12	205	11	.96
41	Gem Village	18	175	15	.85
42	Dry Creek	23	161	21	.93
43	Loma Linda, Upper	33	178	31	.94
	Middle	26	168	23	.89
	Lower	25	177	22	.89
46	McDermott Arroyo	26	197	16	.61
---	Composite for all locations	1218	184	876	.72
---	Composite for all Ojo Alamo units	928	185	653	.70
---	Ave. of mean vectors--Trend 1	---	165	---	---
---	Ave. of mean vectors--Trend 2	---	190	---	---

Explanation of column headings:
Location Number refers to Figure 1.
n = total number of crossbed measurements.
v = mean vector (in degrees) at each location.
R = magnitude of resultant vector rounded to nearest integer.
R/n = consistency ratio.
Note: No crossbed measurements at Locations 9, 10, 44, and 45.

Basin is interpreted to have been deposited by high-energy braided streams, carrying predominantly gravel and coarse sand, in the mid-fan to upper-distal portion of an alluvial plain. Its sediments originated in the area of the present-day Needle Mountains and western San Juan Mountains from the continuation of uplift and alluvial processes that formed the volcaniclastic conglomerate and sandstone lithofacies of the middle and upper Animas Formation near Durango. Ojo Alamo rocks possess many of the sedimentary structures and features described by Reineck and Singh (1980) for sandy braided rivers, including trough crossbedding, coarse-grained sediments, climbing ripples, parallel laminations, convolute bedding, clay and silt interbeds, and organic material. They also possess many characteristics of alluvial fan and plain deposits, including a decrease in sand and conglomerate size down slope (from coarse to upper medium and from 10 cm to 4 cm [4 to 1.6 in], respectively), imbricated conglomerates, laterally discontinuous channels that vary in depth and width, highly oxidized sediments, a limited suite of sedimentary structures, and a dominance of trough crossbeds and increase in overbank deposits downdip (Scholle and Spearing, 1982). However, the deposits are too fine grained and well sorted to be proximal or upper mid-fan deposits. They are also too coarse and too lacking in characteristic thick overbank deposits (except at Mesa Portales) to be the more distal portion of the alluvial plain.

Sediment sources for the western and southern Ojo Alamo are similar to the late-stage Animas deposits. Ojo Alamo braided streams initially carried a high percentage of both volcanic pebbles and volcanic debris as bedload (although decreased from the Animas Formation), which concentrated in the basal portion of the Ojo Alamo Sandstone. The volcanic pebbles carried by these streams appear both fresh and weathered, indicating that some may be eroded and reworked from the older Animas Formation and some may have been eroded from what was left of andesitic flows in the source area. The degree of weathering may also reflect transport distance. A characteristic increase in quartz-bearing pebbles upward in the stratigraphic sections of the Ojo Alamo indicates that new material was also being exposed to erosion by continued uplift in the source area.

The pebbly, trough-crossbedded lithofacies represented by Ojo Alamo and lower Nacimiento rocks on the east side of the San Juan Basin is also interpreted to be braided-stream deposits in an alluvial plain environment. Rocks resulting from sediment deposition in this alluvial system, however, differ in three main ways from Ojo Alamo rocks on the west side of the basin.

First, the eastern Ojo Alamo lithofacies is much finer grained than the western lithofacies. Streams of the eastern alluvial system carried fine to lower coarse-grained sediments and only very fine gravels (less than 0.5 cm—0.2 in) as bedload. These sediments fine southward.

Second, the eastern rocks contain less than 1 percent volcanic pebbles. Some workers attribute the scarcity of volcanics and the small pebble size to instability of volcanics during transport and distance from the source area, respectively (Powell, 1972). However, at Dulce Lake the pebbles are less than 0.5 cm (0.2 in) in diameter, and this is not much farther from a Needle Mountains source area than Pinon Mesa, where the pebbles are as much as 10 cm (3.9 in) in diameter and consist of over 10 percent volcanics. Likewise, south of Llaves (Fig. 1, loc. 31) pebbles do not exceed 0.5 cm (0.2 in) in diameter; yet in rocks farther south there are pebbles as much as 5 cm (2 in) in diameter, composed of 5–30 percent volcanic pebbles (Table 2; Fig. 1, loc. 9–20). The lack of abundant volcanic pebbles in the eastern rocks reflects the paucity of volcanics in the source area rather than the transport instability of the pebbles.

Figure 10. Pebbly, trough-crossbedded facies of Ojo Alamo Sandstone, Animas River section.

Finally, crossbed data from these rocks indicate a dominant flow direction to the southwest, they do not indicate a southeastern transport that would be expected from a Needle Mountains source. Crossbed data to the southeast are only a minor portion of the rose diagrams for eastern locations (Fig. 8, trend 2). These directions indicate a source area of the central to eastern San Juan Mountains.

The major differences in grain size, composition, and transport directions between the eastern and western pebbly, trough-crossbedded lithofacies indicate that the eastern facies rocks in the San Juan Basin were deposited on an alluvial plain by a fluvial system originating in the area of the present-day central to eastern San Juan Mountains. This source region apparently did not contain deposits of abundant andesitic volcanics such as those of the Animas Formation or the La Plata volcanic center to the west. Rocks in the source area may also have been less resistant and thus have shed finer sediment initially.

The channel-form sandstones and shale facies with large-scale, low-angle crossbeds represented by Ojo Alamo rocks in the southeastern outcrop area at Mesa Portales probably represent a transition between the pebbly, trough-crossbedded lithofacies formed by braided streams on an alluvial plain to the north to more isolated, sinuous streams at the lower reaches of the depositional facies. The shale units between the sandstone sheets are

Figure 11. Channel-form sandstones (outlined) and shale facies of Ojo Alamo Sandstone (Fassett, 1966), Mesa Portales East and Arroyo Piedre Lumbre sections.

interpreted to have formed in overbank or other floodplain settings.

Figure 12 shows the relationships between the pebbly, trough-crossbedded lithofacies and the channel-form sandstone facies of the Ojo Alamo Sandstone and other Upper Cretaceous–lower Tertiary lithofacies in the San Juan Basin. Several episodes of uplift and deposition are suggested in these lithofacies.

SUMMARY OF LATE CRETACEOUS-EARLY TERTIARY DEPOSITION IN THE SAN JUAN BASIN

The events that occurred prior to, and which set the stage for, Ojo Alamo deposition began with the regression of the Pictured Cliffs sea from the San Juan Basin and the onset of Laramide tectonism and volcanism in the northern San Juan Basin.

In Late Cretaceous time, as the Pictured Cliffs sea regressed from the area, a wide delta plain formed at the sea's edge. Moderately sinuous to meandering streams flowed northeastward across the low-gradient, poorly drained delta plain landscape to the sea. As the sea regressed to the northeast, the meandering streams lengthened and continued to carry sediment to it. Scattered swamps also migrated northeastward shoreward of the shoreline. Deposition on the delta plain resulted in overbank shales and low-angle, trough-crossbedded, laterally extensive sandstones of the carbonaceous shale and channel-form sandstone lithofacies of the Kirtland Shale and Fruitland Formation in the San Juan Basin. Dinosaurs, crocodiles, turtles, brackish-water invertebrates, fishes, and a variety of flora lived in this environment.

Sometime after the Pictured Cliffs sea retreated from the area, volcanism began in the area of the present-day La Plata Mountains. Bentonitic, varicolored shales with andesitic debris, prevalent in the Kirtland floodplain deposits, indicate that volcanism began during their deposition. The areal extent of these eruptive centers is unknown. According to Eckel (1937), they were probably completely eroded by Paleocene time. Their activity, however, had spectacular effects on the flat, plainlike Cretaceous landscape of the San Juan Basin. Thick expanses of andesitic and rhyolitic rocks and mud spewed from the volcanic center and accumulated as debris flows on the Kirtland surface. At least two such debris-flow episodes are recorded in the volcaniclastic conglomerate facies of the McDermott near Durango (Fig. 4).

Braided streams draining the debris flows transported sand, silt, and mud that was rich in volcanic debris across the northern basin area at least as far south as Farmington. At Pinon Mesa, near Farmington, these streams change from braided to meandering with extensive overbank deposition. The stream deposits still contain abundant volcanic constituents, and the overbank deposits also contain abundant volcanic constituents because of continuing air fall from the volcanic centers.

Subsequent to debris flow and fluvial deposits, an extensive paleosol developed in the northwestern San Juan Basin. The paleosol indicates a period of tectonic inactivity and nondeposition in at least part of the northern basin area.

Uplift in the northern basin area during Laramide tectonism began in the Late Cretaceous (Tweto, 1975). It continued episodically into the Tertiary and resulted in the formation of several alluvial fans. One of the initial fans to spread into the San Juan Basin area originated in the vicinity of the present day La Plata and Needle mountains. Sediments shed southward into the basin from this area resulted in distinctive mid-fan conglomerates of the volcaniclastic conglomerate and sandstone facies of the Animas Formation near Durango. Sediment changes in these conglomerate deposits indicate that sediment was eroded from volcaniclastic deposits and from the paleosol below and that as uplift continued, more quartzose rocks were exposed to erosion in the source area. The cyclic sedimentation and scour surfaces within the volcaniclastic conglomerate and sandstone facies indicate that more than one period of uplift and deposition occurred during this time. Sediments shed southward on these early alluvial fans spread at least as far south as Farmington. If the sediments were deposited farther south than Farmington, they have been eroded by later fluvial deposits of the pebbly, trough-crossbedded facies.

Laramide uplift continued in the area of the present-day Needle Mountains and San Juan Mountains. Drainage of these uplifts by fluvial systems resulted in deposits of the Ojo Alamo Sandstone in the San Juan Basin. Depositional characteristics of the lithofacies forming the Ojo Alamo Sandstone indicate that at

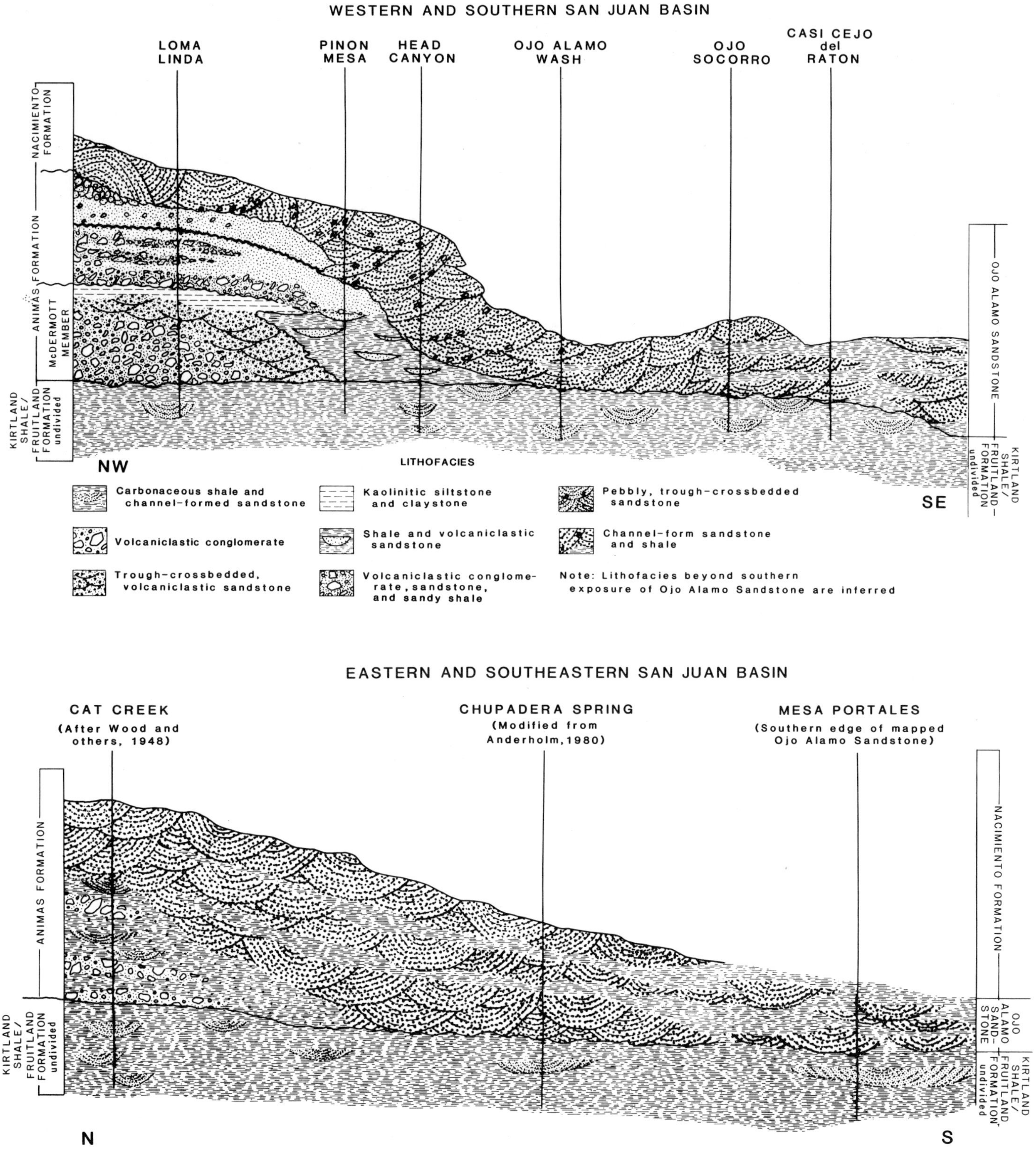

Figure 12. Schematic cross section of the relationship of Late Cretaceous–early Tertiary lithofacies in the San Juan Basin, N. Mex. In line of section, west of Loma Linda, rocks equivalent to the pebbly, trough-crossbedded lithofacies are mapped as Nacimiento Formation. South of Cat Creek, lateral equivalents of the pebbly, trough-crossbedded lithofacies in the upper part of the Animas Formation are mapped as Ojo Alamo Sandstone and Nacimiento Formation.

least two distinct source areas are responsible for the Ojo Alamo sediments.

Sediments of the pebbly, trough-crossbedded lithofacies that compose the western alluvial system originated in the area of the Needle Mountains and western San Juan Mountains. Deposition of these sediments resulted from a continuation of the processes responsible for the mid-fan conglomerates of the volcaniclastic conglomerate facies (Animas Formation) near Durango. High-energy braided streams of this alluvial system flowed southward, carrying sand and quartz-bearing and volcanic pebbles up to 10 cm (4 in) in diameter as bedload. They scoured portions of deposits formed during final regression of the Pictured Cliffs sea from the San Juan Basin—that is, the carbonaceous shale and channel-form sandstone facies consisting of the Kirtland Shale and Fruitland Formation delta plain deposits. They also scoured and reworked deposits formed during volcanic activity in the northwestern San Juan Basin, namely, the volcaniclastic conglomeratic lithofacies and the shale and volcaniclastic sandstone facies of the Animas Formation.

The eastern alluvial system originated in the central to eastern San Juan Mountains. High-energy braided streams carried sand and only very fine gravels (less than 0.5 cm—0.2 in—in diameter) and incorporated few volcanic pebbles as bedload. They scoured portions of the Animas Formation and Kirtland Shale and Fruitland Formation deposits that were exposed during uplift and tilting in the southeastern and/or eastern San Juan Basin prior to Ojo Alamo deposition (Fassett and Hinds, 1971). The streams of this alluvial system underwent a down-dip transition from mid- to upper-distal braided stream deposits in the north to more meandering streams in the south at Mesa Portales. The transitional area is characterized by the channel-form sandstone and shale lithofacies.

The scouring features of the streams in both the eastern and western alluvial systems have important implications to the Cretaceous-Tertiary boundary studies, paleomagnetic studies, iridium anomaly studies, and palynology studies that are presently being conducted in the San Juan Basin. Sediment removed and reworked from underlying Cretaceous sediments and the Cretaceous-Tertiary boundary area affects age-determination studies in both the eastern and western basin.

ACKNOWLEDGMENTS

Many people and organizations have assisted with and greatly improved this study. Dale Carlson, of Triple S. Development Co., Albuquerque, New Mexico, provided uranium drill core and geophysical logs for detailed description and sampling of the mineralized Ojo Alamo Sandstone at Mesa Portales. The Jicarilla Apache, Southern Ute, and Navajo tribes all granted permission to study and sample on their lands.

Field assistance and helpful discussions during this study were provided by J. L. Ridgley, J. A. Campbell, M. W. Green, K. J. Franczyk, and J. E. Fassett, all of the U.S. Geological Survey. Jack Ellingston, Professor of Geology, Fort Lewis College, Durango, pinpointed locations to measure sections in the McDermott and Animas formations.

Special thanks go to Mollie Thompson, Jeanne Davidson, and Sharon Gabel for their help and support as field assistants; to J. A. Campbell, J. L. Ridgley, J. E. Fassett, M. W. Green, J. N. Moore (University of Montana), and B. S. Kues (University of New Mexico) for their critical review of this manuscript.

REFERENCES CITED

Anderholm, S. J., 1980, Hydrogeology and water resources of the Cuba quadrangle [Masters Thesis]: Socorro, New Mexico Institute of Mining and Technology, microfiche (4 cards).

Anderson, R. Y., 1960, Cretaceous-Tertiary palynology, eastern side of the San Juan Basin, New Mexico: New Mexico Bureau Mines and Mineral Resources Memoir 6, 58 p.

Baltz, E. H., 1967, Stratigraphy and regional tectonic implications of part of upper Cretaceous and Tertiary rocks east-central San Juan Basin, New Mexico: U.S. Geological Survey Professional Paper 552, 101 p.

Baltz, E. H., Ash, S. R., and Anderson, R. Y., 1966, History of nomenclature and stratigraphy of rocks adjacent to the Cretaceous-Tertiary boundary, western San Juan Basin, New Mexico: U.S. Geological Survey Professional Paper 524-D, p. D1–D23.

Barnes, H., 1953, Geology of the Ignacio and Pagosa Springs quadrangles, La Plata and Archuleta counties, Colorado: U.S. Geological Survey Oil and Gas Investigations Map 138.

Barnes, H., Baltz, E. H., Jr., and Hayes, P. T., 1954, Geology and fuel resources of the Red Mesa area, La Plata and Montezuma counties, Colorado: U.S. Geological Survey Oil and Gas Investigations Map OM 149.

Bauer, C. M., 1916, Contributions to the geology and paleontology of San Juan County, New Mexico; I. Stratigraphy of a part of the Chaco River valley: U.S. Geological Survey Professional Paper 98-P, p. 271–278.

Brimhall, R. M., 1973, Groundwater hydrology of Tertiary rocks of the San Juan Basin, New Mexico, *in* Fassett, J. E., ed., Cretaceous and Tertiary rocks of the southern Colorado Plateau: Four Corners Geological Society Memoir, p. 197–207.

Brown, B., 1910, The Cretaceous Ojo Alamo beds of New Mexico, with description of the new dinosaur genus *Kritosaurus:* American Museum Natural History Bulletin, v. 28, p. 267–274.

Butler, R. F., Lindsay, E. H., and Jacobs, L. L., 1976, Paleomagnetic polarity stratigraphy through the Cretaceous-Tertiary boundary, San Juan Basin, New Mexico [abs.]: EOS (American Geophysical Union Transactions), v. 57, no. 12, p. 902.

Clemens, W. A., 1973, The roles of fossil vertebrates in interpretation of Late Cretaceous stratigraphy of the San Juan Basin, New Mexico, *in* Fassett, J. E., ed., Cretaceous and Tertiary rocks of the southern Colorado Plateau: Four Corners Geological Society Memoir, p. 154–167.

Dane, C. H., 1936, The LaVentana–Chacra Mesa coal field, *in* Geology and fuel resource of the southern part of the San Juan Basin, New Mexico: U.S. Geological Survey Bulletin 860-C, p. 81–161 (1937).

Davis, J. C., 1973, Statistics and data analysis in geology: New York, John Wiley and Sons, 543 p.

Eckel, E. B., 1937, Mode of igneous intrusion in La Plata Mountains, Colorado: American Geophysical Union Transactions, no. 18, p. 258–260.

Fassett, J. E., 1966, Geologic map of the Mesa Portales quadrangle, Sandoval County, New Mexico: U.S. Geological Survey GQ-590, scale 1:24,000.

—— , 1973, The saga of the Ojo Alamo Sandstone or the rock-stratigrapher and the paleontologist should be friends, *in* Fassett, J. E., ed., Cretaceous and Tertiary rocks of the southern Colorado Plateau: Four Corners Geological Society Memoir, p. 123–130.

PLATE 1.

Petrographic characteristics of Kirtland Shale, Animas Formation, and Ojo Alamo Sandstone (magnified 250×). Qtz = quartz, SRF = sedimentary rock fragment, VRF = volcanic rock fragment, Si = siltstone fragment, Pyr = pyroxene, Plag = zoned plagioclase, Sru = soil ripup, Mi = microcline, Pxq = polycrystalline quartz, Pyx = polycrystalline quartz (quartzite), gar = pyralspite garnet.

Figure A. Sandstone in the Kirtland Shale, Betonnie Tsosie section.

Figure B. Animas Formation (conglomerate matrix), Loma Linda section.

Figure C. Ojo Alamo Sandstone, western outcrop, Kirtland, N. Mex., section.

Figure D. Ojo Alamo Sandstone, southeastern outcrop, Mesa Portales, South section.

PLATE 2.

Figure 1. Erosional bases and trough crossbedding, Mesa Portales East section (loc. 25). Thickness of upper sandstone is approximately 10 m (30 ft).
Figure 2. Trough crossbedding, clay ball, and coal pockets, Tsun Je Zhin section (loc. 12). Ruler to right of coal pocket for scale = 1 ft.
Figure 3. Graded beds, scattered pebbles, and clay ripups, Kirtland, N. Mex. (loc. 3). Scale = 1 ft.
Figure 4. Climbing ripples, symmetrical ripples, and ripple drift, Fork Rock Mesa section (loc. 22). Scale = 1 ft.
Figure 5. Clay ball with preserved plant fragments in matrix, Arroyo Piedre Lumbre section (loc. 26). Hammer for scale is approximately 1 ft.
Figure 6. Contorted bedding in Ojo Alamo Sandstone, Fork Rock Mesa section (loc. 22). Scale = 1 ft.

1

2

3

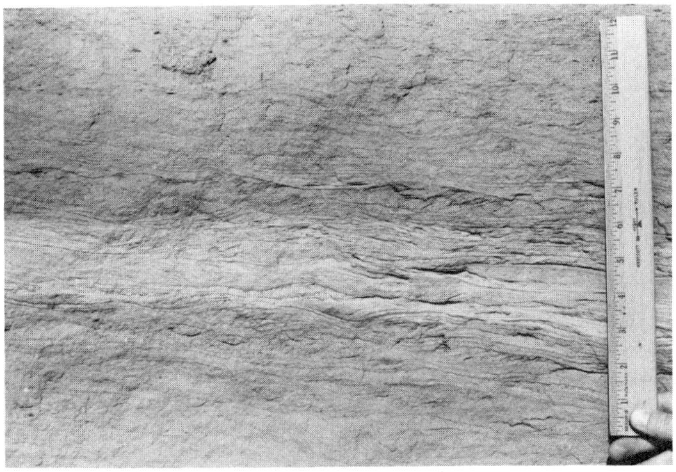

4

5

6

PLATE 3.

Figure 1. Petrified log (outlined on horizon) at Ojo Alamo Trading Post ruins (loc. 8).
Figure 2. Petrified log within Ojo Alamo Sandstone at Ojo Alamo Trading Post ruins (loc. 8). One-foot ruler for scale at arrow. Log diameter is approximately .75 m (2.5 ft).
Figure 3. Volcanic ash beds and bentonitic sandstones between thick sandstone channel deposits, Arroyo Piedre Lumbre section (loc. 26).
Figure 4. Fossil bone fragments of unknown origin weathering from surface of mid to upper Ojo Alamo Sandstone, Gallegos Canyon section (loc. 5).

1

2

3

4

——, 1974, Cretaceous and Tertiary rocks of the eastern San Juan Basin, New Mexico and Colorado, *in* Siemers, C. T., ed., Ghost Ranch (Central-Northern New Mexico): New Mexico Geological Society Guidebook, 25th Field Conference, p. 225–230.

Fassett, J. E., and Hinds, J. S., 1971, Geology and fuel resources of the Fruitland Formation and Kirtland Shale of the San Juan Basin, New Mexico and Colorado: U.S. Geological Survey Professional Paper 676, 76 p.

Gilmore, C. W., 1916, Contributions to the geology and paleontology of San Juan County, New Mexico; II. Vertebrate faunas of the Ojo Alamo, Kirtland, and Fruitland Formations: U.S. Geological Survey Professional Paper 98-P, p. 279–308.

Haynes, D. D., Vogel, J. D., and Wyant, D. G., 1972, Geology, structure, and uranium deposits of the Cortez quadrangle, Colorado and Utah: U.S. Geological Survey Miscellaneous Investigation Series Map I-629, scale 1:250,000.

Hinds, J. S., 1966, Geologic map of the Johnson Trading Post quadrangle, Sandoval County, New Mexico: U.S. Geological Survey GQ-591, scale 1:24,000

Larson, E. S., Jr., and Cross, W., 1956, Geology and petrology of the San Juan region, southwestern Colorado: U.S. Geological Survey Professional Paper 258, 303 p.

Lindsay, E. H., Jacobs, L. L., and Butler, R. F., 1977, Paleomagnetic polarity stratigraphy through the Tertiary-Cretaceous boundary, San Juan Basin, *in* Piatti, G., ed., Paleomagnetic stratigraphy of pelagic carbonate sediments: Società Geologica Italiana Memoir, v. 15, p. 69–72.

Lindsay, E. H., Jacobs, L. L., and Butler, R. F., 1978, Biostratigraphy and magnetostratigraphy of Paleocene terrestrial deposits, San Juan Basin, New Mexico: Geology, v. 6, no. 7, p. 425–429.

Lindsay, E. H., Butler, R. F., and Johnson, N. M., 1981, Magnetic polarity zonation and biostratigraphy of Late Cretaceous and Paleocene continental deposits, San Juan Basin, New Mexico: American Journal of Science, v. 281, p. 390–435.

Lucas, S. G., and Mateer, N. J., 1983, Dinosaurs and the Cretaceous-Tertiary boundary in the San Juan Basin, New Mexico: Geological Society of America Abstracts with Programs, v. 15, no. 5, p. 306.

Manley, K., Scott, G. R., and Wobus, R. A., 1978, Preliminary geologic map of the Aztec 1 × 2 degree quadrangle, northwestern New Mexico and southern Colorado: U.S. Geological Survey Open-File Report 78-466, scale 1:250,000.

Orth, C. J., Gilmore, J. S., Knight, J. D., Pillmore, C. L., Tschudy, R. H., Fassett, J. E., 1981, Iridium anomaly at the palynological Cretaceous-Tertiary boundary in northern New Mexico: Science, v. 214, p. 1341–1343.

O'Sullivan, R. B., and Beikman, H. M., 1963, Geology, structure, and uranium deposits of the Shiprock quadrangle, New Mexico and Arizona: U.S. Geological Survey Miscellaneous Geologic Investigations Map I-345, scale 1:250,000.

O'Sullivan, R. B., Repenning, C. A., Beaumont, E. C., and Page, H. G., 1972, Stratigraphy of the Cretaceous rocks and the Tertiary Ojo Alamo Sandstone, Navajo and Hopi reservations, Arizona, New Mexico, and Utah: U.S. Geological Survey Professional Paper 521-E, p. E1–E65.

Powell, J. S., 1972, The Gallegos Sandstone (formerly Ojo Alamo Sandstone) of the San Juan Basin, New Mexico [Masters thesis]: Tucson, University of Arizona, 131 p.

——, 1973, Paleontology and sedimentation models of the Kimbeto Member of the Ojo Alamo Sandstone, *in* Fassett, J. E., ed., Cretaceous and Tertiary rocks of the southern Colorado Plateau: Four Corners Geological Society Memoir, p. 111–122.

Reeside, J. B., Jr., 1924, Upper Cretaceous and Tertiary formations of the western part of the San Juan Basin of Colorado and New Mexico: U.S. Geological Survey Professional Paper 134, p. 1–70.

Reinick, H. E., and Singh, I. B., 1980, Depositional sedimentary environments with reference to terrigenous clastics, 2d, rev. and updated edition: New York, Springer-Verlag, p. 257–314.

Rigby, J. K., and Lucas, S., 1977, Fossil mammals from the Ojo Alamo Formation: New Mexico Geological Society Guidebook, 28th Field Conference, supplement, p. 55–56.

Scholle, P. A., and Spearing, D., 1982, Sandstone depositional environments: American Association Petroleum Geologists Memoir 31, p. 49–86, 115–137.

Sinclair, W. J., and Granger, W., 1914, Paleocene deposits of the San Juan Basin, New Mexico: American Museum Natural History Bulletin, v. 33, p. 297–316.

Steven, T. A., Lipman, P. W., Hail, W. J., Jr., Barker, F., Luedke, R. G., 1974, Geologic map of the Durango quadrangle, southwestern Colorado: U.S. Geological Survey Miscellaneous Investigations Map I-764, scale 1:250,000.

Tschudy, R. H., 1973, The Gasbuggy core; A palynological appraisal, *in* Fassett, J. E., ed., Cretaceous and Tertiary rocks of the southern Colorado Plateau: Four Corners Geological Survey Memoir, p. 131–143.

Tweto, O., 1975, Laramide (Late Cretaceous–Early Tertiary) orogeny in the southern Rocky Mountains: Geological Society of America Memoir 144, p. 1–74.

Wood, G. H., Kelly, V. C., and MacAlpin, A. J., 1948, Geology of southern part of Archuleta County, Colorado: U.S. Geological Survey Oil and Gas Investigations Preliminary Map 81.

Wyant, D. J., and Olson, A. B., 1978, Geologic map of the Albuquerque 1 × 2 degree quadrangle, northwestern New Mexico: U.S. Geological Survey Open-file Report 78-467, scale 1:250,000.

Manuscript Accepted by the Society March 24, 1986

Printed in U.S.A.

Geological Society of America
Special Paper 209
1987

Geochronologic and geochemical study of volcanic ashes from the Kirtland Shale (Cretaceous), San Juan Basin, New Mexico

Douglas G. Brookins
Department of Geology
University of New Mexico
Albuquerque, New Mexico 87131

J. Keith Rigby, Jr.
Department of Earth Sciences
University of Notre Dame
Notre Dame, Indiana 46556-1020

ABSTRACT

Several volcanic ash units that have recently been discovered in the Kirtland Shale (Cretaceous) in the northwestern part of the San Juan Basin, New Mexico, occur some 20 m (sample 62) to 39 m (sample 54) above the uppermost rocks of the Fruitland Formation. The ashes are of extremely great value in attempting to work out the stratigraphy of the rocks in this area, especially as there is some question concerning the interpretation of faunal assemblages and paleomagnetic signatures; contradicting "dates" of earliest Tertiary to latest Cretaceous have been suggested for the part of the Kirtland Shale sampled. A systematic study of the Potassium-Argon geochronology of some of the ashes has been undertaken with the following results: (1) highest ash (sample 54): sanidine concentrates—72.4 ± 3.1 to 74.4 ± 2.6 Ma; biotite concentrates—73.2 ± 2.7 to 76.1 ± 2.8 Ma; (2) lowest ash (sample 62): sanidine concentrate—75.0 ± 2.7 Ma; and (3) upper middle ash (sample 93): sanidine concentrate—69.8 ± 2.5 Ma.

Petrographic and scanning-electron microscope studies show the biotite and sanidine to be primary phases, containing negligible amounts of detrital material. The rare earth element and other chemical studies of the ashes show them to be slightly different from each other, and this information may allow rare earth element distribution data to be used for purposes of ash correlation. We interpret the Potassium-Argon ages as suggesting a Late Cretaceous age for the part of the Kirtland Shale sampled, a view consistent with recent paleontological and paleomagnetic studies.

INTRODUCTION

During the summers of 1980 and 1981, the Albuquerque District Office of the U.S. Bureau of Land Management, under the direction of J. Keith Rigby, Jr., conducted several field inventories for paleontological materials. These inventories documented the occurrence of several thin, partially reworked air-fall tuffs in the uppermost part of the Fruitland Formation and lower half of the Kirtland Shale. Part of this inventory effort was concentrated in the badlands surrounding the Bisti Chapter House of the Navajo Nation in the drainage of Hunter Wash, San Juan Basin, New Mexico (see Fig. 1). The areas studied included exposures of the Fruitland and Kirtland formations (usage of Lindsay and others, 1981) in 23-24N,R.12-13W.

The ashes evaluated here (Brookins, 1982; O'Sullivan and others, 1972) occur in the basal part of the Kirtland Shale, as defined by the U.S. Geological Survey (Clemens, 1973). Their field occurrence is indicated by concentrated thin layers of red-brown fibrous calcite chips with individual fibers being rarely longer than 5 cm (2 in). The calcite chips form a surface litter on subdued benches formed by differential weathering of slightly more resistant layers within the generally rounded, deeply dissected hills of the Kirtland Shale. Observations from minor surface excavations suggest that the ashes, which are the more resistant material, actually hold up the benches.

There are evidently many "ashes" scattered throughout the

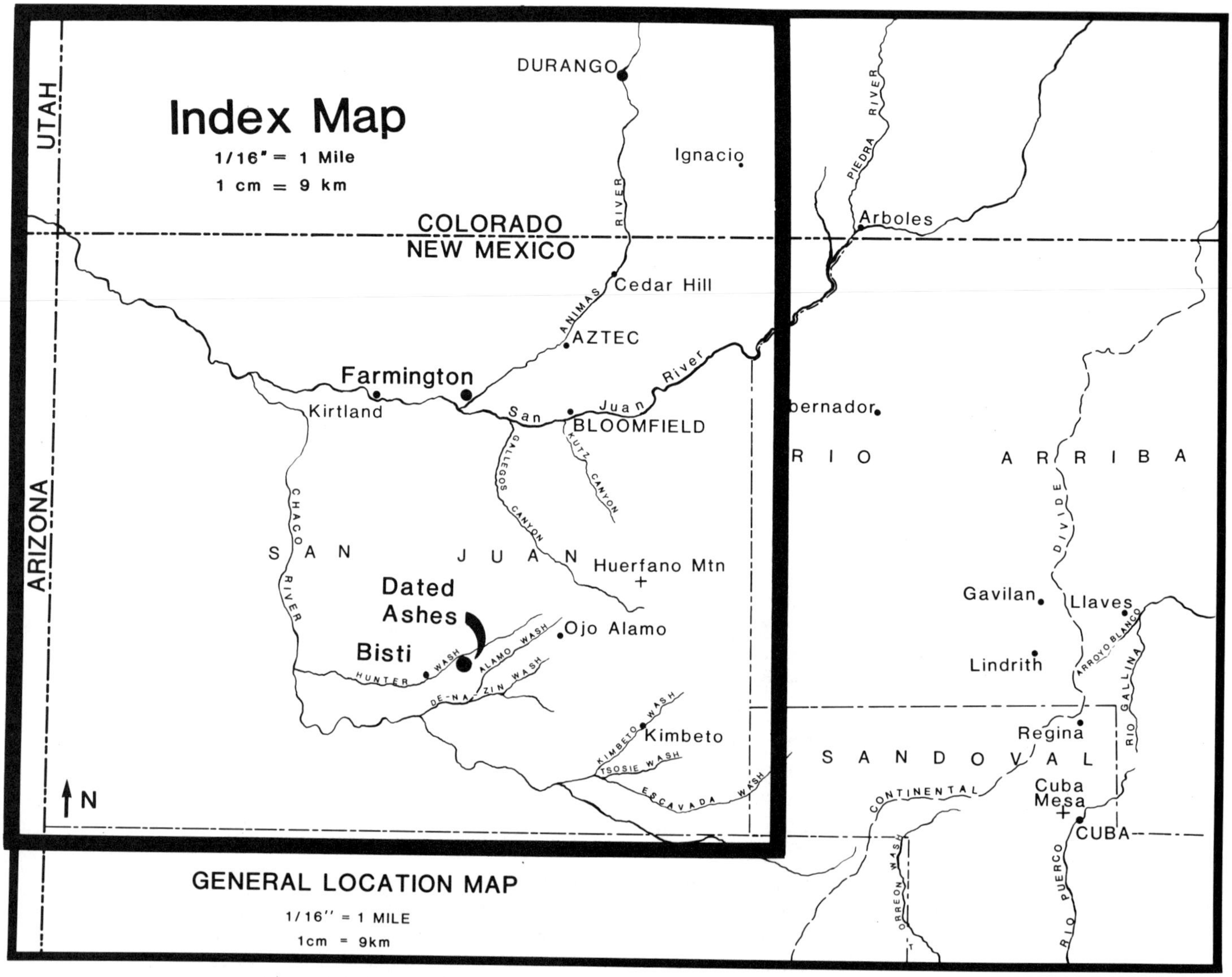

Figure 1. Index map showing location of ash samples, northwestern San Juan Basin.

Kirtland and Fruitland formations. Many are very thin (generally less than 2 cm thick) and are laterally discontinuous. We chose to date radiometrically three of the most widespread (the erosional bench formed can be traced from one side of the Bisti Badlands to the other) and thickest (20–40 cm). These thicker ashes also appear to be the least altered of all ashes observed. Thin sections and chemical analyses later confirmed this field observation.

Stratigraphic Position of the Ashes

Many workers have defined a boundary between the Fruitland and Kirtland formations within the Bisti Badlands and along the south flank of the San Juan Basin. The type sections of the Fruitland and Kirtland formations are along the bluffs of the San Juan River approximately 56.3 km (35 mi) north of the Bisti Chapter House. Bauer (1916) in his original definition of the Fruitland Formation, characterized it as being irregularly bedded and exceptionally variable in composition, including discontinuous fluviatile sandstones, shales, and coal. Bauer characterized the Kirtland Shale as predominantly clayey, but particularly along the San Juan Basin, it included a thick sandstone unit, the Farmington Sandstone Member, which pinches out to the south.

The formational definition of these units is somewhat vague, even in the type section area, and their application to beds farther to the south has at best been problematic. These difficulties are perhaps best summarized by Hunt (1983, and elsewhere in this volume). In the immediate vicinity of our ash sample localities in the Bisti Badlands, the formation distinctions between the Fruitland and Kirtland formations are understandably imprecise, so that authors have defined formational contacts in a variety of ways. For example, Scott and others (1979) most recently mapped the Alamos Mesa West 7½-minute Quadrangle and de-

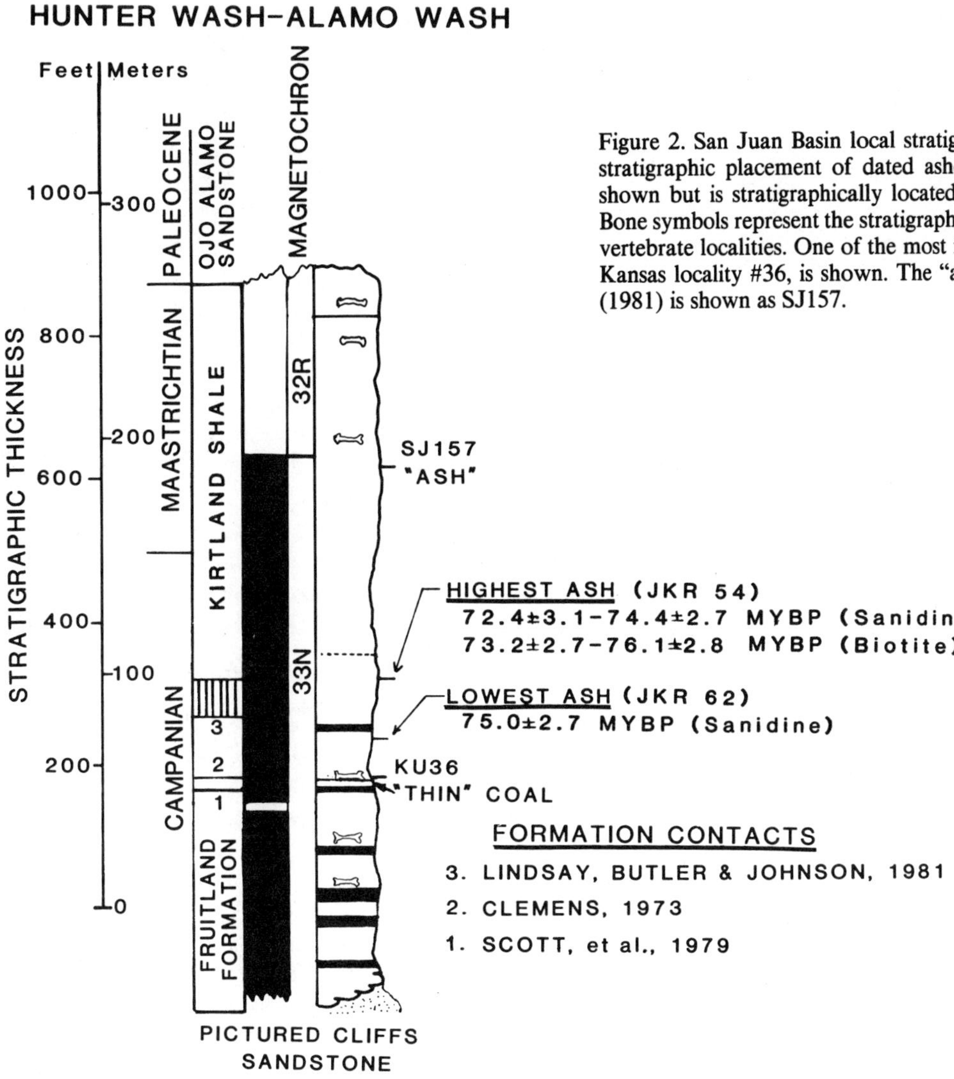

Figure 2. San Juan Basin local stratigraphy (south flank) showing the stratigraphic placement of dated ashes. The upper-middle ash is not shown but is stratigraphically located near the highest ash (JKR 54). Bone symbols represent the stratigraphic placement of a number of fossil vertebrate localities. One of the most important localities, University of Kansas locality #36, is shown. The "ash" dated by Lindsay and others (1981) is shown as SJ157.

fined the top of the Fruitland Formation as being at the top of the most persistent "major" coal bed. This was an attempt to exclude thin lignites that are known to occur quite commonly in the base of the Kirtland Shale. In so defining the boundary, however, they excluded two lenticular, major fluviatile sandstones that, by Bauer's original definition, perhaps should be included within the Fruitland Formation. These two sandstones, with their red-brown–capped subunits, are important because they contain the largest samples of the Hunter Wash local fauna described by Clemens (1973). Clemens defined the formational contact of the Fruitland Formation and Kirtland Shale as at the top of a "thin" persistent coal unit that is present at the base of the upper sandstone. The base of the upper sandstone also contains Kansas University locality 36, which has produced most of the mammalian material described by Clemens (1973). This formational boundary would be placed somewhat above that of Scott and others (1979).

Lindsay and others (1981) have defined the boundary in still a different manner. They include within the Fruitland Formation several carbonaceous and lignitic zones that most authors would consider part of the Kirtland Shale. They cite Fassett and Hinds (1971) as justification for selecting the uppermost persistent "coal" as the contact between the Fruitland Formation and Kirtland Shale. The differences in opinion here rest in the definition of a "coal."

For the benefit of future workers, who may wish to disregard the separation of the Fruitland and Kirtland formations altogether, we think it is necessary to locate our dated ashes stratigraphically in the context of at least the three different definitions above (see Fig. 2). The lowest of our dated ashes is approximately 22.5 m (75 ft) above the "thin," persistent coal of Clemens (1973). Our uppermost ash is approximately 48–51 m (160–170 ft) above the same thin "coal." All ashes are in the basal Kirtland Shale, by Clemens's definition. Using the definition of Scott and others (1979), our dated ashes are approximately 6–9 m (20–30 ft) above the base of the Kirtland Shale. Based on the definition of Lindsay and others (1981), our lowest ash is in the uppermost Fruitland Formation and our highest ash

TABLE 1. K-Ar AGE DATA

Sample	K^+ (%)	$^{40}Ar^*$ (ppm)	$^{40}Ar^*/^{40}Ar$ (total)	Age (Ma)
JKR-54a (sanidine)	3.145 3.184	0.01553 0.01690	0.822 0.758	72.4 ± 3.1
JKR-54b (biotite)	6.450 6.317	0.03437 0.03452	0.775 0.776	76.1 ± 2.8
JKR-54c (biotite)	6.618 6.066	0.03259 0.03082	0.754 0.767	73.2 ± 2.7
JKR-54d (sanidine)	10.227 9.896	0.05240 0.05359	0.886 0.886	74.4 ± 2.6
JKR-62a (sanidine)	9.269 9.474	0.05444 0.04960	0.961 0.917	75.0 ± 2.7
JKR-93a (sanidine)	9.472 9.203	0.04627 0.04596	0.817 0.871	69.8 ± 2.5

Notes: Samples reported as sanidine are sanidine-quartz mixtures, cleaned of all coatings of smectite and carbonate by washing in dilute HF and HNO3 under clean conditions. Average K^+ and $^{40}Ar^*$ values were used for the individual K-Ar age determinations. The ages reported here assume the half life of $^{40}K = 1.25 \times 10^9$ yr. The errors reported here represent the analytical reproducibility of replicate analyses.

is at the top of the zone of gradational contact between the two formations.

Potassium-Argon Age Determinations

Several of the ashes from the Kirtland Shale contain K-feldspar (mainly sanidine) and biotite, although the biotites are commonly weathered. Representative samples of three of these ashes were cleaned and sent to Geochron Laboratories, Cambridge, Massachusetts, for commercial K-Ar study. The early results, summarized in Brookins and Rigby (1982a, 1982b), yielded fairly wide disagreement between K-Ar ages on different aliquots of the same samples. The first set of dates (Brookins and Riby, 1982a) was older than those reported here, and this difference was tentatively explained as resulting from some detrital component. Yet this explanation proved to be inadequate when the second dates were obtained, as they were significantly younger than the first dates (Brookins and Rigby, 1982b). This discrepancy between the two sets of data was presented to Geochron Laboratories, whose personnel investigated the matter at some length. Their conclusion was that the first samples had not been as carefully cleansed prior to analysis as the second set had been, and that in the case of the second set, an old, not well-characterized (for last residual volume) ^{38}Ar spike had been used, which was suspect. They agreed to rerun aliquots of the samples, using refined cleansing techniques (see Brookins and Rigby, 1982b) and a new, well-calibrated ^{38}Ar spike. The results of the rerun are shown in Table 1. The resultant average age of 73.5 ± 2.2 Ma is the preferred date for biotites and sanidine from the volcanic ashes of the basal Kirtland Formation; these data have been presented in Brookins (1982) and Brookins and Rigby (1983).

Sample Treatment

Large samples of ashes nos. JKR-54, JKR-62, and JKR-93 were sent to Geochron Laboratories K-Ar dating. The samples were crushed to −100 mesh, and the 100–200 mesh fraction was used for mineral separation: Biotite was separated magnetically, and a sanidine-rich concentrate was separated by heavy liquids (quartz is a dilutent in the sanidine-rich concentrate). Biotites and the sanidine-rich fraction by washing with dilute HF followed by HNO_3. This treatment removes carbonate and clay mineral coatings as well and does not affect the interiors of the grains.

Analyses

Geochron Laboratories carried out the K and Ar determinations. The K was analyzed by replicate atomic absorption spectrophotometric methods, and Ar was analyzed by standard isotope dilution techniques using a ^{38}Ar-enriched spike. Standard calibration methods were used for the ^{38}Ar tracer, and interlaboratory standards were checked on a routine basis for both K and Ar determinations (H. W. Krueger, personal communication).

Sample Descriptions and Locations

Sample JKR-62. (cen. SE¼SE¼S.21,T.24N.,R.31W.; 108°13′28.6″W.,36°17′43.9″N.) Sample is streaked with some cross laminations. The biotite does not look highly altered, yet is corroded on grain edges and partly altered to chlorite. Some grains are coated by clay minerals on edges and partly along cleavage traces.

The feldspars are euhedral sanidines with well-developed crystals. Minor adularia and Na-plagioclase also appear fresh, although some Na-spar is altered to montmorillonite. Montmorillonite and illite (includes mixed layer illite-montmorillonite as well) and carbonates (calcite) are common in the altered matrix. Quartz crystals are unaltered. Minor opaque minerals (zircon + unidentified phase) are noted, especially along cross laminations.

Sample JKR-54. (center S23,T.24N.,R.13W.;108°11′29.5″W.,36°18′0.35″N.) More fine grained than JKR-62; medium-grained phenocrysts of K-feldspar (sanidine) and altered biotite in altered matrix. Matrix consists predominantly of montmorillonite (minor illite) and calcite. Biotite is altered and partly kinked(?). Alteration is pronounced on grain edges with pseudo-isotropic–appearing nature. Heavy minerals include zircon, hornblende, spinel (?), and unidentified opaques. Kaolinite present on some sanidine surfaces and in surface cracks. The K-feldspar is coated by clay minerals but not infiltrated by them.

Sample JKR-93. (SE¼NW¼SE¼S.32,T.24N.,R21W.; 108°0.8′4.5″W.,36°16′04.8″N.) Medium-grained ash (like JKR-54) with highly altered matrix (calcite and abundant montmorillonite). Biotite is scarce and highly altered, mainly to clay minerals. The sanidines are more altered than JKR-54, although most of the alteration is surficial. Chlorite after biotite is abundant; in places the biotite appears almost isotropic. Some calcite fills cracks in the K-feldspars, but these are minor. Kaolinite is more abundant in this sample than in JKR-54.

TABLE 2. GEOCHEMICAL DATA FOR KIRTLAND SHALE

Element	JKR-54	JKR-62	JKR-93
Na-24	26100 ± 700	25500 ± 700	22400 ± 600
Mg-27	14000 ± 2000	14000 ± 2000	13000 ± 2000
Al-28	139000 ± 5000	172000 ± 6000	158000 ± 5000
Cl-38	370 ± 50	< 200	< 120
K-42	14000 ± 2000	10000 ± 2000	< 4000
Ca-49	21000 ± 2000	22000 ± 2000	13000 ± 2000
Ti-51	3600 ± 300	1700 ± 300	1400 ± 200
V-52	53 ± 5	< 10	< 8
Mn-56	700 ± 30	470 ± 20	250 ± 10
Cu-66	< 400	< 400	< 300
Sr-897	490 ± 70	< 400	600 ± 100
I-128	< 20	< 20	< 20
Ba-139	1200 ± 100	310 ± 80	360 ± 60
Dy-165	5.7 ± 0.4	8.0 ± 0.7	6.4 ± 0.5
U-235	4.8 ± 0.1	21.2 ± 0.6	18.8 ± 0.5
Ga-72	< 70	< 80	< 200
As-76	< 3	< 3	< 4
Br-82	< 3	< 4	< 4
Sb-122	< 1	< 1	< 2
La-140	62 ± 2	60 ± 2	75 ± 2
Sm-153	6.4 ± 0.3	8.9 ± 0.4	11.0 ± 0.5
Yb-175	2.7 ± 0.3	4.1 ± 0.5	3.1 ± 0.4
W-187	< 4	< 5	< 7
Au-198	< 0.02	< 0.02	< 0.02
Sc-46	5.61 ± 0.10	3.16 ± 0.06	3.10 ± 0.06
Cr-51	29 ± 3	< 2	7 ± 1
Fe-59	30700 ± 800	28300 ± 800	27500 ± 800
Co-60	10.0 ± 0.4	2.5 ± 0.1	3.4 ± 0.2
Zn-65	150 ± 20	58 ± 9	28 ± 10
Se-75	< 2	< 0.4	< 2
Rb-86	43 ± 4	< 10	< 8
Zr-95	230 ± 30	180 ± 30	160 ± 30
Ag-110	< 2.1	< 2.1	< 1.8
Sb-124	0.9 ± 0.2	1.5 ± 0.2	1.1 ± 0.2
Cs-134	0.7 ± 0.1	< 0.3	< 0.2
Ba-131	900 ± 100	610 ± 70	440 ± 60
Ce-141	94 ± 3	129 ± 4	143 ± 5
Nd-147	28 ± 5	43 ± 7	55 ± 8
Eu-152	1.26 ± 0.06	0.78 ± 0.05	1.09 ± 0.06
Tb-160	0.5 ± 0.1	0.39 ± 0.10	0.05 ± 0.01
Yb-169	2.3 ± 0.1	3.6 ± 0.2	2.9 ± 0.1
Lu-177	0.30 ± 0.03	0.43 ± 0.05	0.32 ± 0.04
Hf-181	7.0 ± 0.6	6.7 ± 0.6	6.6 ± 0.6
Ta-182	1.1 ± 0.1	4.1 ± 0.4	4.7 ± 0.5
Th-233	19.6 ± 0.8	52 ± 2	56 ± 2

Explanation: Data are reported in ppm. The element and its isotope used for analytical counting are given in the left hand column. All data by instrumental neutron activation analysis (Los Alamos National Laboratory, S. R. Garcia, analyst).

Trace Element Chemistry

Trace elements for three representative whole-rock samples of the ashes were determined by instrumental neutron activation analysis (INAA), using the facilities of the Los Alamos National Laboratory (S. Garcia, analyst). These data are presented in Table 2.

The data are too few to make any wide-sweeping generalities, yet they are of interest in terms of the chemical integrity of the ashes. The Th/U ratios range from 2.45 (JKR-62) to 4.08 (JKR-54), well within the expected range for these rocks. Certainly, despite widely differing U and Th contents, there does not, for example, appear to be preferential loss of U by leaching. The high U contents of ashes JKR-62 and JKR-93 also argue against such leaching. Further, the Fe/Sc ratios ($\times 10^{-3}$) range from 5.48 (JKR-54) to 8.87 (JKR-93), which also argues against any loss of Fe from alteration of biotites.

Rare earth element (REE) patterns are shown in Figure 3, in which REE/CHON (CHON = chondrite) is plotted on a log scale against relative position of the REE in order of increasing atomic number (La to Lu). The REE/CHON curves are very similar for the three ashes; moreover, they are similar to patterns for volcanic rocks elsewhere in the western United States. No perturbations of the REE systematics of the rocks are indicated, although the slight differences in the REE/CHON curves (Fig. 3) suggest that the REE can be used for correlative purposes.

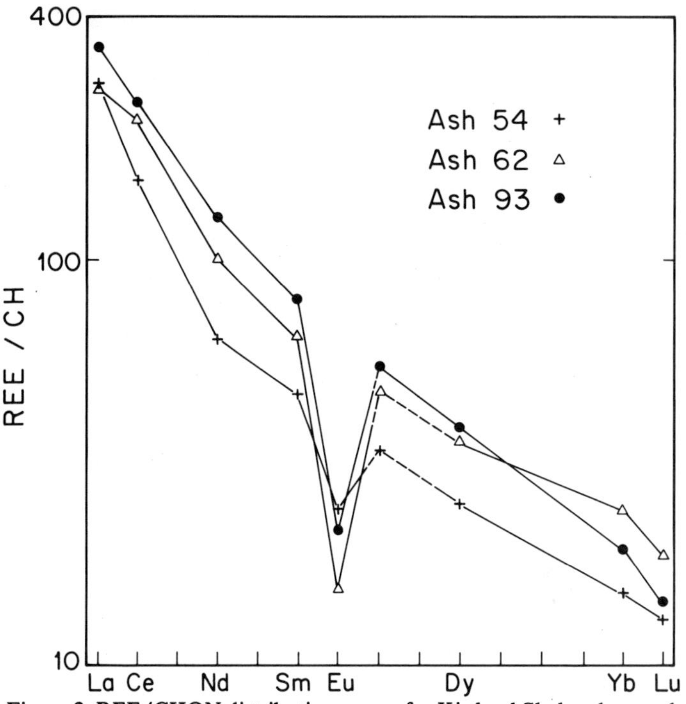

Figure 3. REE/CHON distribution curves for Kirtland Shale ashes used in this study.

Collectively, the data shown in Table 2 argue against these ashes having been affected by extensive postformational weathering and/or leaching. This is important because it indicates that the K-Ar systematics of constituent minerals such as biotite and sanidine in many instances may not have been seriously affected by postemplacement alteration. More data on phenocryst minerals are needed to verify this interpretation, however.

Fission Track Dates

Lindsay and others (1981) have reported a fission track data from a volcanic ash, presumably from the Kirtland Shale; they used a small zircon separate for their age investigation. The information used in their age investigation is given as follows (from Lindsay and others, 1981, p. 401):

number of zircons	13
total spontaneous track count (N_s)	544
total induced track count (N_i)	244
thermal neutron dose (ϕ)	9.44×10^{14}
^{238}U fission constant	6.85×10^{-17}/y
age and error	64.6 ± 3.0 Ma

The authors emphasize the fact that this date of 64.6 Ma is identical, within the error limits given, to the age of the Cretaceous-Tertiary boundary at 65–66 Ma.

One of us (DGB) contacted Dr. Charles Naeser of the U.S. Geological Survey after our own K-Ar work had been initiated and was informed that some of the parameters used by Lindsay and others (1981) in their calculation of age were wrong. Specifically, the thermal neutron dose was actually 9.98×10^{14}, the half life of ^{238}U for spontaneous fission is $7.03 \times 10^{-17}/y$, $N_s = 561$, and $N_i = 490$. Using these parameters, the corrected zircon age by fission track is 68.0 ± 3.0 Ma. In addition, Naeser (personal communication, 1982) and Johnson (written communication, 1981) have stated that all that can really be said about this particular sample is that it is between 60–75 Ma. Lindsay and others (1981) do not appear to be justified in assigning a date of 64.6 ± 3.0 Ma based on the above observations.

DISCUSSION

The purpose of this report is to focus on the K-Ar age determinations of volcanic ash units from the basal Kirtland Shale. The combined K-Ar and geochemical data argue for deposition of the ash units at about 73 Ma, with little subsequent perturbation of the mineral systematics. The nearly identical REE/CHON curves for the ashes, and for similar other trace element contents, suggest a similar origin for these units and, possibly, no extensive reworking of them. Further, the sanidines and biotites in many samples are quite fresh. Where the carbonate and grain-coating smectite content increases, the biotites are commonly chloritized and yield anomalously low K-Ar dates (see Brookins and Rigby, 1982a). Cleansing the samples by HNO_3 wash presumably removes the carbonate and smectite but leaves the sanidines and biotites otherwise undisturbed (H. W. Krueger, personal communication). Quartz is a common diluent for the sanidines, but it usually does not pose any special problem for the K-Ar determinations.

We believe that the average K-Ar date reported here (73.5 ± 2.2 Ma) represents the formational age of the ash units of the basal Kirtland Formation. This age is in good agreement with the invertebrate and vertebrate paleontological evidence reported elsewhere in thhis volume, as well as with the revised paleomagnetic evidence and field relations. The fission track age reported by Lindsay and others (1981) is suspect, and we report a revised age for their samples based on new information provided (Naeser, personal communication). In any event, it is not wise to rely on just one date, regardless of method, to address important problems such as those surrounding the Cretaceous-Tertiary boundary.

We consider that the effect of detrital K-feldspars is negligible. The sanidines are very fresh, and even though elsewhere detrital feldspars contain microcline and orthoclase in addition to sanidine, neither of those were noted in our mineralogic study. The biotites also appear to be newly formed in the ash, rather than detrital, although some of the biotites have been so altered as to yield erroneous K-Ar dates (see Brookins and Rigby, 1982a).

CONCLUSION

We conclude that the time of deposition of the three volcanic ash units in the basal Kirtland Shale from the study area was 73.5 ± 2.2 Ma. This finding supports the other evidence given in this volume of these units being late, but certainly not latest, Cretaceous.

ACKNOWLEDGMENTS

We extend thanks to the U.S. Bureau of Land Management (BLM) for its interest and partial financial support and to all those who participated in the BLM invenstory of the San Juan Basin coal lands. Rigby was supported by a donation of the Hill Foundation to the Dean Clarence Manion Fund of the University of Notre Dame. We also wish to thank M. C. Abashian, M. T. Murphy, H. S. Vogler, R. K. Matheney, and especially R. M. Lowy, all of the University of New Mexico, and H. W. Krueger and R. A. Reesman, Geochron Laboratories, Cambridge, Massachusetts, for their assistance and interest in the project.

REFERENCES CITED

Bauer, C. M., 1916, Stratigraphy of a part of the Chaco River Valley: U.S. Geological Survey Professional Paper, v. 98-P, p. 271–278.

Brookins, D. G., 1982, Revised K-Ar ages for the Kirtland Formation (Cretaceous), San Juan Basin, New Mexico: Isochron/West, n. 35, p. 15.

Brookins, D. G., and Rigby, J. K., 1982a, New K-Ar dates from the Kirtland Formation (Cretaceous), San Juan Basin, New Mexico: Isochron/West, n. 33, p. 17–18.

Brookins, D. G., and Rigby, J. K., 1982b, K-Ar ages for the Kirtland Formation (Cretaceous), San Juan Basin, New Mexico: Isochron/West, n. 34, p. 19.

Brookins, D. G., and Rigby, J. K., 1983, Geochronologic and geochemical study of volcanic ashes from the Kirtland Formation (Cretaceous), San Juan Basin, New Mexico: Geological Society of America Abstracts with Programs, v. 15, no. 5, p. 307.

Clemens, W. A., 1973, The roles of fossil vertebrates in interpretation of Late Cretaceous stratigraphy of the San Juan Basin, New Mexico: Four Corners Geological Society Memoir, p. 154–167.

Hunt, A., 1983, Stratigraphy, sedimentology, taphonomy, and magnetostratigraphy of the Fossil Forest area, San Juan Basin, New Mexico [M.S. thesis]: Socorro, New Mexico Institute of Mining and Technology, 380 p.

Lindsay, E. H., Butler, R. F., and Johnson, N. M., 1981, Magnetic polarity zonation and biostratigraphy of Late Cretaceous and Paleocene continental deposits, San Juan Basin, New Mexico: American Journal of Science, v. 281, p. 390–435.

O'Sullivan, R. B., Repenning, C. A., Beaumont, B. C., and Page, H. G., 1972, Stratigraphy of the Cretaceous rocks and the Tertiary Ojo Alamo Sandstone, Navajo and Hopi Indian reservations, Arizona, New Mexico, and Utah: U.S. Geological Survey Professional Paper 521, 65 p.

Palmer, A. R., 1983, The decade of North American geology 1983; geologic time scale: Geology, v. 11, p. 503–504.

Scott, G. R., O'Sullivan, R., Mytton, J. W., 1979, Reconnaissance geologic map of the Alamo Mesa West quadrangle, San Juan County, New Mexico: U.S. Geological Survey Miscellaneous Field Study Map ME-1074, 1 sheet, scale 1:24,000.

Printed in U.S.A.

Geological Society of America
Special Paper 209
1987

Stratigraphy and depositional environments of the Cretaceous-Tertiary boundary clay and associated rocks, Raton basin, New Mexico and Colorado

Charles L. Pillmore
Romeo M. Flores
U.S. Geological Survey
Box 25046, Denver Federal Center
Denver, Colorado 80225

ABSTRACT

Upper Cretaceous and lower Tertiary rocks exposed in the Raton basin, New Mexico and Colorado, contain a thin, iridium-rich, kaolinitic clay bed that marks the palynologically defined Cretaceous-Tertiary boundary. The clay bed is presumed to have been deposited as the result of a catastrophic event, possibly the impact of an asteroid, at the end of the Cretaceous; it is preserved in sediments laid down in the quiet waters of ponds and coal-forming swamps. Occurring in a conformable sequence of nonmarine rocks, this extraordinary marker bed enables observation of depositional conditions of an instant of geologic time throughout a wide area in the basin. Investigations made at various sites where the boundary clay is present indicate that component elements of a dynamically aggrading fluvial system existed at the close of the Cretaceous. This depositional system was characterized by meandering rivers, broad floodplains, and levees. The levees were breached locally by crevasse splays that periodically invaded the floodplains and associated poorly drained and well-drained swamps.

INTRODUCTION

The Late Cretaceous in the Raton basin was a period of emergence of land and retreat of the Cretaceous epeiric sea. The coastline advanced eastward across New Mexico as broad prograding deltaic and barrier coastal plains were formed seaward of aggrading fluvial systems. This report discusses the marine and marginal-marine formations and the rocks that were deposited by fluvial systems that drained the episodically emergent San Luis Highland (Tweto, 1980) during the Late Cretaceous and Paleocene in the Raton basin. It also describes the Cretaceous and Tertiary boundary interval and adjacent strata.

The Cretaceous-Tertiary (K-T) boundary in the Raton basin, as determined from fossil pollen (Tschudy, 1973; Tschudy and others, 1984), is in a conformable sequence of nonmarine rocks in the lower part of the Raton Formation. At some places, this fossil-pollen break is marked by a thin, light-colored kaolinitic clay bed that is anomalously high in iridium (Ir) concentrations as well as in several other elements (Orth and others, 1981, 1982; Pillmore and others, 1984; Gilmore and others, 1984). This bed, called the boundary clay in this report, is thought to be the result of fallout associated with a global catastrophe (Alvarez and others, 1980; Alvarez and others, 1982), termed the K-T boundary event.

Some floodplain environments of the alluvial plain that developed at the close of the Cretaceous period proved to be ideal for deposition and preservation of Ir-rich fallout material associated with the Cretaceous terminal event (Pillmore and others, 1984; Alvarez and others, 1980; Alvarez and others, 1982). These ideal environments consisted of coal-forming swamps, ponds, and oxbow lakes. The active currents in channels of meandering and braided streams, overbank floodpaths, and crevasse splays would have quickly eroded and dispersed the Ir-rich boundary-clay fallout material. At 15 sites in the Raton-Trinidad area, abundance anomalies of Ir and of several other elements have been found coincident with an abrupt disappearance of several taxa of fossil pollen and a sudden change in the ratio of fern spores to angiosperm pollen (Orth and others, 1981, 1982;

Pillmore and others, 1984; Tschudy and others, 1984). The geochemical anomalies occur in a bed of kaolinitic claystone 1 to 2.5 cm (0.5 to 1 in) thick, that is usually associated with an overlying thin coal bed.

The Raton basin is a large, asymmetric structural depression of about 6500 km^2 (2500 mi^2) extending from Huerfano Park, Colorado, to Cimarron, New Mexico (Fig. 1). The rocks of the lower part of the Raton Formation, which contain the K-T boundary, dip steeply and form hogback ridges along the western margin of the basin but dip more gently along the other margins and are highly dissected. The stratigraphic interval that locally contains the K-T boundary-clay layer crops out along the eastern rim of the Raton basin coal fields from Walsenburg, Colorado, to Cimarron, New Mexico, a distance of about 161 km (100 mi). However, the boundary clay has been found only in the east-central and southern parts of the basin, between Aguilar and Cimarron (Fig. 1).

STRATIGRAPHY

A columnar section of uppermost Cretaceous and Paleocene rocks in the Raton basin is shown in Table 1. The offshore-marine Pierre Shale (Campanian to Maastrichtian) and overlying marginal-marine Trinidad Sandstone (Maastrichtian) underlie the nonmarine Upper Cretaceous and Tertiary rocks in the basin. Nonmarine sedimentary rocks exposed in the Raton basin, from oldest to youngest, include the coal-bearing Vermejo Formation (Maastrichtian) and Raton Formation (Maastrichtian and Paleocene) and the non-coal-bearing Poison Canyon Formation (Maastrichtian and Paleocene), which overlies and intertongues with the Raton Formation.

MARINE AND MARGINAL-MARINE ROCKS

The Pierre Shale

The Pierre Shale consists of dark, silty, noncalcareous shale about 550–580 m (1800–1900 ft) thick, deposited in prodeltaic and offshore-marine environments in the Western Interior Cretaceous epeiric sea (Fig. 2A). The Pierre Shale grades into the overlying Trinidad Sandstone through a tens-of-feet-thick transition zone of interbedded silty shale and bioturbated siltstone and sandstone that contains *Ophiomorpha* trace fossils. Ammonite zone fossils in the upper part of the Pierre below the transition zone are, from youngest to oldest, *Didymoceras cheyennense, Exiteloceras jenneyi, Didymoceras stevensoni,* and *Didymoceras nebrascense* (G. R. Scott, written communication, 1984). Along the margins of the coal fields, the Pierre Shale forms steep slopes that are mantled by soil, talus, and landslide debris beneath cliffs of the Trinidad Sandstone. The Pierre Shale locally intertongues with the overlying Trinidad Sandstone (Harbour and Dixon, 1956; Billingsley, 1978; Wanek, 1963).

The Trinidad Sandstone

The Trinidad Sandstone (Hills, 1888) is tabular in shape, averages about 25–30 m (80–100 ft) in thickness, and was deposited contemporaneously in delta-front and barrier environments during the eastward progradation of the Cretaceous coastline (Fig.2A). Diagnostic fossils have not been reported from the unit, but it is above the Maastrichtian upper Pierre Shale. The Trinidad Sandstone ranges from 0 to 100 m (0 to 300 ft) in thickness along the western margin of the basin where it is locally truncated beneath conglomeratic rocks of the basal part of the Raton Formation (Johnson and Wood, 1956). The Trinidad is composed mostly of very fine grained to fine-grained sandstone that contains *Ophiomorpha, Diplocraterion,* and other trace fossils (Pillmore and Maberry, 1976); it forms persistent, conspicuous, light-colored cliffs along the eastern and southern margins of the basin. *Ophiomorpha* is common to most sections, but *Diplocraterion* is limited to only a few localities in barrier sandstones.

NONMARINE ROCKS

The Vermejo Formation

The Maastrichtian Vermejo Formation consists of interbedded sandstone, siltstone, shale, carbonaceous shale, and coal that together form steep, generally debris-covered slopes above the cliffs of the Trinidad Sandstone. Named by Lee (1917), the Vermejo Formation ranges in thickness from about 115 m (380 ft) at Vermejo Park and along the western border of the Raton basin to 0 m in the eastern part of the basin, east and south of Raton, New Mexico. Coal beds, as much as 4 m (13 ft) thick, occur near the top and bottom of the formation. The Vermejo Formation was deposited conformably on the Trinidad Sandstone in contemporaneous fluvio-deltaic plain and back-barrier environments (Fig. 2A). The upper part of the Vermejo Formation in the west-central part of the basin was probably deposited in an alluviated upper delta plain (Flores and Tur, 1982) in an environment similar to the alluvial plain of the lower coal zone of the Raton Formation (Fig. 2B). The Vermejo Formation intertongues with the Trinidad Sandstone along the southern margin of the basin northeast of Cimarron, New Mexico (Fig. 2A; Wanek, 1963). This intertonguing relationship, along with the intertonguing of the Pierre Shale with the Trinidad Sandstone, possibly represents brief episodes of shoreline transgression or autocyclic (Beerbower, 1964) shifts of depositional systems during the overall regression of the Cretaceous sea.

The Raton Formation

The Raton Formation, defined by Lee (1917), contains rocks of both Late Cretaceous and Paleocene age and includes the K-T boundary interval (Pillmore and others, 1984). The formation unconformably overlies the Vermejo Formation and consists of sandstone, siltstone, mudstone, coal, carbonaceous shale, and conglomerate. It ranges in thickness from more than 640 m (2100 ft) in the west-central part of the basin to about 335 m (1100 ft) in the eastern part. A basal pebble-conglomerate bed in scour-based sandstones commonly marks the lower contact and

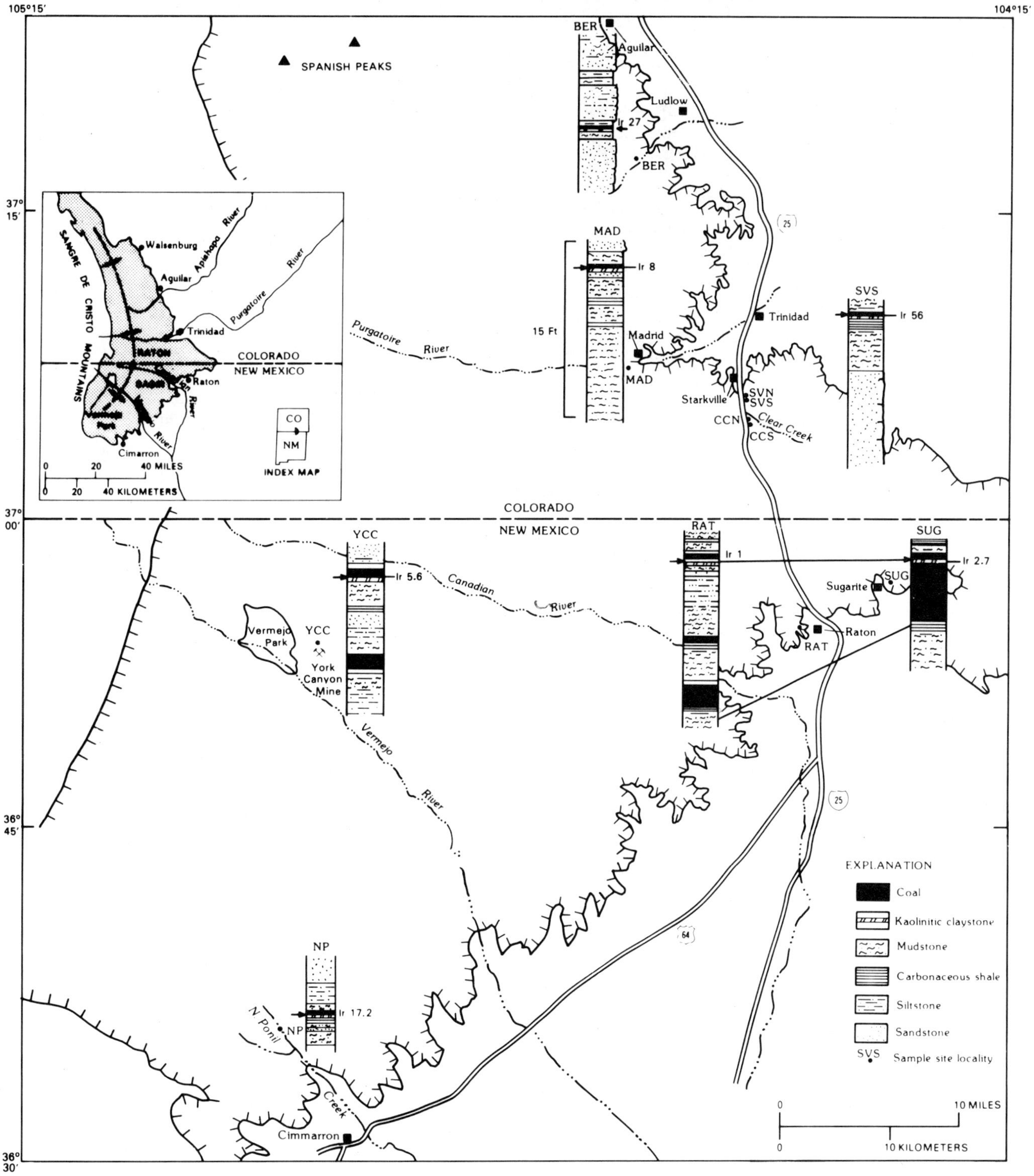

Figure 1. Index map showing location of Raton basin and representative Cretaceous-Tertiary boundary sites in New Mexico and Colorado. Columnar sections at or near top of lower coal zone of Raton Formation show lithology of the boundary interval at the sites: YCC, York Canyon; RAT, Raton; SUG, Sugarite; SVS, Starkville South; BER, Berwind Canyon; NP, North Ponil; and MAD, Madrid. Arrows indicate the K-T boundary. Measured iridium anomalies shown in ng/g (10^{-9} g/g). Hachured line shows top of Trinidad Sandstone. (Modified from Pillmore and others, 1984).

AGE	FORMATION NAME	GENERAL DESCRIPTION	SYMBOL	APPROXIMATE THICKNESS IN FEET
PALEOCENE	POISON CANYON FORMATION	SANDSTONE--Coarse to conglomeratic; beds 15-50 ft thick; interbeds of yellow-weathering, clayey sandstone. Thickens to west at expense of underlying Raton		500+
PALEOCENE	RATON FORMATION	Formation intertongues with Poison Canyon Formation to the west		0?–2100
		UPPER COAL ZONE--Very fine grained sandstone, siltstone, and mudstone with carbonaceous shale and thick coal beds		
		BARREN SERIES--Mostly very fine to fine grained sandstone with minor mudstone, siltstone, carbonaceous shale, and thin coal beds		
UPPER CRETACEOUS		LOWER COAL ZONE--Same as upper coal zone; coal beds mostly thin and discontinuous. Conglomeratic sandstone at base; locally absent		K-T BOUNDARY
UPPER CRETACEOUS	VERMEJO FORMATION	SANDSTONE--Fine- to medium-grained; also mudstone, carbonaceous shale, and extensive, thick coal beds. Local sills		0–380
UPPER CRETACEOUS	TRINIDAD SANDSTONE	SANDSTONE--Fine- to medium-grained; contains casts of *Ophiomorpha*		0–300
UPPER CRETACEOUS	PIERRE SHALE	SHALE--Silty in upper 300 ft. Grades up to fine-grained sandstone. Contains limestone concretions		1800–1900

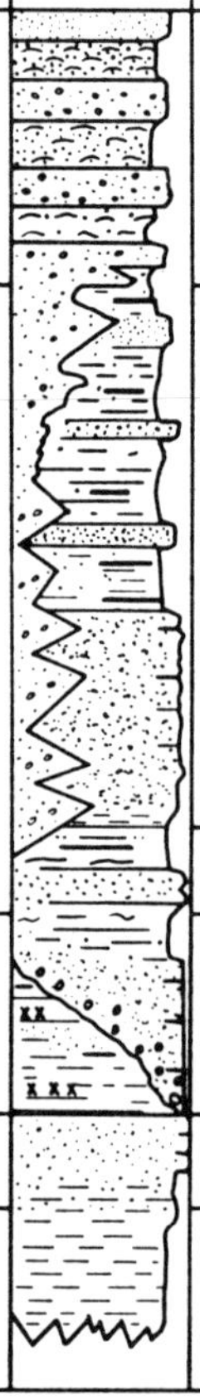

forms a prominent cliff throughout much of the Raton basin. Lee (1917) originally divided the Raton Formation into a basal conglomerate, a lower coal zone, a cliff-forming or barren series, and an upper coal zone. For the purpose of discussing the K-T boundary, and because the conglomerate is absent in some parts of our study area, the basal conglomerate, where present, is included here in the lower coal zone. As used here, this lower coal zone contains all of the Cretaceous rocks of the Raton Formation.

The Raton Formation is herein divided into three lithofacies that are mostly consistent and identifiable throughout the Raton basin. The lithofacies of the lower coal zone generally consists of a ledge-forming basal conglomeratic channel sandstone, locally absent, that grades up into a slope-forming sequence of mudstone, siltstone, carbonaceous shale, and coal beds interbedded with lenticular channel and tabular crevasse-splay sandstone beds. The lithofacies of the cliff-forming barren series consists of a thick sequence of stacked channel sandstones nearly barren of coal beds. The lithofacies of the upper coal zone consists of siltstone, carbonaceous shale, and mudstone beds interbedded with channel and crevasse-splay sandstones and coal beds as much as 4 m (12 ft) thick.

The Lower Coal Zone. The lower coal zone contains the Cretaceous (Maastrichtian) part of the Raton Formation, with the K-T boundary at or near the top of the zone, beneath the cliffs of the barren series. The thickness of the lower zone ranges from greater than 90 m (300 ft) west of Vermejo Park, New Mexico, to about 30 m (100 ft) east of Raton at the abandoned mining town of Sugarite (Fig. 3). Throughout much of the basin, the basal conglomeratic channel sandstone ranges from 0 to 15 m (0 to 50 ft) thick, and local relief beneath the unconformity is as much as 20 m per 0.5 km (60 ft per 0.25 mi). In most places, this basal lithofacies consists of meandering-stream and interchannel deposits of medium- to coarse-grained sandstone with lenses and stringers of conglomerate. Along the western margin of the basin, the conglomeratic channel sandstone beds are as much as several hundred feet thick and represent braided-stream deposits of distal alluvial fans similar to those shown in Figure 2C. These coarse sequences of sandstone are not differentiated from the Poison Canyon Formation in the western Raton basin (Lee, 1917). In the areas near Raton and Trinidad, between the Canadian River and the Purgatoire River (Fig. 1), the basal conglomerate is absent, and no distinct contact between the Vermejo and Raton

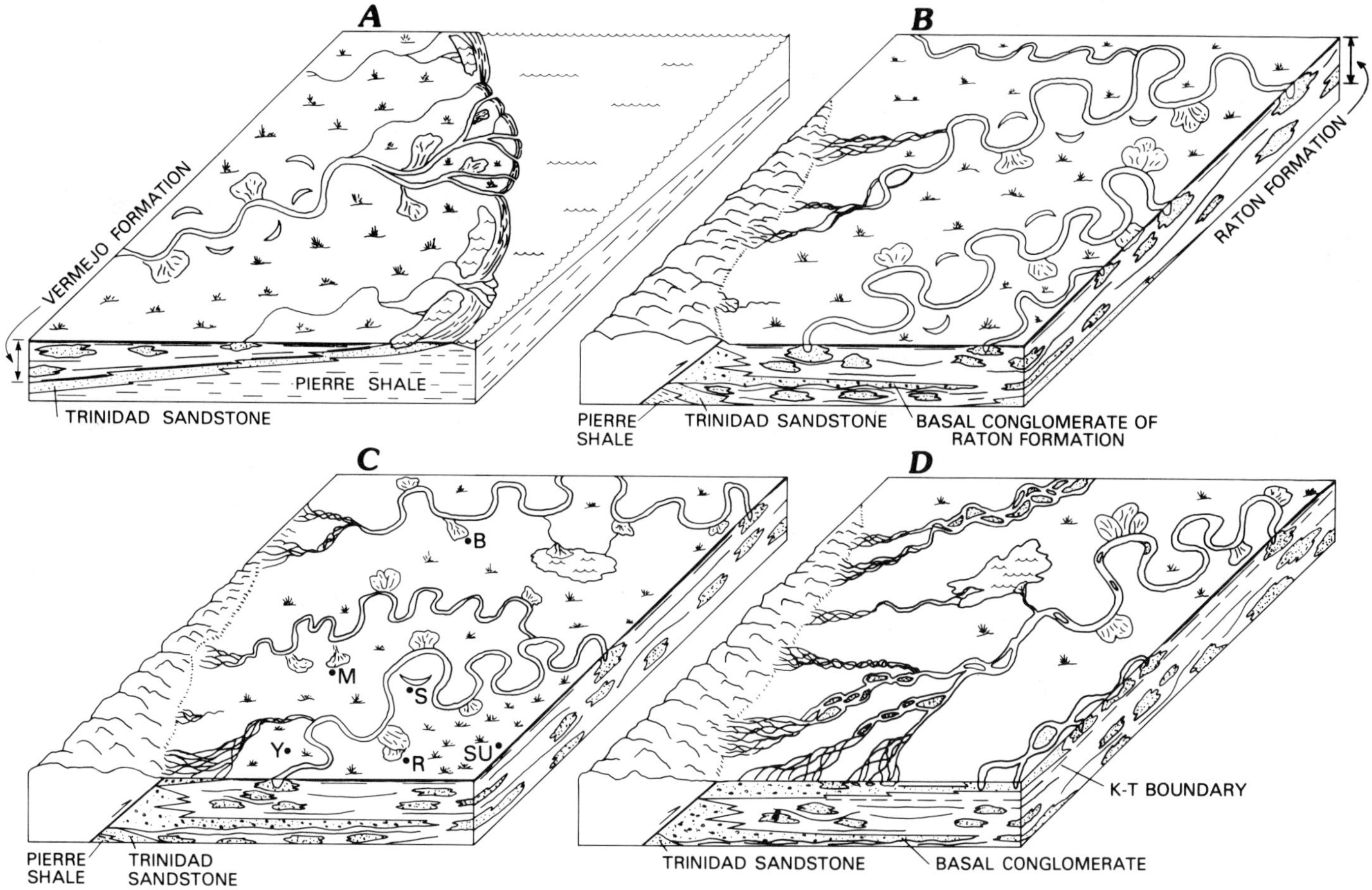

Figure 2. Diagrammatic block diagrams depicting paleoenvironments of Late Cretaceous and Paleocene rocks of the Raton basin: A. The offshore environments of Pierre Shale; the contemporaneous delta front and barrier environments of the Trinidad Sandstone; and the fluvio-deltaic plain of the Vermejo Formation. Diagram shows oxbow lakes, crevasse splays, and swamps related to meandering streams of the lower delta plain. B. The floodplain and fine-grained meanderbelt environments on the alluvial plain that developed during the Late Cretaceous after deposition of the basal conglomerate of the lower coal zone of the Raton Formation. The floodplain was characterized by swamps and oxbow lakes. C. The depositional environments on the alluvial plain of the lower coal zone of the Raton Formation at the end of the Cretaceous Period. The K-T boundary fallout material was deposited on a surface such as that shown here. The letters show representative depositional sites: B, Berwind site (floodplain sequence, dominated by crevasse splays from nearby stream channel); M, Madrid site (channel-floodplain–crevasse splay sequence); S, Starkville sites (floodplain sequence developed on abandoned channel sequence); Y, York Canyon (floodplain–crevasse splay sequence); R, Raton site (floodplain, marginal to major backswamp); SU, Sugarite site (dominated by large swamp). D. The braided stream and coarse meanderbelt environments of the Paleocene barren series of the Raton Formation.

formations is evident. The interval containing the contact locally is occupied by discontinuous quartzose channel sandstones, usually greater than 1.5 m (5 ft) thick, that exhibit a sugary texture due to quartz overgrowths on sand grains (Leighton, 1980). These fine-grained channel sandstones were deposited in fine-grained meanderbelts (Ethridge and others, 1981) developed by mixed-load meandering streams of high sinuosity. These fluvial deposits grade laterally into the overbank-floodplain deposits that constitute the lower coal zone. The remainder of the lower coal zone is made up mostly of beds of siltstone, mudstone, and sandstone with interbeds of carbonaceous shale and thin coal deposited in an alluvial plain containing discrete meanderbelt and floodplain environments (Fig. 2B).

Meanderbelt channel sandstones are bounded by levee, overbank, and crevasse-splay deposits. A crevasse-splay deposit is formed when the levee is breached during floods and a tabular fan-shaped body of sediments is deposited on the floodplain. Enlargement of the breach results in channelized flow that bifurcates around the fan as progradation proceeds. The crevasse-splay sandstones of this area commonly display abundant ripple lami-

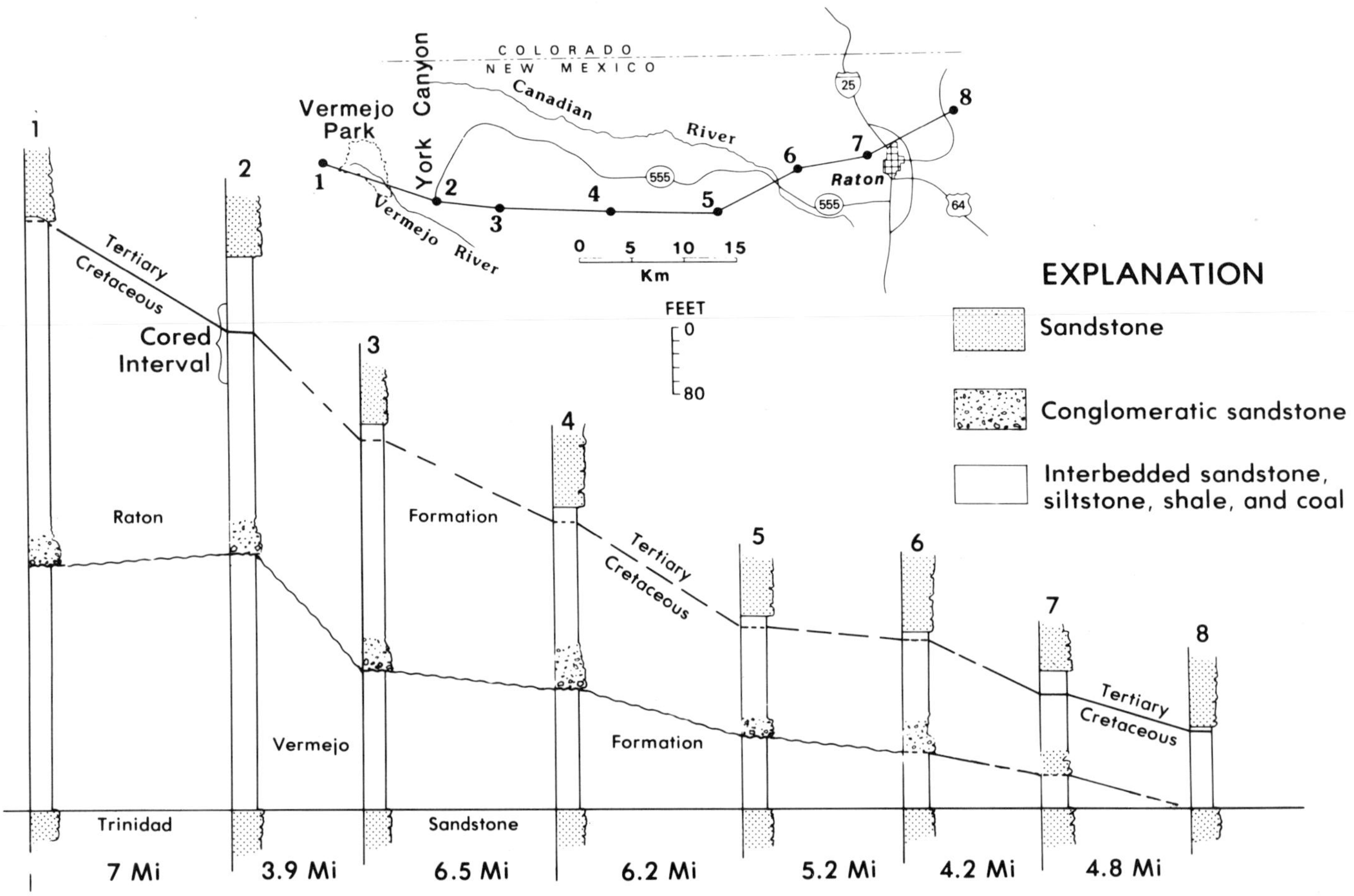

Figure 3. Line of stratigraphic sections across the southern part of the Raton basin showing the eastward thinning of both the Late Cretaceous Vermejo Formation and the lower coal zone of the Raton Formation. The cored interval at York Canyon (column 2) is shown by the bracket.

nations in the lower part and subordinate small-scale cross laminations in the upper part; the overall deposit shows a coarsening-upward sequence. The upper part of the crevasse-splay sandstone is locally scoured and infilled by rooted shale and siltstone representing abandoned crevasse-channel deposits (Saxena and Ferm, 1976). However, the scours commonly are filled by thin, trough-crossbedded sandstone.

The Barren Series. The lithofacies of the barren series is characterized by sparseness of coal beds and is dominated by cliff-forming, multiple-stacked, en echelon channel sandstones that are mostly fine to coarse grained. The channel sandstones are nested to form a series of prominent ledges and cliffs but are locally interbedded with and laterally grade into floodplain deposits of mudstone and siltstone and associated swamp deposits of carbonaceous shale and coal beds that are mostly thin and limited in extent. The barren series ranges from about 180 m (600 ft) thick in the western part of the basin to about 55 m (180 ft) in the eastern part. In the northern part of the basin, this lithofacies thins to about 75 m (250 ft) and forms a less prominent escarpment (Lee, 1917). There, braided streams draining alluvial fans along the western basin margin dominated the upper alluvial plain and deposited Poison Canyon–type channel sandstones. The barren series merges with the cliff-forming conglomeratic channel sandstone beds of the non-coal-bearing Poison Canyon Formation to the west. To the east, meandering sand-rich streams were dominant, and subordinate restricted floodplains were present (Fig. 2D). The rocks in this lithofacies are interpreted as coarse-grained meanderbelt deposits (Ethridge and others, 1981) of high-bedload, low-sinuosity streams.

The Upper Coal Zone. The upper coal zone ranges in thickness from about 335 m (1100 ft) in the central part of the basin near York Canyon to about 180 m (590 ft) near Raton. It consists of floodplain lithofacies of crevasse-splay sandstone, mudstone, siltstone, and shale interbedded with swamp lithofacies of coal and carbonaceous shale. Coal beds in the upper coal zone are commonly lenticular and less than 1 m (3 ft) thick, but several beds that attain thicknesses of 2 m (6 ft) or more are considered economic. These thicker coals outline the principal coal districts of the Raton coal field (Pillmore, 1976). The floodplain and swamp deposits are interbedded with channel sandstones.

According to Flores (1984), lithofacies analysis of this part of the Raton Formation suggests deposition in an upper alluvial plain containing discrete meanderbelt and floodplain environments similar to those of the lower coal zone, shown in Figure 2B. The meanderbelt facies (fine-grained meanderbelt) include channel fills, slumped cutbank deposits, and abandoned-channel clay-plug deposits. These deposits, together with point-bar deposits of channel fills, indicate a pattern of deposition in meandering channels in which point bars formed at convex banks of the loops, slumped cutbank deposits developed at concave banks of meander bends, and abandoned channel clay-plug deposits resulted from neck or chute cutoffs (Fisk, 1944). The meanderbelt facies above some of the thick coal beds in the lower part of this zone are overlain by moderately thick coal beds, suggesting that channel avulsion and aggradation into an adjacent floodplain caused cessation of meanderbelt deposition. This process of meanderbelt abandonment was succeeded by the encroachment of swamp vegetation from outlying floodplains over other abandoned channel deposits that served as a relatively incompactible platform for thick peat accumulation. As in the lower coal zone, the meanderbelt lithofacies grade laterally into floodplain lithofacies of levee, overbank, and crevasse-splay deposits.

The Poison Canyon Formation

The Poison Canyon Formation (Hills, 1888) overlies the Raton Formation throughout most of the Raton basin, but to the west and southwest the Poison Canyon Formation intertongues with the Raton Formation. The lithofacies of the Poison Canyon Formation consists of thick to massive, lenticular, ledge-forming beds of coarse-grained to conglomeratic arkosic channel sandstone intercalated with beds of nonresistant, yellowish-gray- to grayish-orange-weathering, sandy, micaceous mudstone. These fine-grained mudstones represent overbank-floodplain deposits. They grade laterally into and are interbedded with the channel sandstones. The Poison Canyon Formation is barren of coal and is interpreted as deposits of braided streams and coarse-grained meanderbelts (Ethridge and others, 1981) in an upper alluvial plain similar to the upslope conditions represented in the barren-series depositional model in Figure 2D. This setting is characterized by sinuous channel sandstones that pass into non-coal-bearing floodplain deposits. Combined effects of detrital influx and oxidizing conditions in the floodplain prevented accumulation of coals in well-drained swamps. The coarse sediment load of the streams probably reflects intensive erosion of source rocks in the rapidly rising source area to the west.

The Poison Canyon Formation is about 150–200 m (500–650 ft) thick in the central Raton coal field near York Canyon. The contact with the underlying Raton Formation generally is indefinite and gradational through a transition zone as much as 45 m (150 ft) thick. The Poison Canyon Formation is Paleocene in age and is overlain unconformably by the Cuchara Formation of Early Eocene age (Baltz, 1965, p. 2065) in the northern part of the basin. According to Baltz, the Poison Canyon Formation is as much as 760 m (2500 ft) thick in the Colorado portion of the basin.

DEPOSITIONAL HISTORY

The Cretaceous epeiric sea covered the area of the Raton basin during most of Late Cretaceous time. The epeiric sea was filled by a thick sequence of older, calcareous, deep-marine or basinal shales, overlain by the Pierre Shale, which consists of offshore-marine or shelf shales and siltstones. The Pierre Shale was deposited in offshore-marine environments (Fig. 2A) and consists of gray noncalcareous shale that coarsens upward into silty shale and siltstone, reflecting the influx of coarser material as the paleoshoreline retreated to the east across New Mexico and Colorado in late Campanian and early Maastrichtian time. The upper part of the Pierre Shale grades into the overlying Trinidad Sandstone. This transition zone contains increasing-upward amounts of siltstone and sandstone that were deposited in the lower shoreface of coastal barriers, which merged with prodeltaic deposits along the coast.

The Trinidad Sandstone was deposited in delta-front and interdeltaic barrier-bar environments (Fig. 2A) as the sea continued to regress eastward. The delta-front deposits include distributary mouth-bar and distributary channel sandstones (Flores and Tur, 1982). These laterally shifting deltaic systems prograded seaward and supplied sediments that were reworked and redeposited along flanking coastal barriers. The barrier-bar deposits of the Trinidad consist of middle-shoreface, river-estuarine-inlet, and beach sandstones (Leighton, 1980). These delta front–barrier bar deposits grade upward into the fluvial delta-plain deposits of the Vermejo Formation.

The Vermejo Formation was deposited as fluvio-deltaic sediments landward of barrier-bar and delta-front deposits of the Trinidad Sandstone (Flores and Tur, 1982). Coal beds in the lower part of the Vermejo formed in poorly drained back-barrier coastal swamps and in interdistributary swamps of delta plains. The depositional environment of the upper part of the Vermejo Formation was characterized by a lower alluvial plain consisting of meandering streams separated by flood basins (Fig. 2A). In this environment, sandstones were deposited in stream channels, and fine-grained sequences of sandstone, siltstone, and mudstone were deposited in floodplains. Floodplain deposits include crevasse-splay and minor crevasse-channel sandstones. Coal and carbonaceous shale accumulated in poorly drained swamps in the floodplains. These lower alluvial plain deposits grade upward into the more landward deposits of the Raton Formation.

The Raton Formation was deposited on an upper alluvial plain (Flores, 1984). Alluvial plain deposition is characterized by several types of fluvial sedimentation. Following the deposition of fluvio-deltaic plain and back-barrier sediments of the Vermejo Formation, rapid uplift and widespread erosion of the source area to the west (the San Luis Highland, Tweto, 1980) caused braided- and meandering-stream deposition of the coarse sediments of the basal conglomerate unconformably on lower alluvial plain deposits of the Vermejo Formation. After deposition of the conglomeratic sandstone, more stable tectonic conditions returned, and sandstones were deposited by high-sinuosity meandering streams of fine-grained meanderbelts. The sandstones are associated with thin coal, carbonaceous shale, and mudstone that accumulated in floodplains and swamps. Crevasse splays periodically invaded ponds and swamps on low-lying floodplains during river floods (Fig. 2B). Though some thick coal beds were deposited locally in poorly drained swamps of the lower coal zone (mainly the 2-m [6-ft] thick Sugarite coal bed; Lee, 1917; Pillmore, 1976; Pillmore and others, 1984), most coal swamps that formed in the flood basins were well drained, small, and shallow. The well-drained nature of these small swamps restricted peat accumulation, which resulted in coal beds mostly less than 20–30 cm (8–12 in) thick. Thus, at the close of the Cretaceous, the extensive alluvial plain of the lower coal zone, as depicted in Figure 2C, was an ideal environment for deposition and preservation of the fallout associated with the K-T boundary event (Alvarez and others, 1982; Pillmore and others, 1984).

Shortly after the close of the Cretaceous, marked by the K-T boundary event, tectonic conditions in the Raton basin changed as uplift of the source area was reinitiated to the west. Sediment load increased, and a coarse-grained meanderbelt fluvial system (Ethridge and others, 1981) with well-drained floodbasins once again characterized the depositional basin (Fig. 2D). Streams aggraded braid and meander belts across the alluvial plain, resulting in broad belts of vertically stacked channel sandstones that form the prominent, persistent cliffs and ledges of the barren series. The channel sandstones are locally interbedded with and laterally grade into floodplain mudstone and siltstone deposits. Minor associated carbonaceous shale beds and coal zones formed in local, well-drained backswamps; these beds are mostly thin and limited in areal extent.

The barren series was succeeded by deposits of an alluvial plain drained by high-sinuosity meandering streams of a fine-grained meanderbelt (Ethridge and others, 1981), resulting from a decrease in tectonic movement in the source area. This tectonic condition, perhaps combined with basin subsidence, led to aggradation of the alluvial plain by a meandering fluvial system to form the upper coal zone. The system featured floodplains that developed poorly drained swamps in which coals as thick as 3.5 m (12 ft) (York Canyon coal bed; Pillmore, 1976) accumulated. The floodplains were locally filled by overbank and crevasse-splay detritus during episodes of floods. The detrital influx extended into the swamps and caused splits of coal beds where the peat swamps were reestablished after deposition of detrital sediments. The fluvial deposits of the upper coal zone grade upward into the coarser-grained braided-stream deposits of the Poison Canyon Formation.

The Poison Canyon Formation was deposited in an alluvial plain drained by braided and low-sinuosity meandering streams and characterized by non-coal-forming floodplains (Strum, 1984). The high sediment input into the alluvial plain via high-bedload streams probably reflects intensive erosion of a rapidly rising source area. The presence of abundant potash feldspar in the Poison Canyon suggests a granitic core in the source terrane, probably flanked by Paleozoic sedimentary rocks. The Poison Canyon deposits coarsen to the west, where alluvial fans were formed along the flanks of the rising San Luis Highlands.

THE CRETACEOUS-TERTIARY BOUNDARY

Fallout material from the asteroid impact that culminated the Cretaceous has been found at more than 20 sites throughout an area of about 2250 km^2 (868 mi^2) in the Raton basin, northern New Mexico and southern Colorado (Fig. 1). The Cretaceous-Tertiary boundary is in a fine-grained sequence of rocks in the lower part of the Raton Formation (Maastrichtian and Paleocene) and is defined by Tschudy (1973), Tschudy and others (1984), Pillmore and others (1984), and Orth and others (1981) on the basis of the abrupt disappearance of several taxa of fossil pollen. The boundary is also characterized by significant geochemical abundance anomalies of Ir and several other elements (Orth and others, 1981, 1982; Gilmore and others, 1984). The following discussion is modified from Pillmore and others (1984).

Lee (1917) originally placed the K-T boundary at the erosional unconformity at the base of the Raton Formation. He thought that this unconformity represented a major orogenic event at the end of the Cretaceous and that the Raton Formation was Tertiary, as confirmed by Knowlton (Lee, 1917, p. 56). Later, Brown (1962) identified one Cretaceous plant fossil (*Paleoaster inquirenda* Knowlton) in the lower part of the Raton Formation at a site about 5.5 km (3.5 mi) north of Trinidad, Colorado, and indicated that the boundary should be placed at least 15 m (50 ft) above the base of the formation. The K-T boundary was later identified by Tschudy (Tschudy, 1973; Pillmore, 1969; Orth and others, 1981) 81 m (265 ft) above the base of the Raton Formation at York Canyon, New Mexico, on the basis of the abrupt disappearance of several taxa of fossil pollen: namely, *Proteacidites, Tilia wodehousei sensu* Anderson, *Trisectoris,* and *Trichopeltinites.* For discussions of the Raton basin, these taxa are herein called the "*Proteacidites* assemblage." A similar assemblage, which is used to define the K-T boundary in the northern Rocky Mountain region, is characterized by the abrupt disappearance of *Proteacidites* and of most species of *Aquilapollenites* (Tschudy, 1970; Hotten, 1984).

A large group of pollen taxa has been described from coal beds in the Raton and Vermejo Formations (Tschudy, 1973). Though various forms appear and disappear throughout the coal-bearing section, changes comparable to the disappearance of the *Proteacidites* assemblage have not been observed elsewhere in the section (Tschudy and others, 1984). The extinction of these taxa at the end of the Cretaceous is well documented and is significant throughout the Western Interior, as well as in the southern Rocky Mountains. In the Raton basin, the *Proteacidites* extinction horizon is coincident with the top of a 1 to 2.5 cm (0.5 to 1.0 in) thick clay bed termed the boundary clay. This horizon also coincides with the abrupt appearance of shock-metamorphosed minerals (Izett and Pillmore, 1985a, 1985b) and a sudden change in the relative proportion of fern spores to angiosperm pollen (Fig. 4; Tschudy and others, 1984). The boundary clay contains anomalously high geochemical concentrations of Ir and several other elements (Orth and others, 1981, 1982; Gilmore and others, 1984).

The Angiosperm Pollen/Fern Spore Ratio

Fern spores and pollen grains make up the majority of palynomorph assemblages in carbonaceous shale and coal beds of the Raton and Vermejo formations. The proportion of fern spores to angiosperm pollen is usually 15–30 percent fern spores. This proportion shifts to as much as 99 percent fern spores at the top of the K-T boundary clay and then returns to 15–30 percent fern spores within 10–15 cm (4–6 in) above the boundary (Fig. 4). This change implies a significant, though geologically brief, stress on the ecological system caused by the event or events responsible for deposition of the boundary clay (Tschudy and others, 1984; Tschudy and Tschudy, 1986).

Tonsteins

Tonsteins (kaolinitic clay partings thought to result from alteration of volcanic ash beds in acidic coal swamps) occur in coal beds throughout the Western Interior, including the Raton basin, and in many other parts of the world (Williamson, 1970; Bohor and Pillmore, 1976; Bohor and others, 1978; Spears and Kanaris-Sotiriou, 1979). These kaolinite-rich beds commonly weather to a distinctive light-gray "bone" color and are characterized by generally sharp contacts and a blocky ledge-forming appearance in coal outcrops. Most tonsteins in the Raton basin appear smooth and even-textured and contain little or no detectable grit (Bohor and Pillmore, 1976). Petrographic analyses show that they consist largely of coarsely crystalline authigenic vermicules of kaolinite in a fine-grained kaolinitic matrix with varying amounts of quartz and feldspar. In hand specimens, some of the coarser-grained tonsteins resemble siltstones with resinous luster. Tonsteins, mostly less than 5 cm (2 in) thick, occur in many Raton basin coal beds. Some probably represent major volcanic eruptions that blanketed the landscape with ash, but we have not observed a significant change in fern spore–angiosperm pollen ratios in any nonboundary tonstein-bearing coals that we have studied in the Raton basin. However, local concentrations of fern spores, due only to regional ecological conditions, would be expected.

The K-T Boundary Clay Bed

The boundary clay bed resembles tonstein, but it usually weathers to a lighter color and exhibits a fine-grained to amorphous texture and a distinctive hackly to conchoidal fracture. The clay is mostly gray and grayish pink to grayish yellow and commonly contains tiny specks and thin contorted lenses or layers of organic matter, especially near the margins. Small spheroidal structures can be seen on fracture surfaces of some specimens and in thin section. X-ray diffractograms show that, like many tonsteins, the boundary clay is nearly pure well-crystallized kaolinite with lesser amounts of randomly stratified illite-smectite clay and some quartz and feldspar (Pollastro and others, 1983; Pollastro and Pillmore, in press). However, the boundary clay is texturally and chemically different from typical Raton basin tonsteins. As seen in ultrathin section, it is fine grained to amorphous but may exhibit an imbricate fabric and relics of small bubbles in a fine crystallite matrix of kaolinite. Microspherules (40 to 120 microns in diameter) consisting of calcium, aluminum, strontium, cerium, rare earth, and phosphorus (similar to goyazite, a hydrous strontium alumino phosphate in composition; microprobe analysis by Ralph Christian) have been observed in the boundary clay at some sites. These phosphatic spherules have not been seen in other Raton basin tonsteins, but goyazite has been reported from other kaolinitic, altered tuff beds of Cretaceous age (Triplehorn and Bohor, 1983). Jan Smit (1984) refers to similarly shaped grains he has observed in the boundary clay from the Raton basin and other areas as microtektite-like structures, implying an impact origin. The goyazite-like grains from Raton basin sites are mostly spheroidal to subspheroidal and resemble dull to shiny resinous little balls under the microscope. Some are hollow. Under the scanning electron microscope, they have an uneven surface texture and commonly are pitted by irregularly shaped cavities. The spherules may result from the alteration of material blown out during the K-T event, but they probably represent only authigenesis of some phosphatic constituent in the K-T boundary clay layer. The shock-metamorphosed mineral grains are compelling evidence of impact origin (Bohor and others, 1984). They occur at the top of the boundary clay in a layer termed the K-T boundary impact bed (Izett and Bohor, 1986) and consist mainly of quartz, microcline, and plagioclase. The shock-metamorphosed minerals contain as many as nine intersecting sets of closely spaced planar features per grain.

The boundary clay and the tonsteins are high-alumina clays that characteristically contain about 32 percent Al_2O_3 (Gilmore and others, 1984). Ir abundance anomalies as high as 56 ng/g (56 $\times 10^{-9}$ g/g, about 8000 × background) have been measured in samples from the top of the boundary clay collected from the study area (Pillmore and others, 1984). In comparison, background Ir concentrations of 0.004–0.040 ng/g are observed in tonsteins and other beds of coal and shale not associated with the boundary (Gilmore and others, 1984). In addition, titanium, scandium, vanadium, chromium, and antimony in the boundary clay are enhanced by factors of about four or more over their concentrations in all other Raton basin tonsteins that were analyzed by Gilmore and others (1984). Table 2 shows a comparison of these and other elements in the boundary clay with those of tonstein beds in the Raton basin (Gilmore and others,

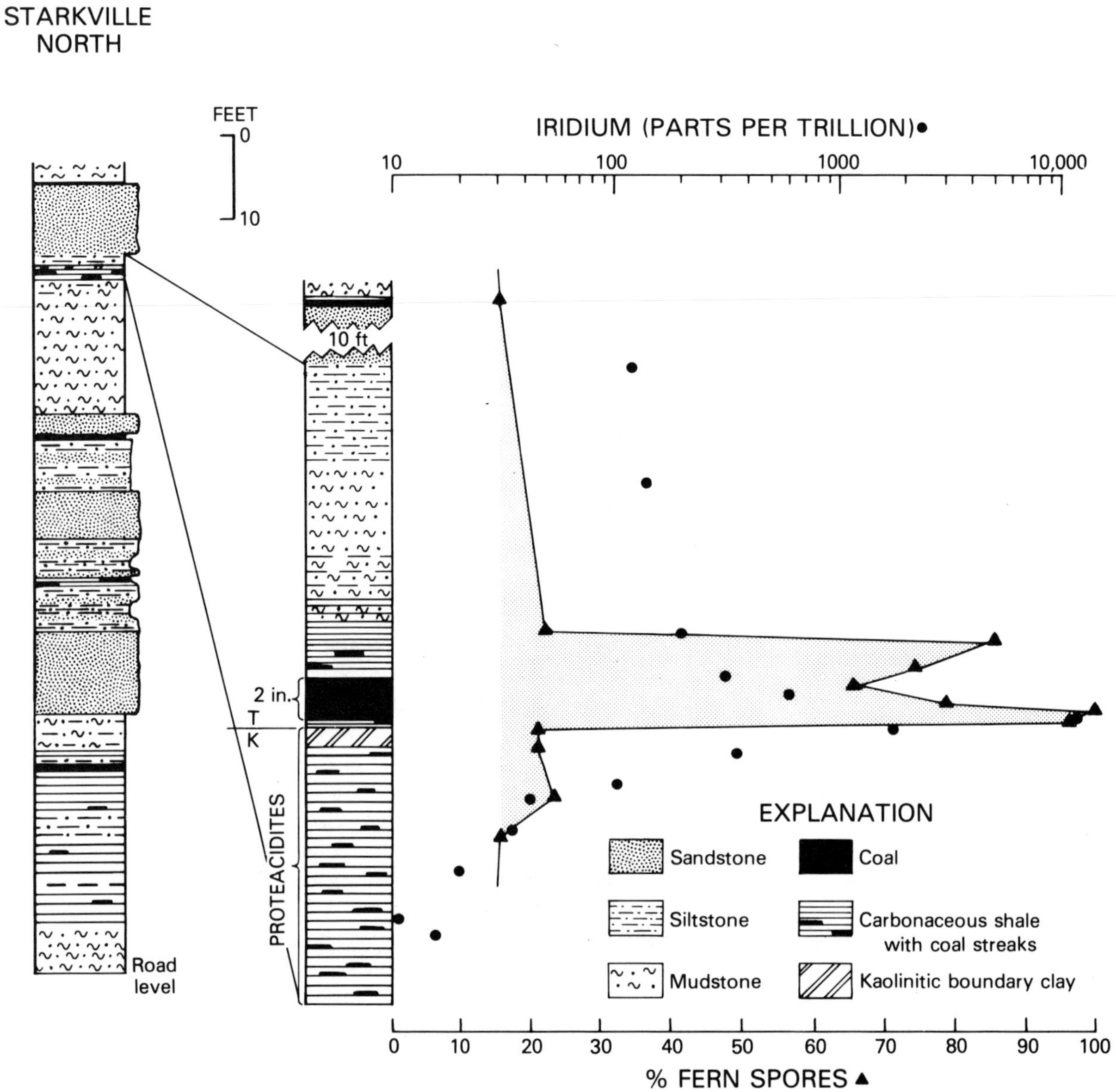

Figure 4. Diagram showing the lithology of the rock sequence containing the K-T boundary at the Starkville North site, 4.8 km (3 mi) south of Trinidad, Colo. The large black dots show the variation in iridium concentration, the solid line and triangles show the fern-spore percentage, and the inset shows the detail of the K-T boundary interval. (Modified from Tschudy and others, 1984).

1984). The Ir abundance anomalies in the boundary clay bed and its unusual textural and chemical character suggest that the clay bed was derived from a different source than were the tonstein beds. However, its relation to the impact event is problematical because of its position directly beneath the K-T boundary impact layer.

The K-T boundary clay bed probably resulted from fallout that rained down on swamps, ponds, oxbow lakes, levees, crevasse splays, and abandoned and active stream channels of the Raton basin alluvial plain. Where the material fell on positive topographic areas, it was soon washed away by rains and dispersed in other sediment. Where the material fell or was washed into streams, it was mixed with other sediment, and the Ir-rich material was diluted. Where the material fell in freshwater ponds or oxbow lakes, it accumulated with detrital clay and silt; however, it is virtually impossible to distinguish in outcrops or cores without detailed chemical and palynological tests. Material that fell in the quiet acidic waters of poorly drained swamps apparently was incorporated into beds of organic-rich clay and peat-forming vegetal matter where, during diagenesis, it was altered to kaolinitic clay. In weathered exposures of coal beds and carbonaceous shale sequences, the boundary clay is usually light-colored and distinguishable. To date, the boundary clay has been found from New Mexico to Alberta, but only where it was deposited in swamps or ponds and preserved in carbonaceous shale and coal beds.

STRATIGRAPHY OF SELECTED K-T BOUNDARY SITES

The K-T boundary clay has been found at more than 20 sites in the Raton basin. Iridium anomalies have been measured at six sites in Colorado and at nine sites in New Mexico (Fig. 1). Berwind Canyon and Starkville North sites are representative of the boundary sites in Colorado; York Canyon, Raton, and Sugarite are representative of the New Mexico sites. Measured stratigraphic sections of the rocks enclosing the K-T boundary at these sites are described in Appendices 1–5.

TABLE 2.

	K-T boundary bed*		Beds above/below K-T boundary bed	
Element**	Range	Average	Range	Average
Al_2O_3 (%)	24-36	32.9	23-36	31.7
K (%)	0.2-1.1	0.52	0.2-2.5	0.81
Sc	21-26	23.3	3-12	6.0
TiO_2 (%)	1.38-2.67	2.00	0.40-0.95	0.757
V	110-187	137	10-67	27
Cr	46-102	67.3	0.9-5.0	3.3
Co	1.2-53	9.8	0.7-4	2.7
As	1-95	36	0.2-46	4.1
Se	2-19	8	<0.1-6	4.8
Sb	0.3-11.5	6.3	0.1-0.8	0.38
La	16-80	43	9-88	28
Yb	0.7-2.2	1.5	0.8-4.1	2.1
Hf	3.3-6.4	4.5	3.2-10.6	7.6
Ir (ng per g)	0.90-2.7	1.7	0.005-0.020	0.010
Th	5-21	7.8	5-34	9.2
La/Yb	15-57	28.7	5-67	13.3
TiO_2/Al_2O_3	0.056-0.074	0.060	0.016-0.027	0.024
Cr/Al_2O_3-relative (ppm/%)	1.64-2.83	2.04	0.036-0.147	0.103

*Boundary data from six sampling sites; non-boundary data from 21 sites.
**Concentrations in ug per g unless noted otherwise.
From Gilmore and others, 1984.

Colorado Sites

Berwind Canyon. The Berwind Canyon site is a road cut reached through Berwind Canyon, 10 km (6 mi) west of Ludlow exit from I-25, 17 km (11 mi) north of Trinidad, Colorado (Fig. 1). This site represents deposition in a floodplain with crevasse-splays and minor swamp deposits. The K-T boundary interval (the stratigraphic interval, common to most sites, of carbonaceous shale and coal that contains the boundary clay) is preserved in a sequence of sandstone, siltstone, mudstone, carbonaceous shale, and coal exposed in a road cut across from the waste pile of the abandoned Toller mine. These rocks are at the top of the lower coal zone of the Raton Formation, which generally forms debris-covered slopes. The Berwind Canyon site shows how fortuitous preservation of the boundary clay can be, as the carbonaceous interval containing the boundary is 25 cm (0.8 ft) thick and is sandwiched between two sandstone beds 1.1 m and 1.4 m (3.7 ft and 4.5 ft) thick, respectively (Fig. 5), that were probably deposited as crevasse splays. Thich channel sandstones are present above and below the sequence of rocks exposed in the Berwind Canyon road cut site. To preserve the boundary clay here, swamp conditions had to form precisely at the time of the fallout in an area dominated by crevasse splays.

The splay sandstones above and below the boundary clay are lenticular to tabular in shape and are rippled and small-scale cross laminated. The coarsening-upward sequences represent overbank-floodplain deposits that resemble crevasse-splay deposits in modern deltaic and fluvial environments. The coal and carbonaceous shale beds are swamp deposits that formed repeatedly and expanded over the floodplain during nonflood conditions. The white clay bed that marks the K-T boundary was deposited in an ephemeral swamp and floodplain. A partial stratigraphic section measured in the road cut at the Berwind Canyon site is in Appendix 1.

The boundary-interval rocks can also be observed in a stream bank about 45 m (50 yd) upstream (west) from the point where these rocks pass beneath road level at the road cut. The total exposure in the stream bank and along the road cut extends only about 65 m (215 ft); updip and east of these exposures, soil cover prohibits observation of the boundary clay.

Starkville North. Starkville North site (Fig. 1) represents deposition on levee-overbank, abandoned-channel, and swamp deposits. The K-T boundary is near the top of the road cut shown in Figure 6. The rocks are exposed along the service road of I-25, 3.2 km (2 mi) south of the Starkville exit, a short distance south of Trinidad. As described in the measured section in Appendix 2, and shown graphically in Figure 4, the carbonaceous interval containing the K-T boundary is at the top of a 7.5-m (25-ft) thick sequence of mudstone with thin lenticular channel sandstone beds in the lower part. Figure 6 shows these channel sandstone beds merging to the right into one channel sandstone bed. About 1000 m (1100 yd) south, at the Starkville South site (Fig. 1), this channel sandstone has thickened to about 5.5–6 m (8–20 ft) and occupies the lower part of the mudstone interval (Fig. 7). The K-T boundary interval at these sites is similar to the other sites in Colorado, such as those along I-25 and the Madrid site (Fig. 1) about 13 km (8 mi) west along the Purgatoire River valley, in that the boundary clay is immediately overlain by a thin coal bed and is on top of a carbonaceous shale zone (Fig. 8). The carbonaceous zone ranges from 25 cm (0.8 ft) thick at Berwind Canyon to 67 cm (2.2 ft) thick at the Starkville North site. The rock sequences that contain this boundary interval show differences in facies association that vary from channel to crevasse-splay deposits of the alluvial plain on which the boundary fallout material accumulated.

The *Proteacidites* pollen assemblage disappears at the top of

Figure 5. Photograph, looking northeast, showing road-cut exposure of the Cretaceous-Tertiary (K-T) boundary stratigraphic sequence in Berwind Canyon. The thin white band crossing the middle part of the photograph above the lower sandstone bed is the K-T boundary clay. Lower sandstone in middle of photograph is about 1.3 m (4 ft) thick. (Photograph by C. J. Orth).

Figure 6. Photograph of Starkville North site, looking east across Raton Creek, showing mudstones and lenticular channel sandstones merging with channel sandstone (at right of picture). The K-T boundary interval is shown by arrow near top of road cut beneath top sandstone ledge.

Figure 7. Photograph, looking east, showing 7-m (21-ft) channel sandstone in road cut of I-25 service road at Starkville South site. The K-T boundary clay is in a carbonaceous interval just a few feet above the sandstone, as shown by the arrow that points to the K-T boundary clay.

unit 12c (Appendix 2), where a 0.64-cm (0.25-in) thick, flaky, dark, carbonaceous shale contains Ir concentrations of 6 ng/g (Pillmore and others, 1984). The flaky shale, which consists mainly of kaolinite, is essentially barren of pollen and marks the change in the ferm spore–angiosperm ratio from 21 percent fern spores commonly observed beneath the boundary clay to more than 99 percent fern spores directly above it. The ratio returns to 22 percent fern spores within 10 cm (4 in) above the shale bed (Fig. 4). The flaky dark shale overlies the distinctive, light-colored layer of kaolinitic clay 2.5 cm (1 in) thick that constitutes the boundary clay (Fig. 9). It grades upward into a thin blocky coal bed about 5 cm (2 in) thick that appears to persist along I-25 road cuts for several miles to the south. This thin coal bed resembles the coal bed overlying the boundary clay at the Berwind and Madrid sites.

At Starkville South (Fig. 1), the mudstone beneath the K-T boundary clay has thinned to only 2.5 m (8 ft) thick, and this mudstone unit overlies a channel sandstone that shows lateral accretion units probably deposited in a meandering stream. This channel sandstone thins toward the Starkville North site where it is partly replaced by the overbank mudstones and siltstones. Thus, prior to accumulation of the K-T boundary clay at these two Starkville localities, the alluvial plain was drained by a meandering stream that overtopped its banks during floods. The stream channel filled and was abandoned, leaving channel-fill and levee-overbank deposits that formed a platform for swamp development and subsequent accumulation of organic-rich shales and the boundary clay. The boundary clay was preserved with the deposition of floodplain detrital sediments over the swamp

Figure 8. Photograph showing the carbonaceous shale interval that contains the K-T boundary clay (white layer near top of dark shaly zone). This zone is near the top of the Starkville North road cut shown in Figure 6. The dark-colored carbonaceous shale interval is about 0.6 m (2 ft) thick.

Figure 9. Photograph of the Cretaceous-Tertiary boundary clay at the Starkville North site (Fig. 6). The pencil points to the fossil-pollen–defined boundary at the top of a 2.5-cm (1-in) thick bed of white-weathering, kaolinitic claystone. A thin flaky dark iridium-rich shale, the K-T boundary impact bed, overlies the Kaoliuitic claystone. The impact bed contain shock-metamorphosed mineral grains.

deposits. The rocks containing the K-T boundary clay at the Starkville North and Starkville South sites are in contrast to the Berwind Canyon site, where they are sandwiched between two splay sandstone beds.

In some depositional sequences, the carbonaceous shale above the thin coal that overlies the boundary clay layer is, in turn, overlain by thick splay sandstone. Near Madrid (Fig. 1), about 13 km (8 mi) west of the Starkville sites, this carbonaceous shale above the coal is only about 7–10 cm (3–4 in) thick and is preserved beneath a prominent 1 to 3-m (3 to 9-ft) thick crevasse-splay sandstone that can be traced for more than 3 km (1.8 mi) along the south side of the Purgatoire River (Fig. 10). A channel sandstone similar to the one at Starkville South is about 4.5 m (15 ft) below the boundary clay. The overlying crevasse-splay sandstone helped preserve the boundary clay, as the base of the sandstone is nearly planar and nonscouring. The Madrid site also serves to illustrate the fortuitous preservation of the boundary clay layer.

New Mexico Sites

The K-T boundary sites in New Mexico exhibit depositional conditions similar to those described for the Colorado sites, but coal-forming environments appear to have played a more important role. The boundary occurs in thicker coal zones at the York Canyon and Raton sites; and at Sugarite (Fig. 1) the boundary clay is in a thick coal bed that was deposited in an extensive coal swamp.

York Canyon. The Ir anomaly at the palynological K-T boundary in continental rocks was first discovered in core from a hole drilled at York Canyon (Fig. 1), about 58 km (36 mi) west of Raton, New Mexico, in 1981 (Orth and others, 1981; 1982). Rocks from this core are described in the stratigraphic section in Appendix 3 and are shown in the columnar section in Shoemaker and others (this volume, Fig. 2).

The cored rocks consist of swamp, floodplain, and channel deposits. The K-T boundary occurs in the lower part of coal bed c (Shoemaker and others, this volume, Fig. 2). Because the boundary clay layer at the base of the coal is high in organics and resembles a thin bony layer, and because it was nearly cut out by a small fault or slump structure (Orth and others, 1981), the clay layer was not immediately recognized in the core and its significance did not become apparent until discovery of the Starkville sites in 1982. The coal bed containing the K-T boundary is underlain by a coarsening-upward sequence of mudstone and interbedded siltstone and sandstone. This sequence probably represents a distal part of a crevasse-splay deposit. The swamp deposit that contains the K-T boundary is overlain by a succession of sandstones, probably of crevasse-splay origin; fine-grained floodplain deposits; and swamp deposits of coal and carbonaceous shale. These successions are, in turn, overlain by a unit interpreted

Figure 10. Photograph, looking south across Purgatoire River, of the Madrid site showing the lower channel sandstone and the tabular crevasse-splay sandstones. These sandstone beds enclose the fine-grained sequence of rocks that contain the K-T boundary clay (shown by arrow), which is immediately beneath the splay sandstone in the upper part of the railroad cut exposure. Splay sandstone is approximately 2 m (6 ft) thick.

as a channel sandstone. A second hole, drilled in 1966, about 140 m (150 yd) north of the 1981 hole, shows floodplain and swamp deposits in the same interval, but the sequence is characterized by fine-grained detritus. Only small pieces of the coal beds were collected from the core at that time, so it is not known whether the K-T boundary clay bed is preserved at the 1966 drill site. The rapid change in types of floodplain sediments in a very short distance at the York Canyon site attests to the variability of the rocks and settings at the time the boundary clay was deposited.

Raton. The Raton site (Fig. 1) is in a road cut along the old Raton Pass scenic road about 1.5 km (1 mi) west of the town of Raton. The rock sequence exposed represents deposition near the margin of a swamp where accumulation of organic material was interrupted periodically by fine-grained floodplain sediments. The boundary clay is in a coal zone considered laterally equivalent to the Sugarite coal (Fig. 1), a correlation that can be made with a high degree of confidence because of the presence of the K-T boundary clay. The coal zone consists of three coal beds in a 3.8-m (12.5-ft) thick sequence of mudstone, siltstone, and carbonaceous shale. The boundary clay is from 5 to 15 cm (2 to 6 in) beneath the base of the 15-cm (6-in) thick upper coal bed; a 41-cm (16-in) thick zone of coal and carbonaceous shale is 1.8 m (6 ft) beneath the upper coal bed, and a 71-cm (28-in) thick coal bed is at the base of the sequence (Fig. 1).

The boundary clay at the Raton site is difficult to distinguish from the enclosing mudstone and shale beds. Rather than being contained in a dark carbonaceous shale zone and lying beneath a thin coal bed, the K-T clay bed occurs as a rusty-weathering 1.3 to 2.5–cm (0.5 to 1–in) thick bed of kaolinitic clay in a sequence of slightly carbonaceous gray mudstone consisting mostly of illite/smectite clay and quartz (R. M. Pollastro, written commun., 1983). It is nearly impossible to distinguish from the enclosing beds on fresh surfaces. The K-T boundary was expected to be directly beneath the upper coal of the zone, as was the case at York Canyon, the discovery site. Pollen recovery was poor, because of the weathered character of the coal, but the clay directly beneath the coal was not kaolinitic. A kaolinite-rich clay bed was finally identified by a combination of microstratigraphy and palynological, X-ray, and scanning electron microscope studies. The tiny goyazite spherules described above were first observed during studies of coarse-fraction splits from samples of the "rusty" bed, about 5 cm (2 in) below the upper coal bed. The sequence of rocks that contains the K-T boundary interval is described in Appendix 4.

The Raton site consists of a succession of floodplain and swamp deposits. The floodplain deposits, which are mostly fine-grained sediments, probably represent accumulation in a distal part of a floodbasin. The sediments may reflect deposition by overbank floods and (or) frontal lobes of crevasse splays that received only mud and silt during floods. Filling of the floodbasin was followed by development of swamps, one of which served as a host environment for the accumulation of the K-T boundary clay.

Sugarite. The Sugarite site (Fig. 1), about 5 km (3 mi) northeast of Raton on highway 526, represents deposition of the boundary clay in a poorly drained swamp environment. A coal termed the Sugarite bed (Lee, 1917) is well exposed in a landslide scar on the east wall of Chicorica Canyon about 150 m (490 ft) above the valley floor near the abandoned town of Sugarite. Because the lower coal zone has thinned (Fig. 3), the Sugarite coal bed is only about 27 m (90 ft) above the top of the Trinidad Sandstone; it is only about 3–4.5 m (10–15 ft) below the base of the cliff-forming channel sandstones of the barren series in the Raton Formation (Fig. 3). At the Sugarite site the coal bed is 1.8 m (6 ft) thick and contains two thin partings: a 2.5-cm (1-in) thick carbonaceous shale 38 cm (15 in) above the base of the coal and the kaolinite-rich boundary clay, 2.5 to 3.8 cm (1 to 1.5 in) thick, 15 cm (6 in) below the top. Iridium concentrations of 2.7 ng/g were measured in the boundary clay (Gilmore and others, 1984; Pillmore and others, 1984), and the characteristic disappearance of the *Proteacidites* assemblage and change in fern spore/angiosperm pollen ratios were also observed (Pillmore and others, 1984; Nichols and others, 1985). The boundary clay has been found at the same position in the Sugarite coal bed at another site about 2.3 km (1.5 mi) south-southwest of Sugarite. The Sugarite area is mostly covered by landslide debris and talus from basalt flows that cap the surrounding mesas, and tracing the coal in the field is very difficult.

The rocks above and below the coal bed at the Sugarite site are mostly covered but appear to be fine-grained mudstones and shale with some sandstone. The sandstones are thin and probably represent deposits of minor crevasse splays that periodically invaded the floodplain. A partial measured stratigraphic section that is representative of the area is shown in Appendix 5.

The Sugarite site, in contrast to the Raton site, represents deposition in a large swamp (roughly 80 km^2 or 30 mi^2) that existed as a poorly drained area for a relatively long time. A stable, poorly drained swamp condition is indicated by the accumulation of a thick vegetal mat that appears nearly uninterrupted by detrital sediments. The small amount of clastic material in the Sugarite bed at this site indicates that this part of the coal-forming swamp was far removed from fluvial incursion. In contrast, the coal zone at Raton probably formed in a swamp that for the most part was well drained but occasionally received waterborne sediments, perhaps from a nearby point source. The coal bed at Sugarite is believed to represent the entire coal zone described at the Raton site (Fig. 1); furthermore, the sequence of rocks adjacent to the coal bed in the Sugarite area also contains a succession of floodplain and swamp deposits similar to those at the Raton site.

SUMMARY

The discovery of the K-T boundary clay and recognition of its significance as a time-stratigraphic unit that records an instant in geologic time permit detailed analysis and comparison of depositional environments throughout a broad area in the Raton basin. The nature and relationships of specific depositional systems that existed on the alluvial plain covering the basin area at the end of the Cretaceous can be determined from the lithofacies types that immediately enclose the K-T boundary interval. These lithofacies include deposits of crevasse splay, swamp, levee-overbank, floodplain, and abandoned channel environments, and they exhibit characteristics very similar to those of modern and ancient fluvial deposits. The presence of the associated boundary clay throughout an alluvial plain that was drained by meandering streams, which were flanked by levees breached by crevasse splays, demonstrates the contemporaneity of these fluvial environments. The meandering streams merged with braided streams and alluvial fans toward the western borderlands of the basin margin. The combined effects of tectonic uplift and input of coarse clastics into the subsiding basin controlled the mode of aggradation by fluvial systems. Increased bedload of streams due to episodic uplift of source areas promoted incursion of alluvial fans, braided streams, and low-sinuosity meandering streams of coarse-grained meanderbelts into the depositional basin. The pulses of detrital influx via these fluvial systems were followed by regional aggradation by highly sinuous meandering streams of fine-grained meanderbelts during periods of tectonic quiescence. The fallout resulting from the K-T boundary impact event took place just before or during the transition from the fine-grained meanderbelt environments of the lower coal zone to the coarse-grained meanderbelt environments of the barren series of the Raton Formation.

ACKNOWLEDGMENTS

We wish to thank C. J. Orth, J. S. Gilmore, and J. D. Knight of Los Alamos National Laboratories for the use of their geochemical and Ir-concentration data and R. H. Tschudy for palynological data used throughout this report. We thank C. K. Molenaar of the U.S. Geological Survey and Barry Kues of the University of New Mexico for their helpful comments and suggestions. We also wish to thank Vera Sable of the U.S. Geological Survey for editorial assistance and preparation of the manuscript.

APPENDIX 1. BERWIND CANYON SECTION

Partial section measured about 8 km (5 mi) southwest of the Ludlow exit from I-25, in a road cut in Berwind (Road) Canyon. The site is near the east edge of the Delagua 7½-minute Quadrangle, Los Animas County, Colorado; SE¼SW¼SW¼Sec.36,T.31S.,R.65W. *(Italics indicate sub-unit thickness.)*

	Thickness (equivalents) Meters	Feet
RATON FORMATION (Tertiary and Cretaceous):		
12. Sandstone, fine-grained, light-gray; weathers yellowish gray to grayish yellow. Forms massive ledge above exposure; top of section	3.+	10.+
11. Coal zone	0.3±	1.±
c. Bluish-gray clay soil above coal. Slumped	*15.2+ cm*	*(6+ in)*
b. Tonstein at top of coal	*3.0 cm*	*(1.2 in)*
a. Coal, thick, blocky, soft, and weathered. Slumped and smeared	*15.2 cm*	*(6 in)*
10. Siltstone, mudstone, and very fine grained sandstone. Sandstone forms beds 15 to 30 cm (0.5 to 1 ft) thick in slope. Slightly carbonaceous mudstone at 1.8 m (6 ft). Top 1 m (3 ft) is rooted siltstone and sandstone	3.1	10.0
9. Sandstone, very fine grained; yellowish gray; weathers grayish yellow and orange to reddish brown; micaceous, lenticular. Characterized by many well-preserved palm impressions	0.5	1.5
8. Mudstone, carbonaceous. 5-cm (2-in) thick carbonaceous shale layer at 0.9 m (3 ft)	1.2	4.0
7. Mudstone and siltstone. 7.6-cm (3-in) carbonaceous shale at base	0.9	3.0
6. Coal, bright, brittle, and blocky. Rests directly on sandstone. 0.6-cm (0.25-in) carbonaceous shale parting 21.6 cm (8.5 in) above base of unit	0.2	0.8
5. Sandstone, very fine grained, light-gray to light-yellowish-gray, micaceous; weathers grayish yellow, grayish orange, yellowish orange and reddish brown. Beds as thick as 1 m (3 ft) interbedded with siltstone. Carbonaceous fragments and plant and root impressions; upper 0.3 m (1 ft) grayish-orange rooted sandstone	1.8	6.0
4. Shale, gray, subfissile, fossiliferous; thin 1.3-cm (0.5-in) coaly stringer in 5–10 cm (2–4 in) carbonaceous shale about 10–15 cm (4–6 in) from base. Shale is in sharp contact with sandstone beneath	0.5	1.5
3. Sandstone and siltstone, interbedded, very fine grained, micaceous. Sandstone beds up to 15.2 cm (6 in), interbeds of nodular-weathering gray siltstone and mudstone. Some thin stringers of coal; carbonized plant and root impressions. Cross laminated	1.1	3.7
2. **K-T boundary interval**	0.24	0.8
d. Shale, gray, fissile, fossiliferous; many carbonaceous impressions of stems, leaves, and reeds	*7.9 cm*	*(3.1 in)*
c. Coal, blocky, bright, brittle, cleated	*4.3 cm*	*(1.7 in)*
b. K-T boundary clay. Claystone, kaolinitic, medium-gray. Thin, fissile, carbonaceous shale at base. Flaky, carbonaceous shale top 0.64 cm (0.25 in); rusty-reddish-brown and yellowish-brown jarosite stain at top	*2 cm*	*(0.8 in)*
a. Mudstone, silty. Grades down into very fine grained sandstone; grades up to carbonaceous mudstone and fissile shale at top	*10 cm*	*(4 in)*
1. Sandstone, very fine grained, light-gray to yellowish-gray, ledge-forming, micaceous; weathers light gray, grayish yellow to grayish orange; some carbonaceous impressions and root remains. Some rooted siltstone and clayey siltstone in top 20.3 cm (8 in), grades up into carbonaceous shale. Laterally to right this zone thickens to 1-m (3-ft) thick sequence of interbedded siltstone and very fine grained sandstone; weathers reddish brown	1.4	4.5
LEVEL OF ROAD: TOTAL THICKNESS MEASURED	14.2	46.8

APPENDIX 2. STARKVILLE NORTH SECTION

Measured about 3 km (2 mi) south of the Starkville exit from I-25, in a road cut on the east side of the service road. The site is in the Starkville 7½-minute Quadrangle, Las Animas County, Colorado; SE¼SE¼SE¼-Sec.2,T.34S.,R.64W. Partial section begins (Unit 1) at service road level, approximately 50 m (170 ft) above top of Trinidad Sandstone.

	Thickness (equivalents) Meters	Feet
RATON FORMATION (Tertiary and Cretaceous):		
13. Sandstone. Lower 0.9 m (3 ft) grades down to siltstone and mudstone	3.0	10.0
12. **K-T boundary interval**	0.67	2.2
e. Shale, carbonaceous. Grades up to overlying mudstone	*0.15*	*0.5*
d. Coal	*0.06 m*	*(0.2 ft)*
c. K-T boundary clay. Kaolinitic claystone, light-grayish-brown; dark organic-rich band at top. Coaly shale at base; flaky dark shale at top	*0.024 m*	*(0.08 ft)*
b. Carbonaceous shale, dark-grayish-brown to dark-gray; thin streaks of bright vitric coal; plant fragments and impressions. Fewer coal streaks in lower part	*0.24 m*	*(0.8 ft)*
a. Dark shale. Tiny coal streaks more common in upper part	*0.18 m*	*(0.6 ft)*
11. Grades down into gray mudstone with scattered carbonaceous fragments, nonfissile, few coal streaks	0.06	0.2
10. Mudstone, gray. Becomes carbonaceous in top 2.5 cm (1 in)	4.6	15.0
9. Sandstone, very carbonaceous. Abundant carbonaceous fragments in lower part, thins to 0.3 m (1 ft); irregular base on coal; discontinuous along road cut	0.8	2.5
8. Coal	0.15	0.5

7. Sandstone and siltstone, interbedded. Lenticular beds as thick as 0.3 m (1 ft); grades up to 0.3-m (1-ft) thick carbonaceous shale bed	1.8	6.0
6. Sandstone, channel-type, fine-grained. Lenses out 18 m (20 yd) north at edge of road cut ...	1.8	6.0
5. Bench in road cut. Partly covered, sandstone grades up to mostly sandstone and siltstone. Thin beds of coaly shale at 1.8 m (6 ft). Grades into thick channel sandstone 18 m (20 yd) to south	3.4	11.0
4. Sandstone, fine- to medium-grained, yellowish-gray. Some contorted beds; mostly even bedding; even base. Ironstone concretions and layers	3.0	10.0
3. Siltstone, shale, and mudstone. Carbonaceous in top 20 cm (8 in)	1.8	6.0
2. Shale and siltstone, carbonaceous. Coaly at 1.7 m (5.5 ft); 10 cm (4 in) coal bed at top ...	5.8	19.0
1. Mudstone	1.8	6.0
LEVEL OF SERVICE ROAD: TOTAL THICKNESS MEASURED	28.8	94.9

APPENDIX 3. YORK CANYON DRILL HOLE LOG

Hole drilled December 29, 1980, to January 1, 1981, approximately 0.16 km (0.1 mi) north of the Santa Fe railhead at the York Canyon Mine, Casa Grande 15′ quadrangle, Colfax County, New Mexico. Latitude 36°52′N; longitude 104°55′30″W. The 1980–81 drilling was funded by Los Alamos National Laboratories in search of the K-T boundary iridium anomaly.

Lithologic Description of Upper Part of Lower Coal Zone of Raton Formation Depth from 243.8 to 274.3 m (800 to 900 ft)

	Thickness (equivalents) Meters	Feet
RATON FORMATION (Tertiary and Cretaceous):		
1. Sandstone, very fine grained to silty, ripple and cross laminated; top few inches grade up to silty shale; bottom half of unit is very fine grained to fine grained, cross laminated sandstone	5.03	16.5
2. Mudstone, dark-gray, carbonaceous	0.91	3.0
3. Coal, bright, banded, cleated, hard	0.38	1.25
4. Mudstone, dark-gray, carbonaceous, sparse coal laminae, very slightly silty in middle part	2.13	7.0
5. Coal, bright, banded, hard, well cleated with calcite on cleat surfaces	0.24	0.8
6. Sandstone, very fine grained to fine-grained, hard, cross-bedded, ripple laminations, hard iron-cemented layer at base underlain by 0.15 m (0.5 ft) silty sandstone	2.87	9.1
7. Mudstone, carbonaceous, coaly streaks and layers	0.21	0.7
8. **K-T boundary interval** Coal, bright, banded, hard, well cleated one direction, slickensided shear plane (40°) cuts lower part of coal. Organic-rich clay layer at base. K-T boundary clay not described in core log	0.21	0.7
9. Mudstone, carbonaceous, coal streaks, plant impressions, dark-gray, slightly silty upper part, curved shear fractures	1.0	3.3
10. Sandstone, very fine grained, silty in upper part	0.61	2.0
11. Mudstone and silty shale, contorted bedding, cross-laminated sandy siltstone to silty sandstone in middle part	0.7	2.3
12. Coal, bright, banded, well-cleated, with shale laminae. At top is 1.02 cm (0.4 in) bony coal, bottom is 1.27 cm (0.5 in) very carbonaceous to coaly shale	0.55	1.8
13. Mudstone, dark-gray	0.61	2.0
14. Siltstone interbedded and interlaminated with very fine grained sandstone, contorted bedding. Erosional contact at base	3.26	10.7
15. Mudstone, dark-gray, silty in upper 0.15 m (0.5 ft)	1.0	3.3
16. Carbonaceous shale, fissil, brittle, coaly in places, coal streaks and laminae. Zone of silty and sandy laminae at base	2.53	8.3
17. Coal, bright, brittle, well-cleated in one direction, calcite on cleat surfaces, carbonaceous shale at base	0.37	1.2
18. Mudstone, dark-gray, slightly carbonaceous, some zones of laminations and streaks of bright coal at 268.8 and 270.6 m (882 and 888 ft). Silty layers at 267.6 m (878 ft) and in lower 0.6 m (2 ft)	5.79	19.0
19. Sandstone, very fine grained to fine-grained, cross-bedded, finely laminated, some irregular silty laminations, silty layers in top 0.6 m (2 ft)	2.13	7.0
TOTAL THICKNESS MEASURED	30.5	100.0

APPENDIX 4. RATON SECTION

Measured in the gully south-southwest of the culvert at the K-T boundary site, west side of saddle on old Raton Pass scenic road about 0.53 km (0.33 mi) directly west of Goat Hill above Raton, in the Raton 7½-minute Quadrangle, Colfax County, New Mexico. Latitude 36°54′ 12″N.; longitude 104°27′4″W. Elevation 2170 m (7120 ft).

	Thickness (equivalents) Meters	Feet
RATON FORMATION (Tertiary and Cretaceous):		
24. Mudstone, thin layers of yellowish-brown weathering siltstone	3.05+	10.0+
23. Carbonaceous shale and mudstone	0.15	0.5
22. Mudstone, like Unit 24	2.13	7.0
21. Shale, carbonaceous; rests on sandstone ..	0.15	0.5
20. Sandstone, very fine grained; grades down to siltstone and mudstone in lower half of unit ...	1.83	6.0
19. Shale, carbonaceous; thin stringers of coal	0.15	0.5
18. Silty mudstone to siltstone	0.15	0.5
17. Mudstone, slightly carbonaceous	0.08	0.25
16. Mudstone, nodular weathering, weathers yellowish gray to grayish yellow; silty about 25 cm (10 in) above base	0.7	2.3
15. Mudstone, weathers yellowish gray; grades down into gray claystone	0.15	0.5

14. Claystone, gray; grades down into carbonaceous shale	0.08	0.25
13. Coal, blocky, brittle, banded; N68W cleat; grades up to carbonaceous shale	0.09	0.3
12. Carbonaceous shale and shaly coal	0.07	0.22
e. Coal, shaly, weathered	*0.02 m*	*(0.08 ft)*
d. Coal, blocky, weathered	*0.01 m*	*(0.04 ft)*
c. Coal, shaly, weathered	*0.01 m*	*(0.04 ft)*
b. Coal, shaly, grades down to very carbonaceous to coaly shale	*0.01 m*	*(0.04 ft)*
a. Shale, carbonaceous, rusty, rests on very carbonaceous, light brown and dark yellowish-orange claystone	*0.006 m*	*(0.02 ft)*
11. **K-T boundary interval**	0.16	0.6
i. Carbonaceous shale; grades down into slightly fissile claystone; many carbonaceous fragments and plant impressions	*0.01 m*	*(0.04 ft)*
h. Claystone, gray, slightly silty; carbonaceous fragments; plant impressions	*0.01 m*	*(0.04 ft)*
g. Claystone, gray, silty, carbonaceous; plant impressions	*0.02 m*	*(0.08 ft)*
f. Claystone, gray, slightly silty, weathers light medium gray; carbonaceous fragments	*0.01 m*	*(0.04 ft)*
e. K-T boundary clay. "Rusty zone," iron-stained kaolinitic claystone, reddish brown to light brown and dark yellowish orange on weathered surface	*0.01 m*	*(0.04 ft)*
d. Claystone, slightly silty	*0.02 m*	*(0.08 ft)*
c. Mudstone, weathers gray to yellowish-gray, carbonaceous fragments	*0.03 m*	*(0.1 ft)*
b. Mudstone, slightly silty, medium gray; indistinct bedding to coarsely bedded; weathers grayish-yellow to pale yellow on bedding surfaces. Carbonaceous fragments scarce	*0.02 m*	*(0.08 ft)*
a. Mudstone like Unit h	*0.03 m*	*(0.1 ft)*
10. Silty mudstone and siltstone. Some reddish-brown iron-cemented concretions; beds up to 15.24 cm (6 in) thick	0.91	3.0
9. Siltstone and silty mudstone, thin 2.54–7.62 cm (1–3 in) beds of siltstone, yellowish-gray to light brownish-gray, weathered grayish-yellow to grayish-orange; root fragments and plant impressions scattered throughout unit; nodular; micaceous	0.88	2.9
8. Coal, 2.5-cm (1-in) bone parting in middle; shaly coal at base grades down into carbonaceous shale	0.24	0.8
7. Shale and mudstone, carbonaceous; grades down into siltstone	0.24	0.8
6. Siltstone and silty mudstone weathers yellowish-gray to grayish-yellow, thin 2.5–5-cm (1–2-in) beds of siltstone about 38 cm (15 in) above base	0.58	1.9
5. Mudstone and silty mudstone, grades down through carbonaceous claystone to carbonaceous shale. Rests on coal bed	0.18	0.6
4. Coal, bright, blocky, banded; strong N65–70W cleat; slight tendency to break into nodules	0.7	2.3
3. Carbonaceous shale; grades down to silty mudstone	0.3	1.0
2. Mostly mudstone, 15-cm (6-in) coaly shale at 0.9 m (3 ft) above base, 15-cm (6-in) coaly shale at 3.35 m (11 ft) above base, top 3 m (10 ft) siltstone and silty mudstone	7.6	25.0
1. Sandstone, fine-grained to very coarse grained channel fill, irregular beds, contorted, iron-cemented stringers, and lenses and concretions; forms massive, persistent ledge; crossbeds and contorted bedding top 3 m (10 ft)	5.8	19.0
TOTAL SECTION MEASURED	26.4	86.7

APPENDIX 5. SUGARITE SECTION

Partial section measured on southwest side of Horse Mesa in a landslide scar exposing the Sugarite coal in the lower coal zone of the Raton Formation (partly covered by soil and talus). The site is at the western edge of the Yankee 7½-minute Quadrangle, Colfax County, New Mexico. Latitude 36°55′22″N; longitude 104°22′57″W. Elevation 2240 m (7350 ft).

	Thickness (equivalents) Meters	Feet
RATON FORMATION (Tertiary and Cretaceous):		
8. Sandstone, fine-grained, forms cliffs of the barren series	3.0+	10.0+
*7. Mudstone	2.4	8.0
*6. Carbonaceous shale and mudstone with 23-cm (9-in) thick coal bed at top	0.9	3.0
*5. Sandstone, very fine grained	0.6	2.0
4. Mudstone and carbonaceous shale with coal streaks	1.8	6.0
3. **K-T boundary interval,** Sugarite coal bed	1.8	6.0
g. Carbonaceous shale, coaly at top	*0.32 m*	*(1.04 ft)*
f. Coal, bright, blocky, shaly top 1 cm (0.03 ft)	*0.15 m*	*(0.5 ft)*
e. K-T boundary kaolinitic clay. Nodular, dark, carbonaceous clay parting, shaly at base, thin shaly layer at top	*0.03 m*	*(0.09 ft)*
d. Coal, bright, blocky	*1.2 m*	*(4.0 ft)*
c. Coal, bony	*0.02 m*	*(0.08 ft)*
b. Coal, bright, blocky	*0.4 m*	*(1.2 ft)*
a. Coal, bony	*0.03 m*	*(0.09 ft)*
*2. Mudstone grading down to sandstone in basal 1.5 m (5 ft). One m (3 ft) below top is 0.3-m (1-ft) thick zone of coaly shale and coal	7.6	25.0
*1. Carbonaceous shale and mudstone, coaly at top	1.5	5.0
TOTAL THICKNESS MEASURED	19.6	65.0

*Description of units from better-exposed section measured on southeast end of Bartlett Mesa, near east-central border of Raton 7½-minute Quadrangle, about 3 km (2 mi) south of the Sugarite site.

REFERENCES CITED

Alvarez, L. W., Alvarez, W., Asaro, F., and Michel, H. V., 1980, Extraterrestrial cause for the Cretaceous-Tertiary extinction: Science, v. 208, no. 4448, p. 1095–1108.

Alvarez, W., Alvarez, L. W., Asaro, F., and Michel, H. V., 1982, Current status of the impact theory for the terminal Cretaceous extinction: Geological Society of America Special Paper 190, p. 305–315.

Baltz, E. H., 1965, Stratigraphy and history of Raton basin and notes on San Luis basin, Colorado–New Mexico: American Association of Petroleum Geologists Bulletin, v. 49, p. 2041–2075.

Beerbower, J. R., 1964, Cyclothems and cyclic depositional mechanisms in alluvial plain sedimentation, *in* Merriam, D. F., ed., Symposium on cyclic sedimentation: Kansas Geological Survey Bulletin 169, v. 1, p. 31–42.

Billingsley, L. T., 1978, Stratigraphy of the Pierre Shale, Trinidad Sandstone, and Vermejo Formation, Walsenburg area, Colorado; A deltaic model for sandstone and coal deposition, *in* Hodgson, H. E., ed., Proceedings of the Second Symposium on the Second Symposium on the Geology of Rocky Mountain Coal: Colorado Geological Survey Resource Series 4, p. 23–34.

Bohor, B. F., and Pillmore, C. L., 1976, Tonstein occurrences in the Raton coal field, Colfax County, New Mexico, *in* Ewing, R. C., and Kues, B. S., eds., Guidebook of Vermejo Park, northeastern New Mexico: New Mexico Geological Society Guidebook, 27th Field Conference, p. 177–183.

Bohor, B. F., Pollastro, R. M., and Phillips, R. E., 1978, Mineralogic evidence for the volcanic origin of kaolinitic partings (tonsteins) in Upper Cretaceous and Tertiary coals of the Rocky Mountain region: Annual Clay Minerals Conference, 15th, Bloomington, Indiana, 1978, Programs and Abstracts, p. 47.

Bohor, B. F., Foord, E. E., Modreski, P. J., and Triplehorn, D. M., 1984, Mineralogic evidence for an impact event at the Cretaceous-Tertiary boundary: Science, v. 224, no. 4651, p. 867–869.

Bohor, B. F., Modreski, P. J., Foord, E. E., 1985, A search for shock-metamorphosed quartz at the K-T boundary: Lunar and Planetary Science Conference, 16th, Houston, March 11, 1985, p. 79–80.

Brown, R. W., 1962, Paleocene flora of the Rocky Mountains and Great Plains: U.S. Geological Survey Professional Paper 375, 119 p.

Ethridge, F. G., Jackson, T. J., and Youngberg, A. D., 1981, Floodbasin sequence of a fine-grained meanderbelt subsystem; The coal-bearing lower Wasatch and upper Fort Union formations, southern Powder River basin, Wyoming: *in* Ethridge, F. G., and Flores, R. M., eds., Recent and ancient nonmarine depositional environments; Models for exploration: Society of Economic Paleontologists and Mineralogists Special Publication, no. 31, p. 191–209.

Fisk, H. N., 1944, Geological investigations of the alluvial valley of the lower Mississippi River: U.S. Army Corps of Engineers, Mississippi River Commission, 78 p.

Flores, R. M., 1984, Comparative analysis of coal accumulation in Cretaceous alluvial deposits, southern United States Rocky Mountain basins, *in* Stott, D. F., and Glass, D., eds., The Mesozoic of Middle North America: Canadian Society of Petroleum Geologists Memoir 9, p. 373–385.

Flores, R. M., and Tur, S. M., 1982, Characteristics of deltaic deposits in the Cretaceous Pierre Shale, Trinidad Sandstone, and Vermejo Formation, Raton basin, Colorado: The Mountain Geologist, v. 19, p. 25–40.

Gilmore, J. S., Knight, J. D., Orth, C. J., Pillmore, C. L., and Tschudy, R. H., 1984, Trace element patterns at a nonmarine Cretaceous-Tertiary boundary, Western Interior: Nature, v. 307, p. 224–228.

Harbour, R. L., and Dixon, G. H., 1956, Geology of the Trinidad-Aguilar areas, Las Animas and Huerfano Counties, Colorado: U.S. Geological Survey Oil and Gas Investigations Map OM-174, scale 1:31,680.

Hills, R. C., 1888, The recently discovered Tertiary beds of the Huerfano River basin, Colorado: Colorado Scientific Society Proceedings, v. 3, pt. 1, p. 148–164.

Hotten, Carol, 1984, Palynofloral changes across the Cretaceous-Tertiary boundary in east central Montana, U.S.A.: Sixth International Palynological Conference, Calgary, 1984, Abstracts, p. 66.

Izett, G. A., and Pillmore, C. L., 1985a, Shock-metamorphic minerals at the Cretaceous-Tertiary boundary, Raton basin, Colorado and New Mexico [abs.]: EOS (Transactions of the American Geophysical Union), v. 66, no. 46, p. 1149.

—— , 1985b, Abrupt appearance of shocked quartz at the Cretaceous-Tertiary boundary, Raton basin, Colorado and New Mexico: Geological Society of America Abstracts with Programs, v. 17, no. 7, p. 617.

Izett, G. A., and Bohor, B. F., 1986, Microstratigraphy of continental sedimentary rocks in the Cretaceous-Tertiary boundary interval in the Western Interior of North America; Geological Society of America Abstracts with Programs, v. 18, no. 6, p. 644.

Johnson, R. B., and Wood, G. H., Jr., 1956, Stratigraphy of Upper Cretaceous and Tertiary rocks of Raton basin, Colorado and New Mexico: American Association of Petroleum Geologists Bulletin, v. 40, p. 707–721.

Lee, W. T., 1917, Geology of the Raton Mesa and other regions in Colorado and New Mexico, *in* Lee, W. T., and Knowlton, F. H., eds., Geology and paleontology of Raton Mesa and other regions in Colorado and New Mexico: U.S. Geological Survey Professional Paper 101, p. 9–221 [1918].

Leighton, V. L., 1980, Depositional environments and petrography of the Trinidad Sandstone and related formations, Raton area, New Mexico [M.S. thesis]: Fort Collins, Colorado State University, 105 p.

Nichols, D. J., Fleming, R. F., Upchurch, G. R., Tschudy, R. H., and Pillmore, C. L., 1985, Paleobotanical changes across the Cretaceous-Tertiary boundary at Sugarite, New Mexico; New data and interpretations [abs.]: Society of Economic Paleontologists and Mineralogists, Midyear Meeting Abstracts, v. 2, p. 68.

Orth, C. J., Gilmore, J. S., Knight, J. D., Pillmore, C. L., Tschudy, R. H., and Fassett, J. E., 1981, An iridium abundance anomaly at the palynological Cretaceous-Tertiary boundary in northern New Mexico: Science, v. 214, p. 1341–1343.

—— , 1982, Iridium abundance measurements across the Cretaceous-Tertiary boundary in the San Juan and Raton basins of northern New Mexico: Geological Society of America Special Paper 190, p. 423–433.

Pillmore, C. L., 1969, Geologic map of the Casa Grande quadrangle, Colfax County, New Mexico, and Las Animas County, Colorado: U.S. Geological Survey Geologic Quadrangle Map GQ—823, scale 1:62,500.

—— , 1976, Commercial coal beds of the Raton coal field, Colfax County, New Mexico, *in* Ewing, R. C., and Kues, B. S., eds., Guidebook of Vermejo Park, northeastern New Mexico: New Mexico Geological Society Guidebook, 27th Field Conference, p. 227–247.

Pillmore, C. L., and Maberry, J. O., 1976, The depositional environment and trace fossils of the Trinidad Sandstone, southern Raton basin, New Mexico, *in* Ewing, R. C., and Kues, B. S., eds., Guidebook of Vermejo Park, northeastern New Mexico: New Mexico Geological Society Guidebook, 27th Field Conference, p. 191–195.

Pillmore, C. L., Tschudy, R. H., Orth, C. J., Gilmore, J. S., and Knight, J. D., 1984, Geologic framework of nonmarine Cretaceous-Tertiary boundary sites, Raton basin, New Mexico and Colorado: Science, v. 223, no. 4641, p. 1180–1182.

Pollastro, R. M., and Pillmore, C. L., 1986, Mineralogy and petrology of the Cretaceous-Tertiary boundary clay bed and adjacent clay-rich rocks, Raton basin, New Mexico and Colorado: Journal of Sedimentary Petrology (in press).

Pollastro, R. M., Pillmore, C. L., Tschudy, R. H., Orth, C. J., and Gilmore, J. S., 1983, Clay petrology of the conformable Cretaceous/Tertiary boundary interval, Raton basin, New Mexico and Colorado: Annual Clay Minerals Conference, 32nd, Buffalo, New York, 1983, Programs and Abstracts, p. 83 (2 p).

Saxena, R. S., and Ferm, J. C., 1976, Lower deltaic sand bodies; Exploration and recognition criteria, *in* Saxena, R. S., ed., Sedimentary environments and hydrocarbons: American Association of Petroleum Geologists short course, New Orleans, Louisiana, p. 22–59.

Smit, Jan, 1984, Evidence for worldwide microtektite strewn field at the

Cretaceous-Tertiary boundary: Geological Society of America Abstracts with Programs, v. 16, no. 6, p. 659.
Spears, D. A., and Kanaris-Sotiriou, R., 1979, A geochemical and mineralogical investigation of some British and other European tonsteins: Sedimentology, v. 26, p. 407–425.
Strum, S. R., 1984, Depositional environments and lithofacics of the Raton Formation, eastern Raton basin, New Mexico [M.S. thesis]: Raleigh, North Carolina State University.
Triplehorn, D. M., and Bohor, B. F., 1983, Goyazite in kaolinitic altered tuff beds of Cretaceous age near Denver, Colorado: Clays and Clay Minerals, v. 31, no. 4, p. 299–304.
Tschudy, R. H., 1970, Palynology of the Cretaceous-Tertiary boundary in the northern Rocky Mountain and Mississippi Embayment regions: Geological Society of America Special Paper 127, p. 65–111.
—— , 1973, The Gasbuggy core: A palynological appraisal, *in* Fassett, J. E., ed., Cretaceous and Tertiary rocks of the southern Colorado Plateau: Four Corners Geological Society Memoir, p. 131–143.
Tschudy, R. H., Pillmore, C. L., Orth, C. J., Gilmore, J. S., and Knight, J. D., 1984, Disruptions of the terrestrial plant ecosystem at the Cretaceous-Tertiary boundary, Western Interior: Science, v. 225, no. 4666, p. 1030–1034.
Tschudy, R. H., and Tschudy, B. D., 1986, Extinction and survival of plant life following the Cretaceous/Tertiary boundary event, Western Interior, North America: Geology, v. 14, no. 8, p. 667–670.
Tweto, O. L., 1980, Summary of Laramide Orogeny in Colorado, *in* Kent, H. C., and Porter, K. W., eds., Colorado geology: Rocky Mountain Association of Geologists Symposium, Denver, Colo., 1980, p. 130.
Wanek, A. A., 1963, Geology and fuel resources of the southwestern part of the Raton coal field, Colfax County, New Mexico: U.S. Geological Survey Coal Investigations Map C45 [1964].
Williamson, I. A., 1970, Tonsteins; Their nature, origins and uses: Mining Magazine, v. 122, p. 203–210.

MANUSCRIPT ACCEPTED BY THE SOCIETY MARCH 24, 1986

Printed in U.S.A.

Geological Society of America
Special Paper 209
1987

Remanent magnetization of rocks of latest Cretaceous and earliest Tertiary age from drill core at York Canyon, New Mexico

Eugene M. Shoemaker
U.S. Geological Survey
Flagstaff, Arizona 86001

Charles L. Pillmore
U.S. Geological Survey
MS 913, Box 25046, Denver Federal Center
Denver, Colorado 80225

Edward W. Peacock
Errol L. Montgomery & Assoc.
1075 E. Fort Lowell Rd.
Tucson, Arizona 85719

ABSTRACT

At the end of 1980, seven complete cores were recovered from a 30-m (100-ft) interval in the Raton Formation at York Canyon, New Mexico. The interval cored spans the palynologically defined Cretaceous-Tertiary boundary, which is marked by a distinctive noble metal–bearing claystone in the Raton basin. Azimuthal orientation of the cores can be recovered both from the average directions of the most stable components of the remanent magnetization, with a root mean square error of 28°, and from the average direction of secondary components of magnetization removed by thermal and alternating field demagnetization, with a root mean square error of 33°. The natural remanent magnetization of about 95 percent of the core is dominated by a secondary normal polarity component. Polarity of the characteristic magnetization of each core, interpreted from 12 to 14 samples per core run, is reversed. No evidence of normal polarity characteristic magnetization was found in the 30-m (100-ft) interval sampled. The characteristic magnetization probably is a depositional remanent magnetization acquired during chron 29r. The noble metal–bearing boundary claystone in the Raton basin is interpreted to be part of a synchronous global deposit laid down at the end of the Cretaceous period.

INTRODUCTION

The present investigation was undertaken to reexamine the remanent magnetization of sedimentary rocks from a core spanning the Cretaceous-Tertiary (K-T) boundary at York Canyon, New Mexico. The rocks cored are part of a relatively thick sequence of continental beds of latest Cretaceous and Paleocene age in the Raton basin of New Mexico and Colorado. In these beds, the K-T boundary has been located on the basis of fossil floras. The precise position of the boundary in the Raton basin has been placed just above the highest occurrence of the *Proteacidites* assemblage of Late Cretaceous palynomorphs (Tschudy, 1973; Pillmore and others, 1984). The upper stratigraphic limit of a related palynomorph assemblage, which includes *Proteacidites,* has also been used to define the K-T boundary in the Powder River Basin of Wyoming (Leffingwell, 1970) and in eastern Montana (Tschudy, 1970; Hotten, 1984) and Alberta (Lerbekmo and others, 1979). In Montana and Alberta, the boundary thus defined is known to occur in a reverse polarity magnetozone (Archibald and others, 1982; Lerbekmo and others, 1979). In contrast to these observations, Payne and others (1982, 1983a, 1983b) concluded from a previous study of the York Canyon core that the rocks spanning the palynological K-T boundary in the Raton basin have a normal polarity primary magnetization.

Inasmuch as the precise boundary based on palynomorphs, both in the Raton basin and in eastern Montana, is locally marked by a thin kaolinitic claystone bed characterized by an unusually high abundance of titanium, noble metals, and other trace elements, it seems likely that the boundary so chosen is isochronous between the Raton basin and eastern Montana. Moreover, a thin claystone with a high content of noble metals has also been found at the K-T boundary in marine beds at numerous localities around the world (Alvarez and others, 1982). Hence, there are fairly strong reasons to suspect that the distinctive boundary claystone in New Mexico and Montana is a chronostratigraphic marker and that it is part of a global deposit formed at the very end of the Cretaceous. If the primary magnetization of the beds enclosing the boundary in the Raton basin were in fact normal polarity, the palynologically defined boundary in the western United States and Canada would appear to be diachronous (Payne and others, 1982, 1983a, 1983b). The distinctive boundary claystone would not then represent a single chronostratigraphic marker.

From a reconnaissance study using fully oriented samples taken from outcrops of the boundary claystone, Shoemaker and others (1983) found that the characteristic magnetization of the boundary claystone in the Raton basin is reversed polarity. All samples measured, however, had a strong secondary normal polarity component of magnetization that could not be completely removed by alternating field (AF) demagnetization. Without adequate knowledge of the orientation of these samples, the polarity of the characteristic or primary magnetization could not have been determined. The azimuthal orientation of the core samples studied by Payne and others (1983b) is not known, and in our view, the polarity of the characteristic magnetization of most of these samples cannot be determined from their reported measurements. Therefore, we have undertaken a new investigation in which a more complete record of the orientation of the core samples was maintained, with the aim of resolving the question of the polarity of the characteristic magnetization of these rocks.

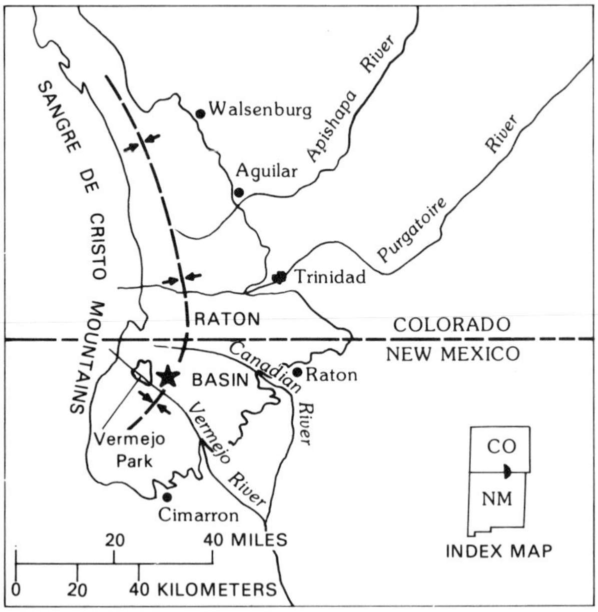

Figure 1. Index map showing Raton basin, Colorado and New Mexico. Star shows location of York Canyon drill hole.

GEOLOGIC SETTING

The Raton basin is a large asymmetric structural depression about 6400 km^2 (2470 mi^2) in area that extends from Huerfano Park, Colorado, south to Cimarron, New Mexico (Fig. 1). The rocks of the lower part of the Raton Formation, which span the K-T boundary, dip steeply along the western margin of the basin on the east flank of the Sangre de Cristo Mountains. These rocks dip more gently inward along the other margins and are highly dissected. The stratigraphic interval that contains the K-T boundary crops out over the irregular eastern limit of the nonmarine rocks of the basin, which extends from Walsenburg, Colorado to Cimarron, New Mexico, a distance of about 160 km (100 mi). Along this outcrop belt the precise position of the K-T boundary claystone bed has been identified at more than 20 sites.

The drill core on which we report here was taken from a hole in the central part of the southern Raton basin, New Mexico. This hole was drilled in the vicinity of the York Canyon coal mines, about 5 km (3 mi) east of Vermejo Park and about 56 km (35 mi) west of Raton, New Mexico. The core was taken specifically to search for an iridium anomaly at the palynologically defined K-T boundary (Orth and others, 1981), and the success of this search led to an extensive study of the boundary along the outcrop (Orth and others, 1982; Pillmore and others, 1984).

REGIONAL STRATIGRAPHY

Sedimentary rocks ranging in age from Permian to Paleocene underlie the Raton basin, but only beds deposited during the Maastrichtian stage of the Late Cretaceous and during the Paleocene are considered here. The marine Pierre Shale (Campanian to Maastrichtian) and overlying Trinidad Sandstone (Maastrictian) underlie and border the nonmarine latest Cretaceous and Tertiary rocks in the basin. Nonmarine sedimentary rocks include coal-bearing strata of the Vermejo Formation (Maastrichtian) and Raton Formation (Maastrichtian and Paleocene) and non-coal-bearing strata of the Poison Canyon Formation (also Maastrichtian and Paleocene), which overlies the Raton and, in part, intertongues with it from the west.

Pierre Shale

The Pierre Shale consists of dark silty noncalcareous shale

more than 600 m (1900 ft) thick deposited in prodeltaic and offshore environments in the Cretaceous Western Interior Seaway (Pillmore and Flores, this volume). No universally accepted criteria have been adopted for determining the Campanian-Maastrichtian boundary in rocks of diverse faunal assemblages in various parts of the world. On the basis of the position of the boundary chosen by Cobban (Obradovich and Cobban, 1975) from ammonite zonation of the Late Cretaceous in the western continental interior, the base of the Maastrichtian probably is 25–30 m (80–100 ft) below the top of the Pierre Shale. In the Raton basin, this position was determined using upper Pierre Shale fossils described by W. A. Cobban that probably are equivalent in age to *Baculites eliasi* and *Baculites reesidei* (Obradovich and Cobban, 1975). The Pierre grades into the overlying Trinidad Sandstone through several meters of transitional beds.

Trinidad Sandstone

The Trinidad Sandstone, generally about 25–30 m (80–100 ft) thick, is composed mostly of very fine to fine grained sandstone deposited in deltaic and contemporaneous barrier-bar marginal marine environments during eastward progradation of the Cretaceous coastline (Pillmore and Flores, this volume).

Vermejo Formation

The Vermejo Formation (Lee, 1917) consists of interbedded sandstone, siltstone, shale, and coal deposited conformably on the Trinidad Sandstone, on fluvio-delta plains. The Vermejo thins from about 114 m (375 ft) at Vermejo Park and along the western border of the basin to 0 m (0 ft) in the eastern part, east of Raton, New Mexico (Pillmore and Flores, Fig. 3, this volume).

Raton Formation

The Raton Formation consists of sandstone, siltstone, mudstone, coal, and a basal conglomeratic sandstone. These beds were deposited on an extensive fluvial plain that formed behind the receding coastline at the close of the Cretaceous (Pillmore and Flores, this volume). The formation ranges in thickness from more than 600 m (2,000 ft) in the west-central part of the basin to about 300 m (1,000 ft) in the eastern part. An unconformity marked by a channeled surface and an overlying conglomeratic sandstone generally is present at the base of the formation, but in the vicinity of Raton and Trinidad the conglomerate commonly is missing and no unconformity is evident.

Lee (1917) originally divided the Raton Formation into four units: (1) the basal conglomerate, (2) a lower coal zone, (3) a cliff-forming or barren series, and (4) an upper coal zone. The Cretaceous-Tertiary boundary is just beneath the barren series, and the subdivision is modified to include all the Cretaceous rocks in the lower coal zone. The three zones are usually identifiable throughout the Raton basin.

Lower Coal Zone. The lower coal zone consists of the basal conglomerate (where present), overlain by mostly fine-grained floodplain deposits of mudstone, shale, and coal that contain splay and channel sandstone deposits. The K-T boundary is at or near the top of the zone. Across the Raton basin, the lower coal zone thins from about 120 m (400 ft) west of Vermejo Park, to about 90 m (295 ft) at York Canyon, and to 30 m (100 ft) east of Raton, at the abandoned coal-mining town of Sugarite, New Mexico (Pillmore and Flores, fig. 3, this volume).

Barren Series. The beds of the barren series are chiefly stacked channel sandstones that consist mostly of fine- to coarse-grained quartz with some feldspar and chert. They are locally interbedded with and laterally grade into minor floodplain deposits of mudstone and siltstone; carbonaceous shale beds and coal zones are present but are mostly thin and limited in extent. The barren series ranges from about 180 m (600 ft) thick in the western part of the basin to about 55 m (180 ft) thick in the eastern part.

Upper Coal Zone. The upper coal zone consists of fine-grained floodplain deposits of mudstone, siltstone, shale, and coal interbedded with splay and channel sandstone beds. It ranges from a thickness of about 300 m (1,000 ft) in the central part of the basin, near York Canyon, to about 180 m (600 ft) near Raton.

Poison Canyon Formation

The Poison Canyon Formation (Hills, 1889) overlies the Raton Formation throughout most of the Raton basin; to the west and southwest, the Poison Canyon intertongues with and thickens at the expense of the Raton Formation. The Poison Canyon consists of thick, massive, lenticular, ledge-forming beds of coarse-grained to conglomeratic arkosic sandstone intercalated with beds of nonresistant, yellow-weathering, sandy micaceous mudstone. This alluvial plain sequence generally is about 200–400 m (650–1300 ft) thick and is barren of coal. Locally it is as thick as 760 m (2,500 ft) in the central part of the basin in Colorado, according to Baltz (1965). The contact with the Raton Formation generally is gradational. The Poison Canyon, together with the barren series and upper coal zone of the Raton Formation, constitutes all of the Paleocene rocks of the southern part of the Raton basin. The Cuchara Formation of early Eocene age unconformably overlies the Poison Canyon Formation in the northern part of the Raton basin in Huerfano Park, Colorado (Baltz, 1965).

CRETACEOUS-TERTIARY BOUNDARY

Lee (1917) originally placed the Cretaceous-Tertiary boundary at the base of the basal conglomeratic sandstone of the Raton Formation. Later, on the basis of plant fossils, Brown (1962) suggested that the boundary should be put at least 15 m (50 ft) above the base of the formation. Tschudy (1973) redefined the boundary on the basis of the disappearance of certain fossil pollen taxa, chiefly *Proteacidites, Tilia woodhousei, Trichopelti-*

nites, and, locally, *Tercissus* (Tschudy and others, 1984). This group of taxa has been termed the *Proteacidites* assemblage. In core samples from a hole drilled in 1966 by Kaiser Steel Corporation at a site in York Canyon, Tschudy noted the disappearance of the *Proteacidites* assemblage about 80 m (265 ft) above the base of the Raton, between two thin coal beds about 3 m (10 ft) apart, approximately 250 m (820 ft) below the surface (Pillmore, 1969; Tschudy, 1973). Samples from the d coal contained the *Proteacidites* assemblage; samples from coal c did not (Fig. 2). The rocks between the coals were not sampled.

At the close of 1980, following the discovery of noble-metal anomalies at the K-T boundary in marine rocks in Europe by Alvarez and others (1980), a new location in York Canyon was cored to test for iridium and other trace-element anomalies at the pollen-defined boundary. The original drill site could not be reoccupied, and a new hole was drilled about 150 m (500 ft) to the south. From the new hole, 30 m (100 ft) of core was collected, starting at a depth of 243.8 m (800 ft) (Orth and others, 1981). Detailed examination of the 1980 drill core showed no particularly significant lithologic break in the rocks near the position of the boundary, as was also the case in the 1966 core. Most of the coal beds near the K-T boundary found in the 1966 core appeared to be present in the 1980 core, as shown by the graphic logs of the two cores (Fig. 2). The boundary was known to lie somewhere between the c and d coals in the 1966 core, but the exact location of the boundary at the base of coal c was not established until detailed examination of the palynomorphs in the 1980 core was completed.

As described by Orth and others (1981), core logs were matched as closely as possible (Fig. 2), and samples from a carbonaceous shale and coal sequence thought to contain the boundary in the 1980 core were analyzed for pollen by Tschudy. At the same time, nearly continuous samples between the a and d coals were analyzed for Ir and other trace elements by neutron-activation methods. The *Proteacidites* extinction was recognized between a coal sample at 255.7 m (838.7 ft) and a carbonaceous shale sample at 256.7 m (842.2 ft), only 1 m (3 ft) apart. An Ir peak was observed in this interval by gamma-ray spectrum analysis of the neutron-irradiated samples (Orth and others, 1981). The interval of core containing the boundary was then resampled in increments of about 2.5 cm (1 in) and analyzed for pollen. The pollen break was found between the depths of 255.79 and 255.81 m (839.09 and 839.27 ft). The highest occurrence of the *Proteacidites* assemblage was just below coal bed c, and an Ir anomaly of 5 ng/g (roughly 1000 x background) was measured in sample YC119 from the lower part of this coal, using radiochemical purification analytical techniques (Orth and others, 1981).

After the discovery of the Ir anomaly at the K-T boundary in the drill core, Ir anomalies were measured at the K-T boundary at 14 additional sites along the margin of the basin; the pollen break was used to pinpoint the boundary. A road cut on the old Raton Pass Scenic Road, about 1.6 km (1 mi) west of Raton, was the site of the first discovery of the boundary in a surface exposure of rocks of fluvial origin. At the Raton site, the boundary claystone is separated from an overlying coal bed by 5–15-cm (2–6 in) of detrital clay. At other sites, discovered later in road cuts along highway I-25 near Starkville, Colorado, about 25 km (15 mi) north of Raton, New Mexico, the boundary is at the top of a 1–3-cm (0.5–1-in) bed of kaolinitic claystone beneath a thin coal bed (Pillmore and others, 1984). In physical appearance this claystone resembles other thin kaolinitic beds, commonly referred to as tonsteins, that are present in coal beds throughout the coal-bearing part of the stratigraphic section. The tonsteins are the product of diagenetic alteration of volcanic ash beds in a coal-swamp environment (Williamson, 1970; Bohor and Pillmore, 1976; Bohor and others, 1978). The boundary claystone has been affected by similar diagenetic alteration (Pollastro and others,

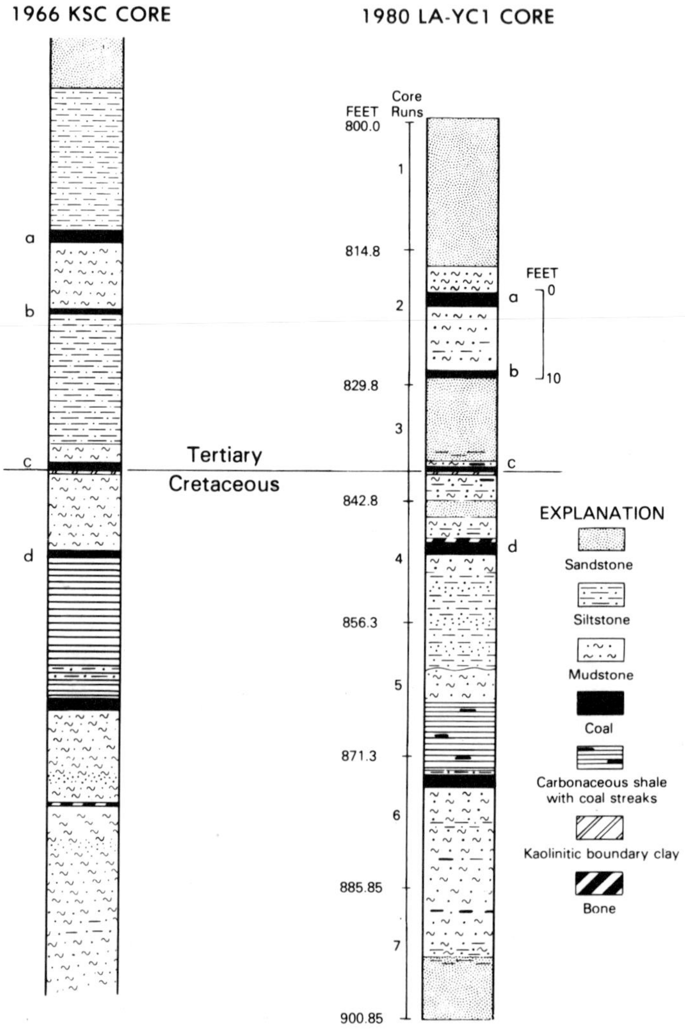

Figure 2. Columnar sections of rocks cored at York Canyon, New Mexico. Core log from hole drilled by Kaiser Steel Corporation in 1966, in which palynological K-T boundary was first discovered, is shown on the left. Core log from hole drilled in 1980-1981 by Los Alamos National Laboratory, in which iridium was first discovered in continental rocks at the palynological K-T boundary, is shown on the right.

1983; Pollastro and Pillmore, in press; Bohor and others, 1984) but can be distinguished from the tonsteins derived from volcanic ash by its trace element composition and, locally, by its texture and the presence of distinctive trace minerals (Pillmore and others, 1984; Pillmore and Flores, this volume).

Discovery of the boundary claystone near Starkville led to recognition of similar high-alumina claystones at all the K-T boundary sites in the Raton basin, including the original York Canyon core. The claystone bed at the base of the coal in the York Canyon core was rechecked and found to be high in aluminum and other elements, including Ti, Cr, Sc, Sb, and V, as well as Ir, that distinctively characterize the boundary claystone observed on the outcrop (Gilmore and others, 1984). The boundary claystone is difficult to differentiate from other layers of dark shale in the York Canyon core because it is dark colored, owing to its high organic content. In outcrop, the boundary claystone weathers to a blocky, light-gray, "boney" bed similar to beds of tonstein, and like most tonsteins, it has little detectable grit.

The boundary claystone has recently been recognized at several sites in a sequence of fluviatile sedimentary rocks in the Hell Creek area, near Jordan, Montana. In the Hell Creek area, which extends over several hundred square kilometers, the boundary is defined by the disappearance of a pollen assemblage characterized by *Proteacidites,* as in the Raton basin region, but also by many species of *Aquilapollenites* (Tschudy, 1970; Pillmore and others, 1984). The extinction occurs at the base of a coal (the "Z" coal) that divides the Tullock Member of the Fort Union Formation from the underlying Hell Creek Formation. A thin bed of kaolinitic claystone, characterized by geochemical anomalies nearly identical to those observed in the Raton basin, lies just beneath the coal, as at most sites in the Raton basin. At the base of the coal, the ratio between angiosperm pollen and fern spores shifts from a dominance of angiosperm pollen to a dominance of fern spores as observed in the Raton basin (Tschudy and others, 1984; Tschudy and Tschudy, 1986; Hotten, 1984).

STRATIGRAPHY OF BEDS CORED IN 1966 AND 1980 AT YORK CANYON

At York Canyon, the Raton Formation is about 640 m (2,100 ft) thick; of this, about 91 m (300 ft) is in the lower coal zone, 131 m (430 ft) in the barren series, and 418 m (1370 ft) in the upper coal zone. York Canyon is approximately on the structural axis of the basin (Fig. 1) where the thickest section of the Raton Formation was deposited. The conglomerate at the base of the lower coal zone occurs as stringers and layers containing pebbles as large as 2 cm (1 in) in diameter in a fining-upward sequence of very coarse to medium-grained sandstone. The first thin coal is 53 m (173 ft) above the base, followed by a 38-m (125-ft) sequence of fine-grained floodplain deposits that contain several coal beds interbedded with shale, mudstone, and sandstone. The K-T boundary occurs in this sequence of fine-grained rocks. Thick sandstones of the barren series first appear about 20 m (65 ft) above the boundary. The base of the barren series is picked above the highest coals of the lower zone where sandstone becomes dominant. The stratigraphic position of the base is inexact. As shown in Figure 2, a sandstone that may represent the base of the barren series occupies the top 4.8 m (16 ft) of the core drilled in 1980, from 244 to 274 m (800 to 900 ft). The hole drilled by Kaiser Steel Corporation at York Canyon in 1966 was begun in rocks of the upper coal zone of the Raton Formation about 340 m (1,115 ft) above the base. The hole was continued below the Raton Formation through 97 m (318 ft) of Vermejo Formation into the Trinidad Sandstone. Unfortunately, the core from the hole drilled by Kaiser Steel Corporation in 1966 was not boxed and was left to be scattered on the ground, except for coal and carbonaceous shale samples that were collected for pollen analysis and a few samples of clay and mudstone collected for geochemical analysis. Hence, it was necessary to drill a new hole for detailed investigation of the K-T boundary at this site.

The core drilled in 1980 by Los Alamos National Laboratory was taken entirely within the lower part of the Raton Formation. The stratigraphic section describing the rocks of the lower coal zone that were core drilled December 29, 1980, to January 1, 1981, is included in Pillmore and Flores (this volume). The cored interval lies mainly beneath the thick sandstone beds of the barren series and above sandstone beds that constitute the lower 50 m (165 ft) of the lower coal zone. The beds of fine-grained material in this interval appear to represent a history of nearly continuous deposition on floodplains and in back swamps that was episodically interrupted by crevasse splays that deposited tabular beds of sandstone.

The pollen-defined K-T boundary is 81 m (265 ft) above the base of the Raton Formation at the York Canyon drill site, based on the combined results from the 1966 and 1980 drill cores, and it lies 175 m (574 ft) above the top of the Trinidad Sandstone.

SAMPLING OF YORK CANYON CORE FOR PALEOMAGNETIC MEASUREMENTS

The core obtained from the 1980 drill hole is 7.6 cm (3 in) in diameter and was taken with a 4.57-m (15-ft) core barrel. The entire 30 m (100 ft) of core was pulled in seven core runs (Fig. 2), each of which yielded a continuous stick of core. Despite the presence of fractured claystone and coal beds, the driller, who is highly skilled and has had very extensive experience drilling these beds for coal exploration, managed to recover a complete core. The core was captured in a 7.6-cm (3-in) diameter steel split barrel that was taped closed during the coring process. After each core barrel was pulled, the tape was cut and the barrel opened, exposing the core. At this point, the core was cleaned and a continuous line marked on the core to insure that the exact alignment of the core could be recovered in case of accidental disturbance. The core was then placed carefully in 1.52-m (5-ft) long split plastic tubes and has remained stored in these tubes at Los Alamos National Laboratory. In order to assure the integrity of the core at various stages of sampling, each 1.5-m (5-ft) core segment was marked with chalk lines of two different colors along its length by Dr. C. J. Orth, of Los Alamos National

Laboratory, who has closely supervised or carried out each stage of the sampling.

As a result of Orth's precautions, it was possible to establish with confidence the relative azimuthal orientation of most of the pieces of core within each core run. With few exceptions, orientation had been lost only in the cases of small fragments, chiefly claystone and coal, that did not carry one or the other of the chalk lines. In general, the gaps in the core were small and were the result chiefly of the samples taken for a previous paleomagnetic study by Payne and others (1982, 1983a, 1983b). Before taking any further samples, all pieces of core in each tube were lined up and fitted together, and the consistency of the chalk lines was checked. In addition, the alignment of continuous grooves left in the surface of the core by the core catcher was checked. Next, the ends of the cores from each tube from a given core run were fitted together. This established the orientation of the chalk lines from one tube to the next; a new line was then drawn for the entire core run, which served as a reference line for the azimuthal orientation of the samples taken for this study.

An attempt was made to fit the ends of cores between core runs. Some of these ends could be fitted, but others could not. We decided to treat each core run as a separate problem in estimating the full orientation of the core from the paleomagnetic measurements. One sample from the lower end of core run 1 was aligned by fitting it with the top of core run 2, because it was felt that a more secure relative orientation was attained thereby. This sample has subsequently been grouped with core run 2 to which it was fitted.

Within core runs, the relative orientation of all samples taken is secure except for small possible rotational errors in core runs 1, 3, and 5. Owing to some inexactness in fitting and to possible rotation across soft beds, there may be relative rotational errors of as much as 10° between samples from different tubes in these runs. As will be seen, these errors are smaller than the likely errors of estimating the full orientation of the core from each run from the paleomagnetic results.

An initial study of 21 samples was made to evaluate demagnetization characteristics and to test the suitability of various rock types for paleomagnetic measurements. Coal was found to be very weakly magnetized and to give unreliable paleomagnetic results. Somewhat to our surprise, we found that the paleomagnetic results from sandstone were indistinguishable from the results from claystone and siltstone. This may be because the sandstones in the York Canyon core are poorly sorted; most have a significant component of clay, which probably contains much of the principal magnetic phase that carries the observed remanent magnetization. Similar results were obtained for the various rock types from a core across the Cretaceous-Tertiary boundary from a hole drilled near the city of Raton. Inasmuch as it was necessary to acquire a sufficient number of samples to derive a statistically useful orientation of the core from each core run, we elected to sample sandstones, where necessary, as well as fine-grained clastic rocks.

A total of 92 pieces of core, about 5 cm (2 in) in length, were taken for paleomagnetic investigation. The pieces were taken at relatively uniform intervals, on the order of 30 to 35 cm (12 to 13 in), with preference given to fine-grained rocks. On average, 13 pieces were taken from each core run. The minimum number was 12 pieces, taken from core run 4.

At the U.S. Geological Survey paleomagnetics laboratory at Flagstaff, Arizona, each piece of 7.6-cm (3-in) diameter core was oriented vertically in a jig, and three 2.5-cm (1-in) diameter plugs were cut parallel with the axis of the original core with a diamond bit in a drill press. The azimuthal orientation of each plug sample relative to the reference line for the respective core run was marked and preserved. The plugs were then trimmed to 2.5-cm (1-in) lengths for measurement. We estimate that the cumulative orientational errors of the plug samples probably do not exceed 15°, except in rare cases.

LABORATORY PROCEDURES

Measurements of the remanent magnetization of the 2.5-cm (1-in) diameter plug samples were made with an SCT superconducting magnetometer at the Flagstaff paleomagnetics laboratory. Inasmuch as secondary components of magnetization were only partially removed by AF demagnetization methods from fully oriented surface samples, we tried thermal demagnetization on the initial set of 21 plug samples. In this set, thermal demagnetization at 300°C (572°F) appeared to remove a major fraction of the strong secondary normal polarity component of magnetization from nearly all samples. Accordingly, two plug samples from each core interval were selected for thermal demagnetization successively at 225°C (437°F) and 300°C (572°F). The third plug sample from each interval was progressively AF demagnetized at steps of 10, 20, and 40 millitesla (mT).

Thermal demagnetization was done in a specially designed oven enclosed in a large μ-metal can that is surrounded, in turn, by Ruebens coils tuned to cancel the Earth's field. Cooling of the samples takes place in a second μ-metal can that is slid into place in contact with the can around the oven at the time of transfer of the samples from the oven. AF demagnetization was done with a shielded two-axis Schonstedt demagnetizer that is operated in a space enclosed by Ruebens coils next to the superconducting magnetometer. All samples were maintained in a low field region throughout the demagnetization procedure.

INTERPRETATION OF MAGNETIC MEASUREMENTS

All rocks measured from the York Canyon core were found to be relatively weakly magnetized. Intensity of the natural remanent magnetization (NRM) ranges from 1.3×10^{-5} to 2.3×10^{-3} Am^{-1} (1.3×10^{-8} to 2.3×10^{-6} gauss). The NRM intensity was found to be about equally strong, on average, for claystones, siltstones, and sandstones. The mean intensity of the remaining magnetization drops to 57 percent of the NRM intensity after demagnetization at 225°C and to 55 percent after demagnetiza-

TABLE 1. MEAN DIRECTIONS OF REMANENT MAGNETIZATION AND FISHER STATISTICS FOR CORE RUNS 1 THROUGH 7.*

	Declination of mean	Inclination of mean	Kappa	α95
Core Run 1				
NRM	136.1°	+65.8°	1.92	42.5°
225°C	140.6°	-13.7°	1.68	49.1°
300°C	set to 163.9°	-10.8°	2.15	38.1°
Core Run 2				
NRM	27.2°	+65.9°	3.27	27.3°
225°C	67.7°	+35.1°	1.34	68.5°
300°C	set to 163.9°	+ 1.5°	1.31	71.2°
Core Run 3				
NRM	315.7°	+69.6°	2.75	31.0°
225°C	222.7°	+50.3°	1.16	83.0°
300°C	set to 163.9°	+20.4°	1.29	73.3°
Core Run 4				
NRM	69.4°	+80.3°	7.81	16.6°
225°C	172.1°	+22.9°	2.39	36.4°
300°C	set to 163.9°	+ 4.9°	2.54	34.6°
Core Run 5				
NRM	0.1°	+75.7°	6.29	17.2°
225°C	106.0°	+52.0°	1.71	46.3°
300°C	set to 163.9°	-5.1°	1.73	49.7°
Core Run 6				
NRM	281.6°	+74.7°	6.10	18.3°
225°C	189.3°	+26.1°	1.56	54.1°
300°C	set to 163.9°	-0.5°	2.37	35.0°
Core Run 7				
NRM	27.6°	+47.4°	2.64	30.7°
225°C	78.5°	-30.9°	1.40	60.9°
300°C	set to 163.9°	-52.0°	1.82	43.1°

*Azimuthal orientation of each core run is based on setting the declination of the mean direction of the remanence after demagnetization at 300°C to 163.9°.

tion at 300°C. With AF demagnetization, the mean intensity drops to 43, 25, and 16 percent of the NRM intensity at peak AF demagnetization fields of 10, 20, and 40 mT, respectively. In many specimens, the dipole moment after demagnetization at 300°C or 40 mT peak alternating field approached the threshold of reliable measurement with the superconducting magnetometer. In a few cases, the magnetization of the 2.5-cm (1-in) plug specimens became too weak to measure at the highest demagnetization steps used (300°C or 40 mT).

Thermal Demagnetization

Our principle objective in this study is to evaluate the polarity of the characteristic magnetization of the rocks as a function of their stratigraphic position. To this end, the shift in directions of magnetization, as the less stable components of magnetization are removed with progressive demagnetization steps, will be examined for each core run. The final directions of magnetization are interpreted as closest to, but not necessarily identical with, the characteristic or primary direction of magnetization.

The inclination of the NRM of all but a few percent of the samples measured is positive. The average NRM direction for each core run is fairly steeply down; mean inclinations, obtained by vector summation of the NRM directions for each core run, range from +47° to +80° (Table 1). In every core run, the mean inclination decreased after demagnetization at 225°C and in two cases became negative (Table 1). At 300°C, the mean inclination is negative for four core runs, and the directions are shallow to nearly horizontal in the remaining core runs (Table 1). It is clear from these results that the NRM of the majority of the samples is dominated by a steep downward component of magnetization, which is partially or largely removed by thermal demagnetization. In nearly all samples, the magnetization remaining after heating to 300°C has a direction much different from the NRM direction, and it lies in the upper hemisphere in the majority of cases.

For purposes of illustration and of combining the results of all the core runs, we have utilized the mean direction of magnetization at the 300°C demagnetization step to estimate the azimuthal orientation of each core run. The characteristic direction of magnetization of the rocks from each core run corresponds to reversed polarity, as will be shown, and we have aligned the declination of the mean direction at the 300°C demagnetization step obtained for each core run with an estimated declination for the mean reversed polarity direction of the geomagnetic field in earliest Tertiary time. This estimated reversed polarity declination, 163.9°, is derived from the observed directions of both normally and reversely magnetized rocks reported by Butler and Taylor (1978) from the Nacimiento Formation of Paleocene age in the San Juan Basin, New Mexico. Although most or all of the Nacimiento studied by Butler and Taylor (1978) is probably somewhat younger than the rocks in the York Canyon core, the reversed polarity declination obtained from their study probably represents the best available estimate of the declination of the average reversed polarity geomagnetic field at the beginning of the Paleocene in the Raton basin. Correction for the difference in longitude of the two basins is negligible. The confidence with which the azimuthal orientation of the core can be estimated will be discussed for each core run.

Core Run 1. Thirteen segments of core distributed in depth from 244.1 to 247.7 m (800.7 to 812.8 ft) were taken from core run 1. As may be seen from Figure 3, the NRM directions from core run 1 are exceptionally scattered. Directions measured from the two specimens from each horizon have been averaged and are shown as a single point in Figure 3. Most directions are in the lower hemisphere, but two are in the upper hemisphere. At 225°C, inclinations at eight horizons become negative, and the mean inclination for the core run becomes shallow and negative. At 300°C, the mean direction remains in the upper hemisphere, at a shallow inclination, but shifts slightly in declination. Directions from five horizons remain in the lower hemisphere but are more tightly clustered around the mean direction than at 225°C.

With the declination of the mean direction for the core run at 300°C aligned with the estimated early Paleocene reversed polarity declination, the mean NRM direction lies close to the path between the present axial dipole field direction (PADFD), at an inclination of 56°, and the Paleocene reversed polarity direction. This suggests that the NRM of most samples is a composite of a normal polarity component of magnetization, probably a viscous component acquired during the Brunhes normal polarity epoch, and a reversed polarity component of magnetization.

The presence of a secondary normal polarity component of magnetization is verified by examination of the direction of the components removed during demagnetization (Fig. 4). The components of magnetization removed at 225°C are clustered in two directions: At 10 horizons the components removed point steeply down, and at three horizons the components removed point steeply up. A similar result is found for the total components removed at 300°C. The means of the components removed at 225°C and 300°C are dominated by the downward cluster and lie close to the PADFD (Fig. 4; Table 2).

In cases where the NRM points downward, all directions of remaining magnetization shifted away from the PADFD during thermal demagnetization. Those directions that remain downward at 300°C have southerly declination and are shifted strongly toward the mean of the reversed polarity directions. Hence, there can be little doubt that all horizons sampled in core run 1 have reversed polarity characteristic magnetization. It is also clear, however, that thermal demagnetization at 300°C failed to remove all of the secondary components of normal polarity magnetization that dominate the NRM of most specimens. Without information about the relative azimuthal orientation of the samples in the core run, the polarity of the characteristic magnetization would be ill determined at several horizons.

Confidence limits for the azimuthal orientation adopted for each core run are given by the half angle of the 95 percent cone of confidence of the mean direction at 300°C, divided by the cosine of the inclination of the mean (Table 1). The 95 percent confidence limits for the orientation of core run 1 are ±38.8° and the estimated error, expressed as a standard deviation, is about ±19°.

Core Run 2. NRM directions for 13 horizons (depths from 248.1 to 252.8 m—814.0 to 829.3 ft) in core run 2 are more tightly clustered than was the case in core run 1 (Fig. 3). Using the mean direction of magnetization at 300°C for orienting the core, one horizon has an NRM direction with southerly declination and very small negative inclination, whereas most of the other NRM directions are fairly well clustered around the PADFD. At 225°C, the mean direction becomes almost horizontal. Directions of magnetization become badly scattered at 225°C but are clustered somewhat better at 300°C. The directions of the remaining magnetization shift strongly away from the present dipole field direction during thermal demagnetization. This shift is also away from the early Paleocene normal polarity direction and, in most cases, fairly clearly toward a reversed polarity direction. We conclude that the characteristic direction of magnetization at all horizons probably is reversed polarity. From the 95 percent confidence cones of the mean direction at 300°C (Table 1), the error of orientation of the core run is estimated to be about ±36° (1σ). With one exception, the components of magnetization removed at 225°C and 300°C (Fig. 4) show fairly good clustering around the PADFD, which confirms the approximate solution for the orientation of the core run.

Core Run 3. NRM directions from 13 horizons (depths from 252.9 to 256.6 m—829.8 to 841.9 ft) are fairly well clustered in core run 3, except for one direction in the upper hemisphere and one that is nearly horizontal (Fig. 3). At 225°C and 300°C, seven directions are in the upper hemisphere and six in the lower hemisphere. The mean NRM direction is close to but steeper than the PADFD, whereas the mean direction at 300°C is shallower but still down. Again, the general shift of directions with demagnetization is away from both the PADFD and the mean Paleocene normal polarity direction and toward a reversed polarity direction. On average, a larger secondary normal polarity component remains in these samples at 300°C than is found in the other core runs. The direction at one horizon, at 256.3 m (840.8 ft), remains close to but steeper than the PADFD at 300°C. AF demagnetization to 40 mT of the third specimen from this horizon, however, successfully removed most of the normal polarity component. The directions of the components removed by thermal demagnetization are rather scattered (Fig. 4), but the 95 percent confidence cones of the mean directions of these components at the 225°C and 300°C steps include the PADFD. Hence, the orientation of the core run, on the interpretation that the characteristic magnetization of all horizons is reversed polarity, is confirmed by the mean direction of the components removed. The error of azimuthal orientation is about ±39° (1σ).

Core Run 4. In this run, NRM directions from all 12 horizons (depths from 256.9 to 260.9 m—843.0 to 856.0 ft) are clustered in the lower hemisphere. The mean NRM direction is nearly straight down (Fig. 3). At the 225°C step the mean direction is southerly, shallow, and down, and at 300°C the mean is still down but nearly horizontal. The magnetization of the plug samples from one horizon became too weak to measure at the

TABLE 2. MEAN DIRECTIONS AND FISHER STATISTICS FOR COMPONENTS OF MAGNETIZATION REMOVED BY DEMAGNETIZATION AT 225°C AND 300°C FOR CORE RUNS 1 THROUGH 7*

	Declination of mean	Inclination of mean	Kappa	α95
Core Run 1				
NRM - 225°C	11.1°	+70.8°	1.95	41.9°
NRM - 300°C	31.4°	+59.5°	1.82	45.0°
Core Run 2				
NRM - 225°C	22.2°	+65.0°	4.05	23.6°
NRM - 300°C	23.9°	+57.6°	4.52	22.0°
Core Run 3				
NRM - 225°C	318.7°	+58.9°	2.83	30.4°
NRM - 300°C	316.1°	+57.2°	2.89	29.9°
Core Run 4				
NRM - 225°C	13.6°	+70.5°	10.11	14.4°
NRM - 300°C	2.7°	+63.9°	7.29	17.3°
Core Run 5				
NRM - 225°C	324.4°	+70.8°	9.15	13.9°
NRM - 300°C	354.8°	+64.6°	15.63	10.4°
Core Run 6				
NRM - 225°C	332.2°	+65.6°	5.25	20.0°
NRM - 300°C	321.4°	+62.9°	6.06	18.4°
Core Run 7				
NRM - 225°C	2.3°	+55.8°	3.11	27.2°
NRM - 300°C	31.4°	+48.8°	3.64	24.3°

*Azimuthal orientation of each core run is based on setting the declination of the mean direction of the remanence after demagnetization at 300°C to 163.9°.

TABLE 3. MEAN DIRECTIONS AND FISHER STATISTICS FOR COMPONENTS OF MAGNETIZATION REMOVED BY THERMAL AND BY AF DEMAGNETIZATION FOR ALL SAMPLES FROM THE YORK CANYON CORE*

	Declination of mean	Inclination of mean	Kappa	α95
Thermal demagnetization				
NRM - 225°C	350.5°	+67.4°	3.85	8.7°
225°C - 300°C	19.7°	+36.4°	1.80	16.4°
NRM - 300°C	1.4°	+62.7°	3.83	8.8°
AF demagnetization				
NRM - 10 mT	24.5°	+67.7°	1.97	15.1°
10 mT - 20 mT	286.3°	+71.7°	2.15	14.0°
20 mT - 40 mT	24.1°	+72.5°	1.67	18.2°

*Azimuthal orientation of each core run is based on setting the declination of the mean direction of the remanence after demagnetization at 300°C to 163.9°.

300°C step. At the other horizons, directions of magnetization are fairly well clustered at 300°C, with one exception, and it is clear that the interval sampled has reversed polarity characteristic magnetization. Components of magnetization removed at 225°C and 300°C are fairly well clustered, with means close to the PADFD (Fig. 4). Error of the azimuthal orientation of the core run is about ±17° (1σ).

Core Run 5. NRM directions from 12 horizons (depths from 261.2 to 265.2 m—857.0 to 870.1 ft) are well clustered in the lower hemisphere about a mean somewhat steeper than the PADFD (Fig. 4). At 225°C the directions scatter, and the mean direction shifts to the southeast; at 300°C most of the directions are clustered about a mean that is in the upper hemisphere but nearly horizontal. The components of magnetization removed at 225°C and 300°C are tightly clustered around mean directions close to the PADFD (Fig. 4). This confirms that the NRM is dominated by a secondary component of magnetization of recent origin and also provides a good confirmation of the approximate azimuthal orientation adopted for a core run. The estimated error of orientation of the core run is ±25° (1σ). The shift of directions with the progressive thermal demagnetization to the upper hemisphere or to the southeast shows that the characteristic magnetization in this core run is reversed.

Core Run 6. NRM directions from 13 horizons (depths from 265.6 to 270.0 m—871.3 to 885.8 ft) are somewhat more scattered in core run 6 than in core run 5 (Fig. 3). The mean NRM direction is steeply down. Mean directions of the remanence remaining after 225°C and 300°C thermal demagnetization shift south and then east. After 300°C demagnetization, the mean direction is nearly horizontal. Directions for all but one horizon, at 269.3 m (883.4 ft) shift to the southeast with progressive demagnetization, about half to the upper hemisphere. In the one exception, the inclination becomes shallow at 300°C; AF demagnetization to 40 mT of the third specimen from this horizon produced a strong shift of direction to the southeast, consistent with the demagnetization behavior at other horizons. Components of magnetization removed at 225°C and 300°C, from all horizons in core run 6, are clustered around a mean fairly close to the PADFD (Fig. 4), as expected. The characteristic magnetization over the interval sampled in this core run is clearly reversed polarity; error of orientation of this core run is about ±18° (1σ).

Core Run 7. Remanent magnetization at 14 horizons sampled (depths from 270.2 to 274.3 m—886.5 to 900.0 ft) from this core run generally is not as strongly dominated by the secondary normal polarity component as is the magnetization at most higher stratigraphic levels. Half of the NRM directions from core run 7 have shallow inclination and two are in the upper hemisphere (Fig. 3). The mean NRM direction is close to, but somewhat shallower than, the PADFD. Demagnetization at 225°C removed secondary components with a mean nearly coincident with the direction of the PADFD (Fig. 4); directions of the remaining remanence at nine horizons are in the upper hemisphere. At the 300°C step, the mean direction of magnetization is fairly steeply negative with an inclination close to that expected for the reversed polarity latest Cretaceous to early Paleocene field (inclination = 51°, Butler and Taylor, 1978). One direction that remained down to the southeast, found at 272.9 m (895.4 ft, was shifted only slightly by thermal demagnetization; the magnetization of the third specimen from this horizon was shifted to a shallower inclination, away from the Paleocene normal polarity field direction by AF demagnetization. Although the directions at individual horizons are broadly scattered after 300°C demagnetization, there can be little doubt that the characteristic magnetization for this interval is reversed. Because of the fairly large scatter of directions at 300°C and the steep inclination of the mean direction, the orientation of this core run has an estimated error of ±35° (1σ).

Summary of Thermal Demagnetization Results. As described above, the characteristic magnetization of the rocks from each core run was found to be reversed polarity on the basis of the shift of the directions of magnetization with progressive thermal demagnetization. A strong, completely independent check of the derived azimuthal orientation of each core run is provided by the directions of the components removed at the 225°C step of thermal demagnetization. For each core run, it was found that the mean direction of the component removed at 225°C lies close to the PADFD.

The approximate alignment of the most labile component of magnetization with the average direction of the recent geomagnetic field demonstrates that much of this component has been acquired at a time or times of normal polarity, very probably during the Brunhes normal polarity epoch. As the rocks studied were deeply buried and unweathered, a major part of this recently acquired component is almost certainly a viscous magnetization. The mean directions of the components removed by AF demagnetization of the alternate set of specimens from the York Canyon core, at the 10 and 40 mT steps, are also approximately aligned with the PADFD (Table 3). The mean of the softest component, that removed at 10 mT, is closely aligned with the

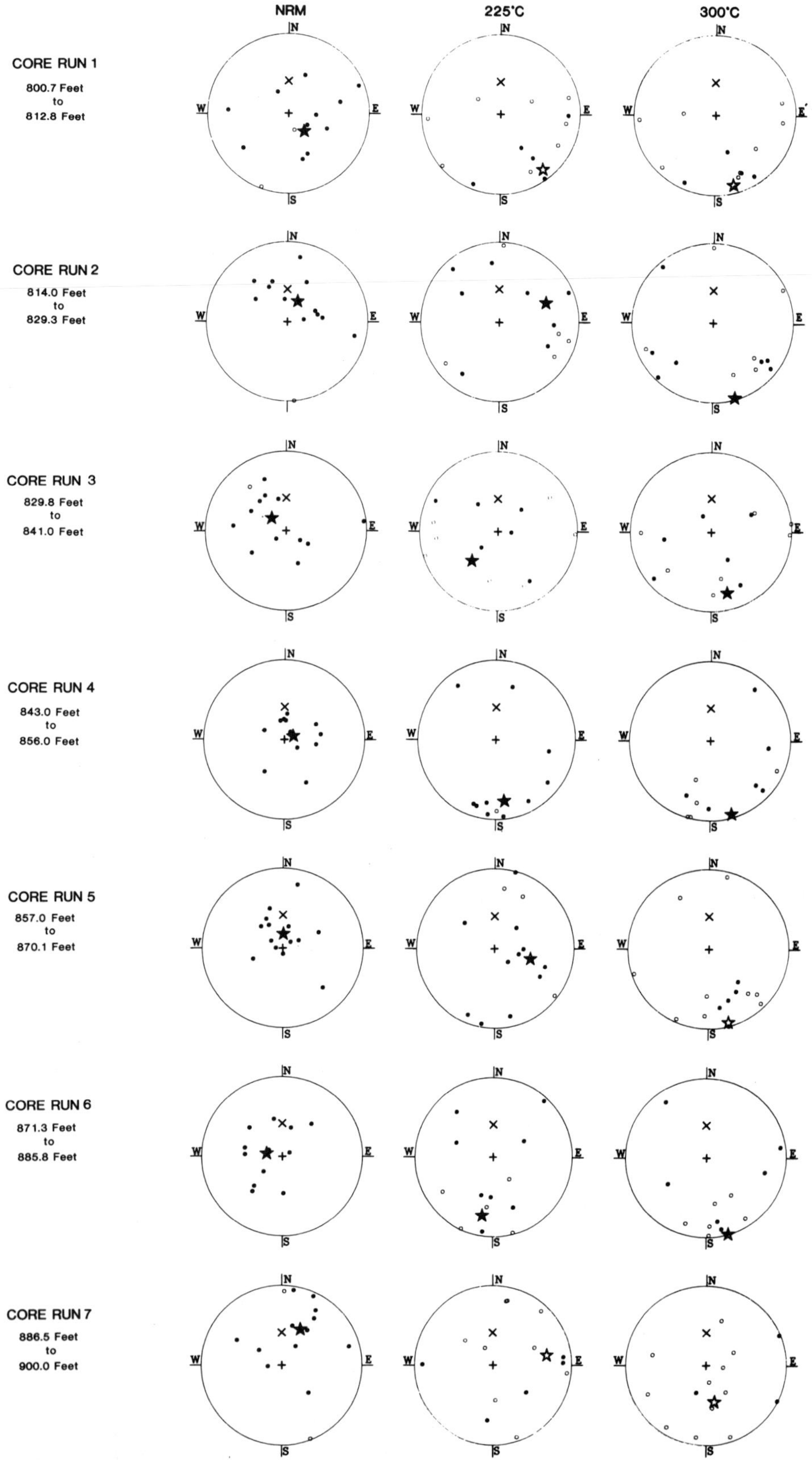

Figure 3. Stereographic equal-area projections showing directions of natural remanent magnetization (NRM) and the directions of remanent magnetization after demagnetization at 225°C (437°F) and 300°C (572°F) for samples from core runs 1 through 7 of the York Canyon core. Each point is the vector mean of the directions measured from two 2.5-cm (1-in) plugs cut from each original sample. Solid dots are in the lower hemisphere (positive inclination), and open circles are in the upper hemisphere (negative inclination). Stars (solid in lower hemisphere, open in upper hemisphere) are the mean directions of magnetization for each core run. The azimuthal orientation shown for each core run is based on setting the declination of the mean direction at the 300°C demagnetization step to 163.9° (earliest Tertiary mean reversed polarity declination; Butler and Taylor, 1978). The present axial dipole field direction at York Canyon is shown with an X.

present field direction, which supports the interpretation that it is dominantly a viscous magnetization.

The accuracy of azimuthal orientation of the individual core runs, based on the mean directions at the 300°C demagnetization step, can be assessed from the mean directions of the components removed at 225°C. Deviations from north of the declination of the mean component removed at 225°C range from 2° in core run 7 to 41° in core run 3 (Table 4). The large deviation found in core run 3 is comparable to the 1σ error of orientation for that core run. The rms deviation from north for all seven core runs is 25.5°, which is slightly smaller than the 28.3° rms error of orientation based on magnetization remaining at 300°C. It can be seen that the azimuthal orientation of individual core runs could be determined almost as well by setting the mean declinations of the secondary components of magnetization removed at 225°C to north, as by the method we used. The rms of the 1σ errors of orientation for the core runs based solely on the secondary components of magnetization removed at 225°C would be 33°, as compared with 28° for the method we adopted (Table 4). The core runs could also have been oriented using the total component removed by demagnetization at 300°C. The rms error of orientation by this method would be 25° (Table 4), but inspection of the components removed between 225°C and 300°C indicates that a significant reversed polarity component is removed in many samples. We conclude that each core run has been correctly oriented to a precision ranging between 17° and 39°, as estimated in the preceding sections. If the results from all seven core runs are combined, the increment to the error in the declination of the mean directions of magnetization introduced by misalignment of the individual core runs should not exceed about 4° (1σ).

The combined results from all seven core runs are illustrated in Figure 5. The grand mean NRM direction lies approximately between the PADFD and the vertically down direction. Although the NRM is clearly dominated by the normal polarity secondary component of magnetization, the displacement of the mean from the PADFD is significant. This displacement may be the consequence either of superposition of the strong components of normal polarity viscous magnetization on weaker components of reversed polarity characteristic magnetization or of the pres-

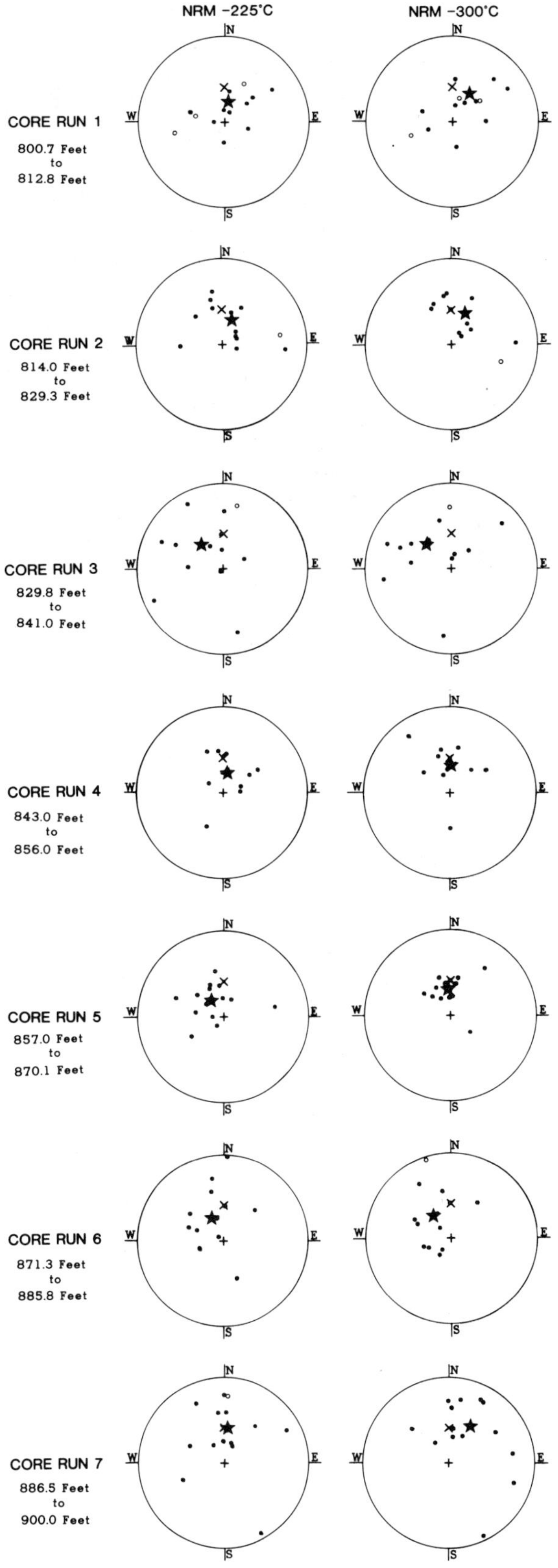

Figure 4. Stereographic equal-area projections showing the directions of the magnetization removed by demagnetization to 225°C (437°F) and 300°C (572°F) for samples from core runs 1 through 7 of the York Canyon core. The directions of magnetization removed from each specimen are found by vector subtraction of the remanence remaining after the 225°C and 300°C demagnetization steps from the NRM. Each point represents the vector sum of the results for two 2.5-cm (1-in) plugs cut from each sample. Solid dots are in the lower hemisphere (positive inclination), and open circles are in the upper hemisphere (negative inclination). The azimuthal orientation shown for each core run is based on setting the declination of the mean direction of remanence at the 300°C demagnetization step to 163.9°, as in Figure 3. Stars (all in the lower hemisphere) are the mean directions of the components of magnetization removed for each core run. The present axial dipole field direction at York Canyon is shown with an X.

ence of components of isothermal remanent magnetization acquired during drilling. As the secondary component or components are progressively removed by thermal demagnetization, the mean direction shifts approximately along a great circle path toward the latest Cretaceous–early Paleocene reversed polarity direction (Fig. 5). After 300°C demagnetization, the mean direction of the remaining remanence is in the upper hemisphere, but there is still a substantial secondary component of magnetization in most specimens. Even though there is a large scatter in the directions, it is noteworthy that no specimens have a direction after 300°C demagnetization that is very close to the latest Cretaceous–early Paleocene normal polarity direction. The region around the early Paleocene normal polarity direction is the emptiest part of the equal-area diagram (Fig. 5). In specimens with NRM directions close to the early Paleocene normal polarity direction, the direction of magnetization shifts away with progressive thermal demagnetization. There is no evidence, at any of the 92 horizons sampled, of a normal polarity component of primary magnetization.

The grand average directions of the components removed by demagnetization are listed in Table 3. Deviation from north of the mean components removed by thermal demagnetization is -9.5 ±11.3° at 225°C and +1.4±9.6° at 300°C.

AF Demagnetization

AF demagnetization was carried out on one specimen from each horizon sampled, as a check on the results obtained from thermal demagnetization. Azimuthal orientations of the core runs based on the directions of the remanence after 300°C thermal demagnetization have been adopted for analysis of the AF demagnetization results.

The mean NRM direction for the set of specimens subjected to AF demagnetization is essentially identical with the mean obtained for the two sets of specimens subjected to thermal demagnetization (Fig. 6). A larger dispersion in the NRM directions is seen in Figure 7, owing to the fact that each point represents the measured direction of a single specimen rather than the average of two specimens as shown in Figure 5. Demagnetization at 10 mT and 20 mT peak AF increases the dispersion of directions of magnetization and shifts the mean direction moderately. At 40 mT the majority of directions shift to the upper hemisphere, and the inclination of the mean is -35°. The 95 percent cone of confidence of the mean at the 40 mT demagnetization step includes the estimated early Paleocene reversed polarity direction. Although individual directions of magnetization are widely scattered at 40 mT, none lie close to the PADFD or the early Paleocene normal polarity direction (Fig. 6). As found in thermal demagnetization, there is a general shift of directions of the remanence during AF demagnetization away from a normal polarity and toward a reversed polarity direction. There is no evidence of normal polarity characteristic magnetization at any horizon. We conclude that the entire 30-m (100-ft) stratigraphic section cut by the core taken in 1980 has a reversed polarity characteristic magnetization.

TABLE 4. DEVIATIONS FROM NORTH OF THE MEAN DIRECTIONS OF COMPONENTS OF MAGNETIZATION REMOVED BY THERMAL DEMAGNETIZATION FOR CORE RUNS 1 THROUGH 7

Components of magnetization removed by demagnetization to 225°C

Core Run	Declination of mean	Deviation from North	$\sigma_{300°}$*	$\sigma_{NRM-225°}$**
1	11.1°	+11.1°	19.4°	63.9°
2	22.1°	+22.1°	35.6°	27.9°
3	318.7°	-41.3°	39.1°	29.4°
4	13.6°	+13.6°	17.4°	21.6°
5	324.4°	-35.6°	25.0°	21.3°
6	332.2°	-27.8°	17.5°	24.2°
7	2.3°	+ 2.3°	34.9°	24.2°
rms error		25.5°	28.3°	33.4°

Components of magnetization removed by demagnetization to 300°C

Core Run	Declination of mean	Deviation from North	$\sigma_{300°}$*	$\sigma_{NRM-300°}$***
1	31.4°	+31.4°	19.4°	44.3°
2	23.9°	+23.9°	35.6°	20.5°
3	316.1°	-43.9°	39.1°	27.7°
4	2.7°	+2.7°	17.4°	19.8°
5	354.9°	-5.1°	25.0°	12.1°
6	311.4°	-38.6°	17.5°	20.2°
7	31.4°	+31.4°	34.9°	18.5°
rms error		29.3°	28.3°	25.2°

Note: Azimuthal orientation of each core run is based on setting the declination of the mean direction of the remanence after demagnetization at 300°C to 163.9°.

*$\sigma_{300°}$ is estimated 1σ error of azimuthal orientation adopted for each core run. The error is estimated from

$$\sigma_{300} = \frac{\alpha_{95} \text{ (for mean direction of remanence at 300°C)}}{2 \cos I \text{ (for mean direction of remenance at 300°C)}}$$

*$\sigma_{NRM-225°}$ is 1σ error of azimuthal orientation if the core were oriented by setting the declination of the mean component of magnetization removed at 225°C to north. Error is estimated from

$$\sigma_{NRM-225°} = \frac{\alpha_{95} \text{ (for mean component removed at 225°C)}}{2 \cos I \text{ (for mean component removed at 225°C)}}$$

***$\sigma_{NRM-300°}$ is 1σ error of azimuthal orientation if the core were oriented by setting the declination of the mean component of magnetization removed at 300°C to north. Error is estimated from

$$\sigma_{NRM-300°} = \frac{\alpha_{95} \text{ (from mean component removed at 300°C)}}{2 \cos I \text{ (for mean component removed at 300°C)}}$$

For purposes of comparison with the results of Payne and others (1983b), the inclinations measured at the 300°C and 40 mT demagnetization steps are plotted as a function of stratigraphic position in Figure 8. Average inclination for two specimens from each horizon, obtained by vector summation of the measured directions, is shown for the thermally demagnetized rocks, whereas only one measurement of inclination was obtained from each horizon for the AF demagnetized specimens. Hence, there is

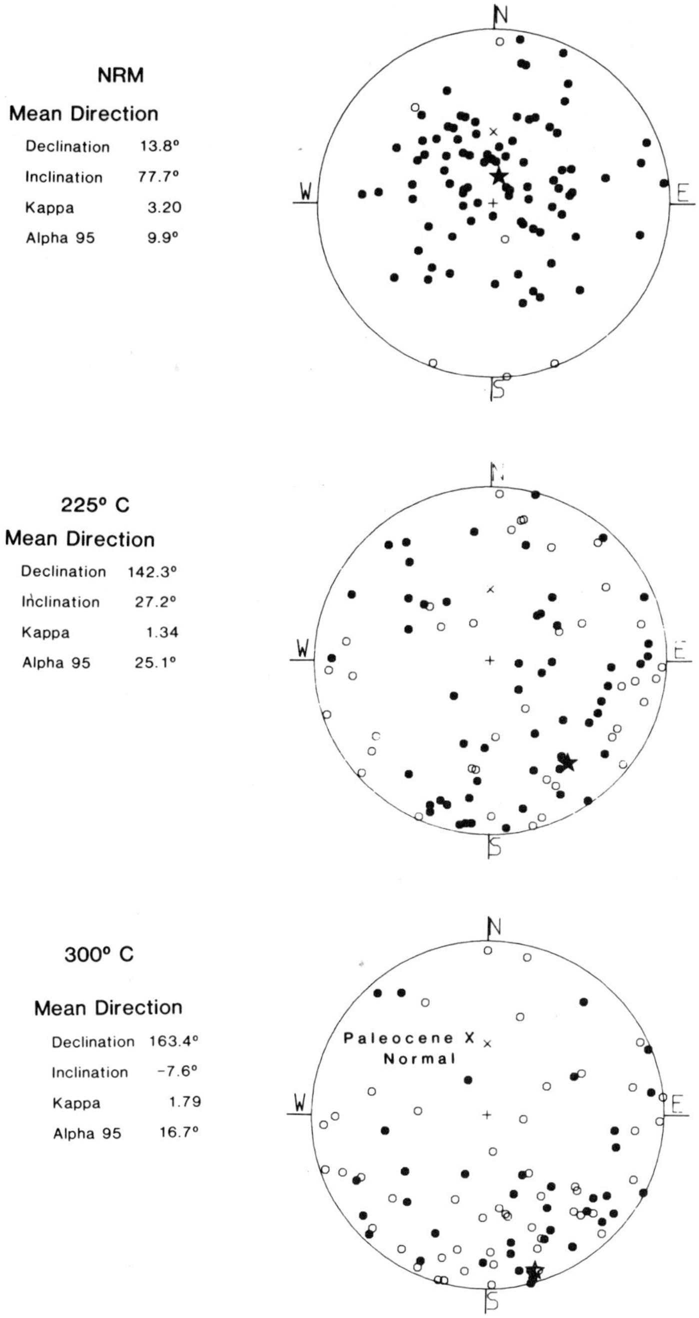

Figure 5. Stereographic equal-area projections showing directions of natural remanent magnetization (NRM) and the directions of magnetization after demagnetization at 225°C (437°F) and 300°C (572°F) for all samples taken from the York Canyon core. Each point is the vector mean of the directions measured from two 2.5-cm (1-in) plugs cut from each original sample. Solid dots are in the lower hemisphere (positive inclination), and open circles are in the upper hemisphere (negative inclination). The azimuthal orientation adopted for each core run is the same as that shown in Figure 3. Stars (solid in the lower hemisphere, open in the upper hemisphere) are the mean directions of magnetization for all samples taken from the core. The present axial dipole field direction at York Canyon is shown with an X.

a larger dispersion in inclination for the AF demagnetized rocks. It can be seen from Figure 8 that the observed inclinations vary in a highly irregular way from the base to the top of the section. For rocks in which the directions of the most stable component of magnetization are as badly scattered as observed in the York Canyon core, the polarity of the characteristic magnetization cannot be determined from observations of the inclination alone.

The observed variations in inclination from one horizon to the next do not correspond to magnetostratigraphic polarity structure. Indeed, there is only a weak correlation between the inclination measured after 300°C thermal demagnetization and the inclination measured after 40 mT demagnetization. There is no correlation between the inclinations shown on Figure 8 and those reported from a different set of samples by Payne and others (1983b). The observed variation in inclination simply reflects the variation from specimen to specimen in the proportion of two or more components of magnetization in the observed remanence. We have seen that the strong component of secondary normal polarity magnetization found in about 95 percent of the samples is incompletely removed at the highest demagnetization steps. At these steps, varying proportions of the normal polarity secondary component superimposed on a reversed polarity characteristic component of magnetization probably is responsible for part of the observed dispersion in inclination.

Large dispersion in the directions of magnetization away from the great circle path between the PADFD and the reversed polarity Paleocene direction is not satisfactorily explained simply by varying the proportion of two components of magnetization, however. Other possible sources of dispersion include components of magnetization acquired during drilling and during mild regional metamorphism. The coals of the Raton basin are high rank Bituminous A coking coals; this classification has been the basis for their industrial use (Lee, 1924). Metamorphism of the coals to coking grade may be related to local igneous activity chiefly of Miocene age (Pillmore, 1976). Partial remagnetization during an episode of increased thermal gradient might account for much of the observed dispersion in directions of magnetization.

MAGNETIZATION OF CLAYSTONE NEAR THE CRETACEOUS-TERTIARY BOUNDARY AT RATON, NEW MEXICO

Payne and others (1983b) reported on paleomagnetic measurements of three samples from 25 cm (10 in) below, 10 cm (4 in) below, and 75 cm (30 in) above the Cretaceous-Tertiary boundary near the city of Raton. Azimuthal orientations of these samples were lost during sample preparation, and only the inclinations were reported. Positive inclinations were measured after AF demagnetization of these samples. This fact was used by Payne and others (1983b) to support their interpretation that the palynologically defined K-T boundary in the Raton basin occurs in a normal magnetic polarity zone.

The site sampled by Payne and others (1983b) is the Raton K-T boundary site found by Pillmore in a road cut on the old

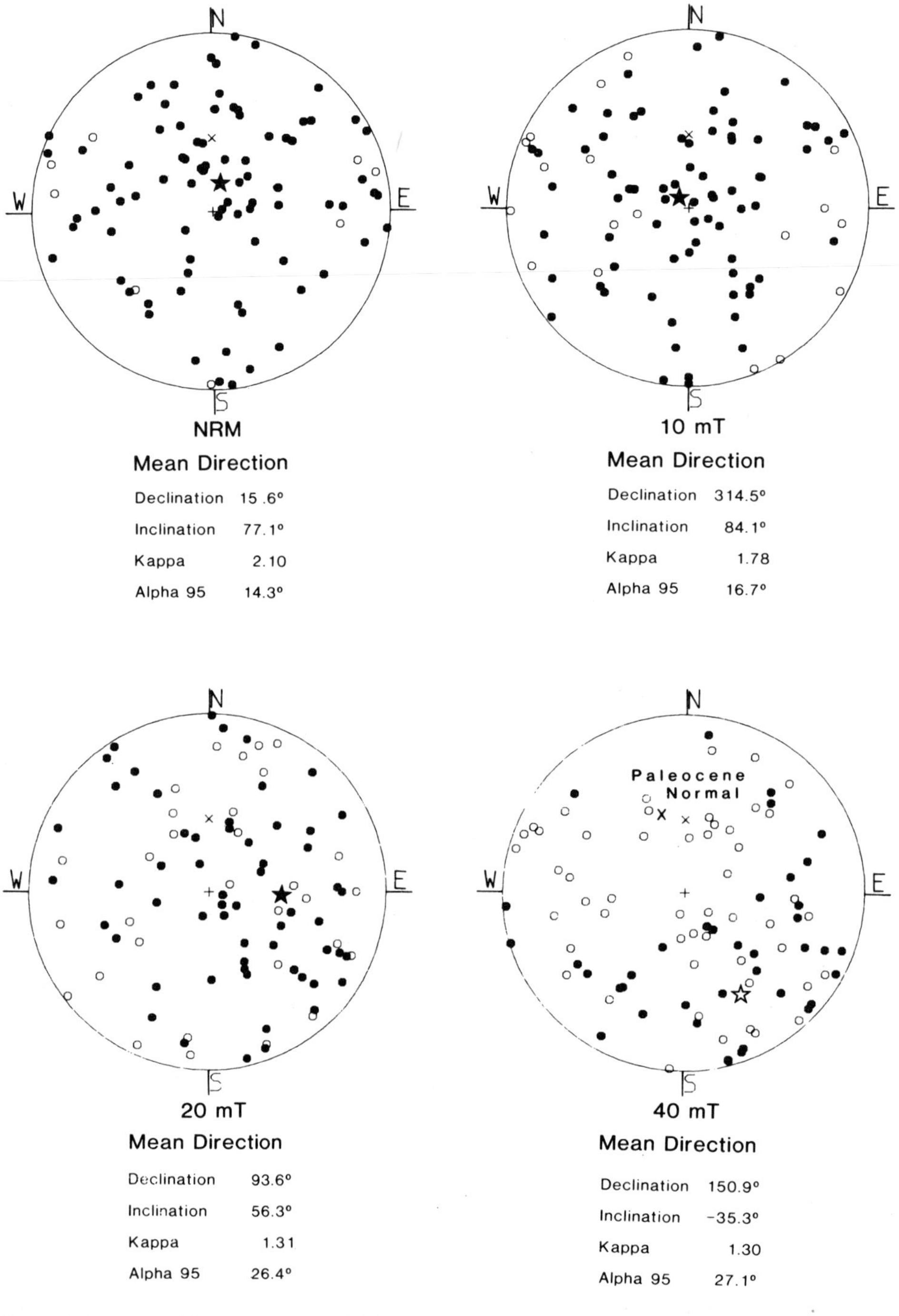

Figure 6. Stereographic equal-area projections showing directions of natural remanent magnetization (NRM) and the directions of magnetization after demagnetization to 10, 20, and 40 millitesla (mT) peak alternating fields for all samples taken from the York Canyon core. Each point represents the direction measured from a single 2.5-cm (1-in) plug taken from each sample. Solid dots are in the lower hemisphere, and open circles are in the upper hemisphere. The azimuthal orientation of each core run is the same as that shown in Figure 4. Stars (solid in the lower hemisphere, open in the upper hemisphere) are the mean directions of magnetization for all samples taken from the core. The present axial dipole field direction at York Canyon is shown with an X.

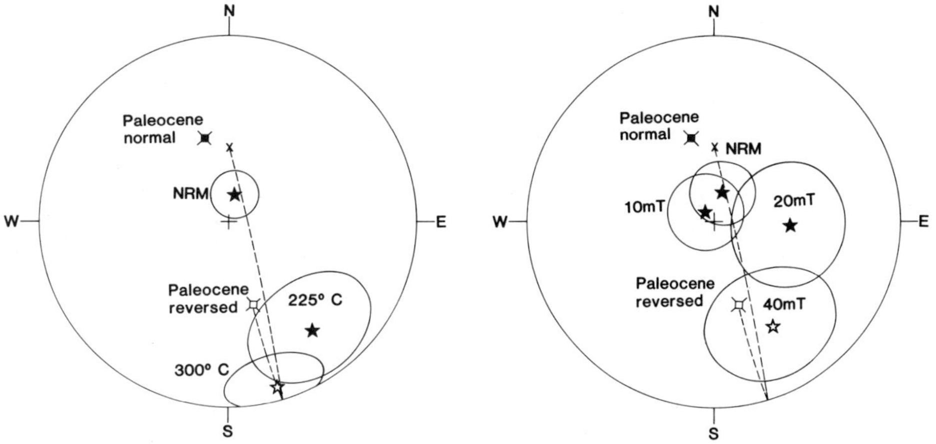

Figure 7. Stereographic equal-area projections showing the shift in mean direction of magnetization with progressive thermal and alternating field demagnetization for all samples taken from the York Canyon core. Ellipses represent 95 percent cones of confidence. Paleocene normal and reversed polarity directions shown are from Butler and Taylor (1978). Present axial dipole field direction is shown with an X. The great circle path from the present axial dipole field direction to the Paleocene reversed direction is shown with a dashed line. It should be borne in mind that azimuthal orientation for each core run was derived by setting the declination of the mean direction of magnetization at the 300°C (572°F) demagnetization step to the declination of the Paleocene reversed polarity direction. If the core runs had been oriented by setting the declination of the mean component of magnetization removed at 225°C (437°F) to north, the declinations of the mean directions would be increased about 10°.

Raton Pass scenic road (Pillmore and others, 1984). The site was revisited by Shoemaker and Orth in 1982, and 10 carefully oriented specimens were collected from a claystone about 10 cm (4 in) below the boundary. The fully oriented samples were thermally demagnetized and measured at the Flagstaff paleomagnetics laboratory, using methods identical to those used for the samples from the York Canyon core.

NRM directions of the claystone near the K-T boundary at Raton are tightly clustered around a mean direction that is indistinguishable from the present field direction, which has a declination of 11° and an inclination of +64° (Fig. 9). At the 95 percent confidence level, the mean NRM direction is different from the PADFD, however, and it is even further removed from the estimated Paleocene normal polarity direction. Alignment of the mean NRM with the present field suggests that the NRM consists predominantly of viscous magnetization that tracks the secular variation of the geomagnetic field. Weak components of reversed polarity characteristic magnetization could also account for the displacement of the mean NRM direction from the PADFD.

After demagnetization at 225°C, the directions of magnetization of all the samples remain fairly tightly clustered, and the mean shifts to steeper inclination (Fig. 9). This shift, although small, is away from the Paleocene normal polarity direction. After demagnetization at 300°C, the directions become widely scattered. Two directions of magnetization shift to the upper hemisphere, and several shift to shallow inclination. None shift toward the Paleocene normal polarity direction. The mean direction at the 300°C demagnetization step has a southwesterly declination, but we conclude that the polarity of the characteristic magnetization at this site is ill defined.

The demagnetization behavior of the samples suggests that the characteristic magnetization of the claystone near the boundary may be reversed polarity. There is no evidence that it is normal polarity. The magnetization appears to be insufficiently stable to preserve a clear record of a primary polarity. Any attempt to determine a characteristic or primary magnetization solely from the magnetic inclination of these samples would lead to a spurious result.

DISCUSSION

From the combined evidence from thermal and AF demagnetization of the York Canyon core, we have found that the characteristic magnetization of the 30 m (100 ft) of beds represented by the core is reversed polarity. We interpret the characteristic magnetization as chiefly a primary depositional magnetization, although the possibility that the magnetization was entirely reset during mild regional metamorphism cannot be completely excluded. The demagnetization characteristics of the rocks are consistent with the hypothesis that very fine grained detrital titaniferous magnetite or possibly titanohematite (Butler and Lindsey, 1983) is the principal carrier of the observed remanence. However, a detailed study of the magnetic phase or phases present has yet to be carried out. A principal future task will be to demonstrate the presence of both normal and reversed polarity

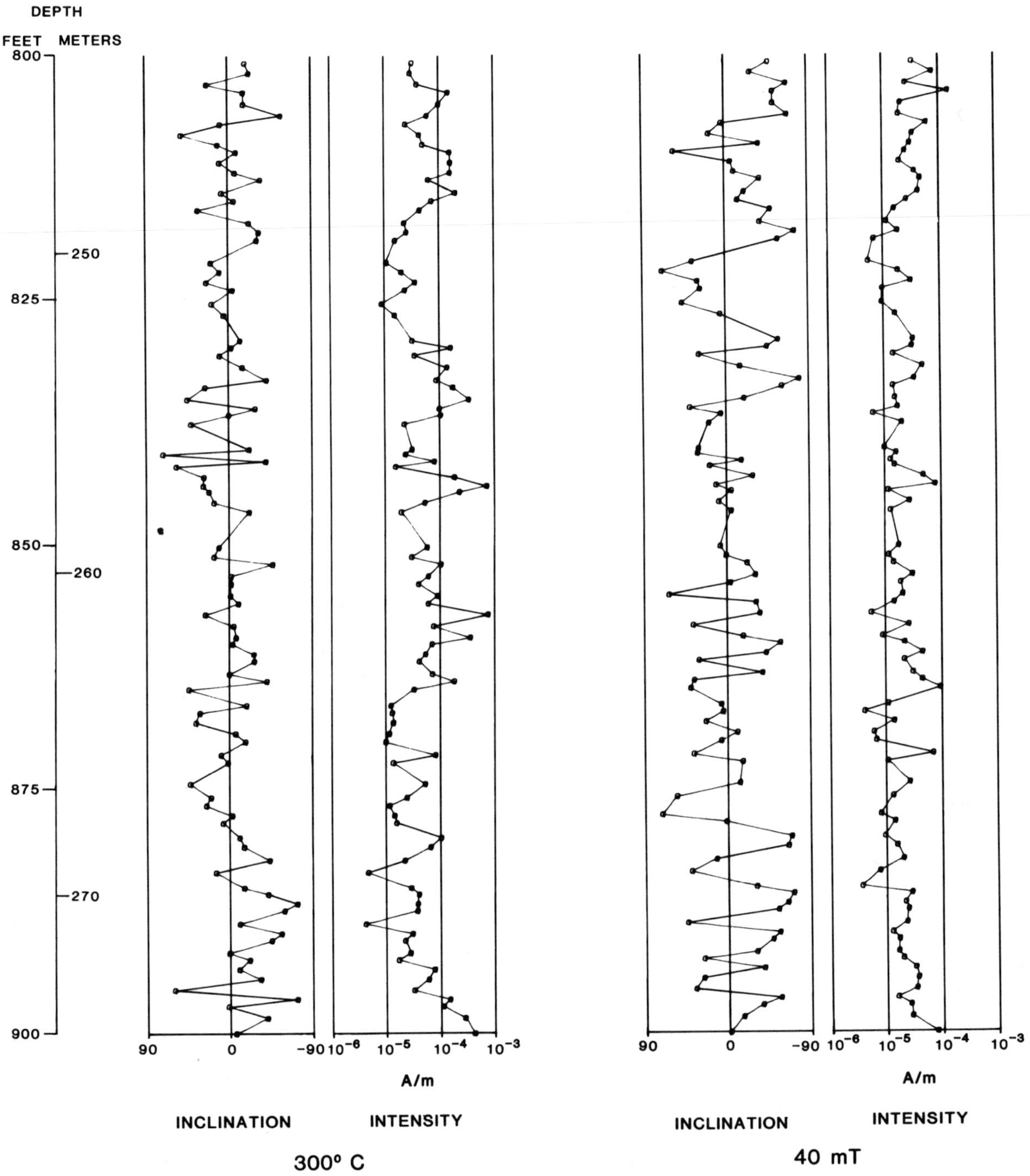

Figure 8. Plot of inclination and intensity of magnetization at the highest demagnetization steps as a function of stratigraphic position in the York Canyon core. Left-hand diagrams show remanent magnetization after demagnetization at 300°C (572°F); all points are averages of measurements from two 2.5-cm (1-in) plugs. Right-hand diagrams show remanent magnetization after demagnetization at 40 millitesla (mT) peak alternating field; all points represent measurements made on a single 2.5-cm (1-in) plug.

zones in the Raton Formation. In the present study it remains for us to show that the observed thickness of beds with reversed polarity characteristic magnetization is consistent with the minimum thickness expected from the known history of the geomagnetic field in latest Cretaceous and early Tertiary time.

The principal control for the polarity history of the magnetic field near the end of the Cretaceous comes from combined study of sections of deep-water marine beds, chiefly a section at Gubbio, Italy (Alvarez and others, 1977; Lowrie and Alvarez, 1977) and the marine (or seafloor spreading) magnetic anomalies (LaBrecque and others, 1977). As shown by Lowrie and Alvarez (1977), the end of the Cretaceous defined by marine fossils occurred in the reversed polarity chron 29r. The duration in time of chron 29r has been estimated from the distance between anomalies 29 and 30 on the sea floor, on the basis of various calibration points for the ages of the marine magnetic anomalies. LaBrecque and others (1977) derived a duration of 0.47 m.y., Ness and others (1980) obtained 0.61 m.y., Lowrie and Alvarez (1981) found 0.44 m.y., and Harland and others (1982) found a duration of 0.53 m.y. for 29r. The duration of 29r evidently is of the order of 0.5 m.y.

In order to find the expected thickness of beds occupying magnetozone 29r at York Canyon, we need an independent estimate of the rate of deposition of the fluviatile sediments of the Raton Formation spanning the K-T boundary. The total thickness of Maastrichtian beds above the base of Cobban's *Baculites reesidei* zone (Obradovich and Cobban, 1975) probably is about 230–245 m (750–800 ft), as determined from the 1966 core hole and from surface exposures of marine rocks. The duration of this interval of the Maastrichtian (post basal *B. reesidei*) is estimated at 4.5 to 5 m.y. (J. D. Obradovich, personal commun., 1986; Obradovich and Cobban, 1975). Averaged thickness (235 m; 770 ft) and time intervals (4.75 m.y.) yield an estimated minimum deposition rate of about 50 m (165 ft)/m.y. for this portion of the Maastrichtian at York Canyon. As there probably were hiatuses in deposition in the nonmarine part of the section, represented by unconformities beneath fluvial channel sandstones and at the base of the Raton Formation, the actual rate of deposition of beds in the lower coal zone of the Raton may have been substantially higher.

An alternative estimate of the minimum deposition rate for rocks in the Raton Formation can be made from the total thickness of the Paleocene section near York Canyon. The uppermost part of the section is not preserved in the York Canyon area; however, the Paleocene part of the Raton Formation and the noneroded beds of the Poison Canyon Formation combine to form a total thickness of at least 770 m (2525 ft). Based on this minimum thickness figure (the total thickness deposited probably was much greater) and the 10.1 m.y. duration of the Paleocene (as estimated by Harland and others, 1982), the minimum average Paleocene deposition rate in the York Canyon area was about 76 m (250 ft)/m.y. We realize that deposition of these rocks was probably episodic and characterized by relatively short periods of very rapid deposition followed by long periods of nondeposition

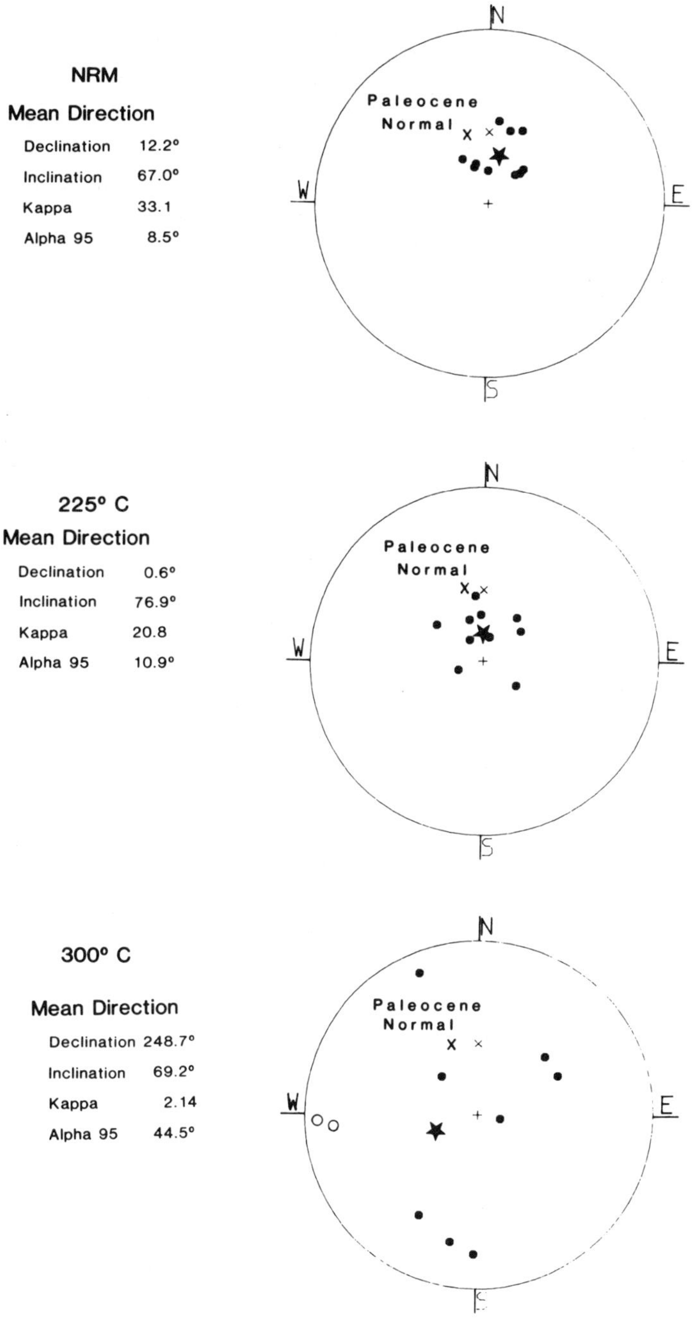

Figure 9. Stereographic equal-area projections showing directions of natural remanent magnetization (NRM) and the directions of magnetization after demagnetization at 225°C (437°F) and 300°C (572°F) for 10 samples from a clay layer about 10 cm (4 in) below the Cretaceous-Tertiary boundary in a road cut on the old Raton Pass Scenic Road near Raton, New Mexico. Solid dots are in lower hemisphere, and open circles are in upper hemisphere. Stars (all in lower hemisphere) are the mean directions of magnetization. Present axial dipole direction is shown with an X.

and erosion. The paleomagnetic record at any one site may not be complete.

Applying the two estimates of the minimum deposition rate, the minimum thickness of beds deposited during the estimated 0.5 m.y. duration of chron 29r would be ~25 to ~38 m (~82 to ~125 ft). The minimum expected thickness of beds with reversed polarity characteristic magnetization, in other words, is comparable to the observed thickness of 30 m (100 ft) in the York Canyon core. Lerbekmo and others (1979) and Lerbekmo and Coulter (1985) found that the reversed polarity zone spanning the palynological K-T boundary in Red Deer Valley, Alberta, is about 30 m (100 ft) thick, and Archibald and others (1982) found that the reversed polarity zone spanning the boundary in northeastern Montana is about 50 m (165 ft) thick. The beds spanning the boundary at the sites in Alberta and Montana are similar in lithology and environment of deposition to the lower coal zone of the Raton Formation. The thickness of beds with reversed polarity characteristic magnetization cut by the York Canyon core, therefore, is consistent both with likely minimum rates of deposition of the Raton Formation at York Canyon and with the thickness of the reversed polarity zone in beds of similar lithology that span the boundary elsewhere in western North America.

Considerable debate has taken place as to whether the K-T boundary determined in continental deposits from fossil floras and land-dwelling vertebrates is synchronous with the boundary defined by fossils of marine organisms and also whether the boundary picked from palynomorphs or vertebrate fossils might be diachronous. From a detailed study of the latest Cretaceous and early Tertiary deposits of the San Juan Basin, New Mexico, Butler and others (1977) suggested that the K-T boundary determined from vertebrate fossils was younger than the boundary determined from planktonic fossils at Gubbio, Italy. Butler and Lindsay (1983) found, however, that a spurious normal polarity magnetozone had been identified in their earlier work in the San Juan Basin. They now place the K-T boundary in the San Juan Basin section in chron 29r, consistent with the boundary at Gubbio (Butler and Lindsay, 1983; Butler, 1984). Although the apparent discrepancy has been removed, there is still debate about whether a major hiatus occurs in the stratigraphic sequence in the San Juan Basin (Alvarez and Vann, 1979; Fassett, 1979, 1982, 1984; Fassett and others, this volume; Lindsay and others, 1979; Lindsay and others, 1979a, 1979b; Lucas and Rigby, 1979; Lucas and Schoch, 1982; Lindsay and others, 1982; and Shoemaker and others, 1984). From combined magnetostratigraphic studies of Paleocene deposits in the Clark's Fork Basin, Wyoming (Butler and others, 1981), the San Juan Basin (Butler and others, 1977; Lindsay and others, 1978; and Lindsay and others, 1981), and the Fort Peck Reservoir area, Montana (Archibald and others, 1982), a plausible correlation can be made between the standard magnetic polarity chrons and the Paleocene North American Land Mammal Ages that suggests the base of the Paleocene determined from mammalian faunas lies in chron 29r. Lerbekmo and Coulter (1985) report that geomagnetic polarity zones in Late Campanian to Early Paleocene fluvial-deltaic rocks of the Edmonton Group in Red Deer Valley, Alberta, correlate to sea-floor magnetic anomalies 29–33 and that the Maastrichtian-Paleocene (K-T) boundary falls near the top of 29r. The identification of the reversed polarity zone containing the last of the dinosaurs in Alberta and Montana and the last of the *Proteacidites* assemblage of palynomorphs rests chiefly on the position of the pollen-defined boundary in North Dakota just below marine beds of the Cannonball Formation that have been dated as earliest Paleocene on the basis of planktonic foraminifera (Lerbekmo and Coulter, 1984).

The best evidence that the pollen-defined K-T boundary in the Raton Basin and in Wyoming, Montana, and Canada is synchronous with the K-T boundary in marine beds is provided by the presence of the distinctive noble metal–bearing claystone at the boundary in both the marine and continental deposits. The claystone has been found at five sites in northeastern Montana where it is coincident with the pollen-defined boundary at the contact of the Tullock Formation with the Hell Creek Formation. It has also been found at the pollen-defined K-T boundary at several sites in Canada, including Morgan Creek, Saskatchewan, and at Coal Valley and Red Deer Valley, Alberta (Nichols and others, 1986; Lerbekmo and Coulter, 1985; and Lerbekmo and St. Louis, 1986). The most extensive collections of latest Cretaceous dinosaurs have been obtained in these areas; the highest dinosaur bone fragments generally are found about 3 m (10 ft) below the pollen-defined boundary (W. A. Clemens, oral commun., 1983; Baadsgaard and Lerbekmo, 1980). Recently, however, Sloan and others (1986) reported the presence of dinosaur remains with Paleocene pollen and mammal teeth 1.3 m (4.3 ft) above the Ir-rich layer in the Ferguson Ranch channel at the base of the Tullock Formation, which suggests to them that some dinosaurs survived the terminal Cretaceous event.

Archibald and others (1982) reported that the base of the Tullock at Billy Creek, Montana, occurred in a normal polarity magnetozone. Reexamination of this locality during a field conference in 1983 revealed that the paleomagnetic sites had been misplaced within the lithostratigraphic section (R. F. Butler and E. H. Lindsay, oral commun., 1983). The noble metal–bearing boundary claystone was found at the base of the Tullock at Billy Creek during the course of the field conference, and it lies in the reversed polarity Fort Peck B- magnetozone described by Archibald and others (1982). No hard evidence has been found that the distinctive boundary claystone, here or elsewhere, is diachronous.

There has been some confusion about the possible presence of short normal polarity intervals in chron 29r. Lerbekmo and others (1979) report three possible short normal polarity intervals in the reversed polarity zone spanning the pollen-defined K-T boundary in Red Deer Valley, Alberta. As noted by Butler and Lindsay (1980), there is no evidence of normal polarity in chron 29r either in several marine stratigraphic sequences in Italy or in the marine magnetic anomaly record. Lerbekmo and others (1979) recognized that some of their reported normal polarity intervals could be due to secondary normal polarity magnetization, which may account for the normal polarity magnetization observed by them in the reversed polarity zone spanning the boundary. In a more recent paper, Lerbekmo and Coulter (1985, p. 574) again reported normal polarity zones in 29r in Red Deer Valley. The best documented of these normal polarity zones occurs in the Scollard Canyon portion of the valley, in two sections, about 3 km (1.9 mi) apart. The authors draw attention to the fact that the normal polarity zone at −18m in the East Scollard Canyon section is at the same relative position as a short normal polarity interval seen by Keating and others (1975) in samples from DSDP sites 207A and 208. Within the portion of the reversed zone included in the York Canyon core, we see no evidence of normal characteristic magnetization.

ACKNOWLEDGMENTS

We wish to thank Dr. C. J. Orth of the Los Alamos National Laboratory for his help and encouragement. This study was made possible as a result of the care he exercised in previous sampling and storing of the York Canyon core.

Dr. D. P. Elston kindly made facilities available at the U.S. Geological Survey paleomagnetic laboratory at Flagstaff, Arizona, and provided important support for this work.

We also wish to thank Dr. D. E. Champion of the U.S. Geological Survey and Dr. R. F. Butler of the University of Arizona for their technical review, and Ve Sable of the U.S. Geological Survey for her assistance in editing and preparation of the manuscript.

REFERENCES CITED

Alvarez, L. W., Alvarez, W., Asaro, F., and Michel, H. V., 1980, Extraterrestrial cause for the Cretaceous-Tertiary extinction: Science, v. 208, no. 4448, p. 1095–1108.33.

Alvarez, W., and Vann, D. W., 1979, Comment (*on* Lindsay, E. H., Jacobs, L. L., and Butler, R. F., 1978, Biostratigraphy and magnetostratigraphy of Paleocene terrestrial deposits, San Juan Basin, New Mexico: Geology, v. 6, no. 7, p. 425–429): Geology, v. 7, no. 2, p. 66–67.

Alvarez, W., Arthur, M. A., Fischer, A. G., Lowrie, W., Napoleone, G., Premoli Silva, I., and Roggenthen, W. M., 1977, Upper Cretaceous–Paleocene magnetic stratigraphic at Gubbio, Italy—Part V, Type section for the Late Cretaceous–Paleocene geomagnetic reversal time scale: Geological Society of America Bulletin, v. 88, no. 3, p. 383–389.

Alvarez, W., Alvarez, L. W., Asaro, F., and Michel, H. V., 1982, Current status of the impact theory for the terminal Cretaceous extinction: Geological Society of America Special Paper 190, p. 305–316.

Archibald, J. D., Butler, R. F., Lindsay, E. H., Clemens, W. A., and Dingus, L., 1982, Upper Cretaceous–Paleocene biostratigraphy and magnetostratigraphy, Hell Creek and Tullock Formations, northeastern Montana: Geology, v. 10, no. 3, p. 153–159.

Baadsgaard, H., and Lerbekmo, J. F., 1980, A Rb/Sr age for the Cretaceous-Tertiary boundary (Z coal), Hell Creek, Montana: Canadian Journal of Earth Sciences, v. 17, p. 671–673.

Baltz, E. H., 1965, Stratigraphy and history of Raton Basin and notes on San Luis Basin, Colorado–New Mexico: American Association of Petroleum Geologists Bulletin, v. 49, p. 2041–2075.

Bohor, B. F., and Pillmore, C. L., 1976, Tonstein occurrences in the Raton coal field, Colfax County, New Mexico, *in* Ewing, R. C., and Kues, B. S., eds., Guidebook of Vermejo Park, northeastern New Mexico: New Mexico Geological Society Guidebook, 27th Field Conference, p. 177–183.

Bohor, B. F., Pollastro, R. M., and Phillips, R. E., 1978, Mineralogic evidence for the volcanic origin of kaolinitic partings (tonsteins) in upper Cretaceous and Tertiary coals of the Rocky Mountain region: Annual Clay Minerals Conference, 15th, Bloomington, Indiana, Oct. 8–12, 1978, p. 47.

Bohor, B. F., Foord, E. E., Modreski, P. J., and Triplehorn, D. M., 1984, Mineralogic evidence for an impact event at the Cretaceous-Tertiary boundary: Science, v. 224, no. 4651, p. 867–869.

Brown, R. W., 1962, Paleocene flora of the Rocky Mountains and Great Plains: U.S. Geological Survey Professional Paper 375, p. 24.

Butler, R. F., 1984, Paleomagnetic and rock-magnetism of continental deposits, San Juan Basin, New Mexico: Geological Society of America Abstracts with Programs, v. 16, no. 4, p. 217.

Butler, R. F., and Lindsay, E. H., 1980, Discussion (*of* Lerbekmo, J. F., Evans, M. E., and Baadsgaard, H., 1979, Magnetostratigraphy, biostratigraphy and geochronology of Cretaceous-Tertiary boundary sediments, Red Deer Valley: Nature, v. 279, no. 5708, p. 26–30): Nature, v. 284, no. 5754, p. 375.

——, 1983, Magnetic mineralogy of continental deposits, San Juan Basin, New Mexico [abs.]: EOS (American Geophysical Union Transactions), v. 64, no. 45, p. 683.

Butler, R. F., and Taylor, L. H., 1978, A middle Paleocene paleomagnetic pole from the Nacimiento Formation, San Juan Basin, New Mexico: Geology, v. 6, no. 8, p. 495–598.

Butler, R. F., Lindsay, E. H., Jacobs, L. L., and Johnson, N. M., 1977, Magnetostratigraphy of the Cretaceous-Tertiary boundary in the San Juan Basin, New Mexico: Nature, v. 267, p. 318–323.

Butler, R. F., Gingerich, P. D., and Lindsay, E. H., 1981, Magnetic polarity, stratigraphy, and biostratigraphy of Paleocene and Lower Eocene continental deposits, Clark's Fork Basin, Wyoming: Journal of Geology, v. 89, p. 299–316.

Fassett, J. E., 1979, Comment (*on* Lindsay, E. H., Jacobs, L. L., and Butler, R. F., 1978, Biostratigraphy and magnetostratigraphy of Paleocene terrestrial deposits, San Juan Basin, New Mexico: Geology, v. 6, no. 7, p. 425–429): Geology, v. 7, no. 2, p. 69–70.

——, 1982, Dinosaurs in the San Juan Basin, New Mexico, may have survived the event that resulted in creation of an iridium-enriched zone near the Cretaceous/Tertiary boundary: Geological Society of America Special Paper 190, p. 435–447.

——, 1984, Significance of the rate of deposition of uppermost Upper Cretaceous rocks of the San Juan Basin, New Mexico and Colorado: Geological Society of America Abstracts with Programs, v. 16, no. 4, p. 221.

Gilmore, J. S., Knight, J. D., Orth, C. J., Pillmore, C. L., and Tschudy, R. H., 1984, Trace element patterns at a nonmarine Cretaceous-Tertiary boundary, Western Interior: Nature, v. 307, p. 224–228.

Harland, W. B., Cox, A. V., Llewellyn, P. G., Pickton, C. A., Smith, A. G., and Walters, R., 1982, A geologic time scale: Cambridge, England, Cambridge University Press, 131 p.

Hills, R. C., 1889, The recently discovered Tertiary beds of the Huerfano River basin, Colorado: Colorado Scientific Society Proceedings, v. 3, pt. 1, p. 148–164.

Hotten, C., 1984, [abs.]: International Palynological Conference, 16th, San Francisco, California, 1984, Programs and Abstracts.

Keating, B., Helsley, C. E., and Pessagno, E. A., Jr., 1975, Late Cretaceous reversal sequence: Geology, v. 3, no. 2, p. 73–76.

LaBrecque, J. L., Kent, D. V., and Cande, S. C., 1977, Revised magnetic polarity time scale for Late Cretaceous and Cenozoic time: Geology, v. 5, no. 6, p. 330–335.

Lee, W. T., 1917, Geology of the Raton Mesa and other regions in Colorado and New Mexico, *in* Lee, W. T., and Knowlton, F. H., Geology and paleontology of Raton Mesa and other regions in Colorado and New Mexico: U.S. Geological Survey Professional Paper 101, p. 9–221 [1918].

——, 1924, Coal resources of the Raton coal field, Colfax County, New Mexico:

U.S. Geological Survey Bulletin 752, 254 p.

Leffingwell, H. A., 1970, Palynology of the Lance (Late Cretaceous) and Fort Union (Paleocene) Formations of the type Lance Area, Wyoming: Geological Society of America Special Paper 127, p. 1–64.

Lerbekmo, J. F., and Coulter, K. C., 1984, Magnetostratigraphic and biostratigraphic correlations of Late Cretaceous to Early Paleocene strata between Alberta and North Dakota: Canadian Society of Petroleum Geologists Memoir 9, p. 313–317.

——, 1985, Late Cretaceous to early Tertiary magnetostratigraphy of a continental sequence: Red Deer Valley, Alberta, Canada: Canadian Journal of Earth Sciences, v. 22, no. 4, p. 567–583.

Lerbekmo, J. F., and St. Louis, R. M., 1986, The terminal Cretaceous iridium anomaly in the Red Deer Valley, Alberta, Canada: Canadian Journal of Earth Sciences, v. 23, p. 120–124.

Lerbekmo, J. F., Evans, M. E., and Baadsgaard, H., 1979, Magnetostratigraphy, biostratigraphy and geochronology of Cretaceous-Tertiary boundary sediments, Red Deer Valley: Nature, v. 279, no. 5708, p. 26–30.

Lindsay, E. H., Jacobs, L. L., and Butler, R. F., 1978, Biostratigraphy and magnetostratigraphy of Paleocene terrestrial deposits, San Juan Basin, New Mexico: Geology, v. 6, no. 7, p. 425–429.

Lindsay, E. H., Butler, R. F., and Johnson, N. M., 1979, Reply (*to* Lucas, S. G., and Rigby, J. K., Jr., 1979, Comment *on* Biostratigraphy and magnetostratigraphy of Paleocene terrestrial deposits, San Juan Basin, New Mexico: Geology, v. 7, no. 7, p. 323–325): Geology, v. 7, no. 7, p. 326–327.

Lindsay, E. H., Butler, R. F., Johnson, N. M., and Jacobs, L. L., 1979a, Reply (*to* Alvarez, W., and Vann, D. W., 1979, Comment *on* Biostratigraphy and magnetostratigraphy of Paleocene terrestrial deposits, San Juan Basin, New Mexico: Geology, v. 7, no. 2, p. 66–67): Geology, v. 7, no. 2, p. 68–69.

——, 1979b, Reply (*to* Fassett, J. E., 1979, Comment *on* Biostratigraphy and magnetostratigraphy of Paleocene terrestrial deposits, San Juan Basin, New Mexico: Geology, v. 7, no. 2, p. 69–70): Geology, v. 7, no. 2, p. 70–71.

Lindsay, E. H., Butler, R. F., and Johnson, N. M., 1981, Magnetic polarity zonation and biostratigraphy of Late Cretaceous and Paleocene continental deposits, San Juan Basin, New Mexico: American Journal of Science, v. 281, no. 4, p. 390–435.

——, 1982, Reply (*to* Lucas, S. G., and Schoch, R. M., 1982, Discussion of magnetic polarity zonation and biostratigraphy of Late Cretaceous and Paleocene continental deposits, San Juan Basin, New Mexico: American Journal of Science, v. 282, no. 6, p. 920–927): American Journal of Science, v. 282, no. 6, p. 928–933.

Lowrie, W., and Alvarez, W., 1977, Upper Cretaceous–Paleocene magnetic stratigraphy at Gubbio, Italy; Part III, Upper Cretaceous magnetic stratigraphy: Geological Society of America Bulletin, v. 88, p. 374–377.

——, 1981, One hundred million years of geomagnetic polarity history: Geology, v. 9, p. 392–397.

Lucas, S. G., and Rigby, J. K., Jr., 1979, Comment (*on* Lindsay, E. H., Jacobs, L. L., and Butler, R. F., 1978, Biostratigraphy and magnetostratigraphy of Paleocene terrestrial deposits, San Juan Basin, New Mexico: Geology, v. 6, no. 7, p. 425–429): Geology, v. 7, no. 7, p. 323–325.

Lucas, S. G., and Schoch, R. M., 1982, Discussion (*of* Lindsay, E. H., Butler, R. G., and Johnson, N. M., 1981, Magnetic polarity zonation and biostratigraphy of Late Cretaceous and Paleocene continental deposits, San Juan Basin, New Mexico: American Journal of Science, v. 281, no. 4, p. 390–435): American Journal of Science, v. 282, no. 6, p. 920, 927.

Ness, G., Levi, S., and Couch, R., 1980, Marine magnetic anomaly timescales for the Cenozoic and Late Cretaceous; A precis, critique, and synthesis: Reviews of Geophysics and Space Physics, v. 18, no. 4, p. 753–770.

Nichols, D. J., Jarzen, D. M., Orth, C. J., and Oliver, P. Q., 1986, Palynological and iridium anomalies at Cretaceous-Tertiary boundary, south-central Saskatchewan: Science, v. 231, no. 4739, p. 714–717.

Obradovich, J. D., and Cobban, W. A., 1975, A time-scale for the Late Cretaceous of the Western Interior of North America: Geological Association of Canada Special Paper, no. 13, p. 31–54.

Orth, C. J., Gilmore, J. S., Knight, J. D., Pillmore, C. L., Tschudy, R. H., and Fassett, J. E., 1981, An iridium abundance anomaly at the palynological Cretaceous-Tertiary boundary in northern New Mexico: Science, v. 214, p. 1341–1343.

——, 1982, Iridium abundance measurements across the Cretaceous/Tertiary boundary in the San Juan and Raton basins of northern New Mexico: Geological Society of America Special Paper 190, p. 423–433.

Payne, M. A., Wolberg, D. L., and Hunt, A. A., 1982, Magnetostratigraphy Raton and San Juan Basins, New Mexico; Implications for synchroneity of Cretaceous-Tertiary boundary events: Geological Society of America Abstracts with Programs, v. 14, no. 7, p. 584.

——, 1983a, Magnetostratigraphy Raton Basin, New Mexico; Implications for magnetostratigraphy of the San Juan Basin and synchroneity of Cretaceous-Tertiary boundary events: Geological Society of America Abstracts with Programs, v. 15, no. 5, p. 308–309.

——, 1983b, Magnetostratigraphy of a core from Raton Basin, New Mexico; Implications for synchroneity of Cretaceous-Tertiary boundary events: New Mexico Geology, v. 5, no. 3, p. 41–44.

Pillmore, C. L., 1969, Geologic map of the Casa Grande quadrangle, Colfax County, New Mexico, and Las Animas County, Colorado: U.S. Geological Survey Geologic Quadrangle Map GQ-823, scale 1:62,500.

——, 1976, Commercial coal beds of the Raton coal field, Colfax County, New Mexico, *in* Ewing, R. C., and Kues, B. S., eds., Guidebook of Vermejo Park, northeastern New Mexico: New Mexico Geological Society Guidebook, 27th Field Conference, p. 227–247.

Pillmore, C. L., Tschudy, R. H., Orth, C. J., Gilmore, J. S., and Knight, J. D., 1984, Geologic framework of nonmarine Cretaceous-Tertiary boundary sites, Raton basin, New Mexico and Colorado: Science, v. 223, no. 4641, p. 1180–1182.

Pollastro, R. M., and Pillmore, C. L., 1986, Mineralogy and petrology of the Cretaceous–Tertiary boundary clay bed and adjacent clay-rich rocks, Raton basin, New Mexico and Colorado; Journal of Sedimentary Petrology (in press).

Pollastro, R. M., Pillmore, C. L., Tschudy, R. H., Orth, C. J., and Gilmore, J. S., 1983, Clay petrology of the conformable Cretaceous/Tertiary boundary interval, Raton Basin, New Mexico and Colorado [abs.]: Annual Clay Minerals Conference, 32nd, Buffalo, New York, 1983, Programs and Abstracts, p. 83 (2 pages).

Shoemaker, E. M., Pillmore, C. L., Tschudy, R. H., and Orth, C. J., 1983, Characteristic magnetization of Cretaceous/Tertiary boundary claystone in Raton Basin is reversed: Geological Society of America Abstracts with Programs, v. 15, no. 5, p. 309.

Shoemaker, E. M., Steiner, M., Fassett, J. E., and Tschudy, R. H., 1984, Magnetostratigraphy of Upper Cretaceous rocks at Mesa Portales, New Mexico: Geological Society of America Abstracts with Programs, v. 16, no. 4, p. 255.

Sloan, R. E., Rigby, J. K., Jr., Van Valen, L. M., and Gabriel, Diane, 1986, Gradual dinosaur extinction and simultaneous ungulate radiation in the Hell Creek Formation: Science, v. 232, p. 629–633.

Tschudy, R. H., 1970, Palynology of the Cretaceous-Tertiary boundary in the northern Rocky Mountain and Mississippi Embayment regions: Geological Society of America Special Paper 127, p. 65–111.

——, 1973, The Gasbuggy Core; A palynological appraisal, *in* Fassett, J. E., ed., Cretaceous and Tertiary rocks of the southern Colorado Plateau: Four Corners Geological Society Memoir, p. 131–143.

Tschudy, R. H., Pillmore, C. L., Orth, C. J., Gilmore, J. S., and Knight, J. D., 1984, Disruption of the terrestrial plant ecosystem at the Cretaceous-Tertiary boundary, Western Interior: Science, v. 225, no. 4666, p. 1030–1034.

Tschudy, R. H., and Tschudy, B. D., 1986, Extinction and survival of plant life following the Cretaceous–Tertiary boundary event, Western Interior, North America: Geology, v. 14, no. 8, p. 667–670.

Williamson, I. A., 1970, Tonsteins; Their nature, origins, and uses: Mining Magazine, v. 122, p. 203–210.

Manuscript Accepted by the Society March 24, 1986

Printed in U.S.A.

Geological Society of America
Special Paper 209
1987

Biostratigraphic correlation of Cretaceous-Tertiary boundary rocks, Colorado to San Juan Basin, New Mexico

Karl R. Newman
Department of Geology
Colorado School of Mines
Golden, Colorado 80401

ABSTRACT

Biostratigraphic zone data compiled from outcrop and cored sections in Colorado and New Mexico are used for correlation of rock units below and above the Cretaceous-Tertiary boundary. Data localities are in the Sand Wash basin, Williams Fork Mountains, Grand Hogback, and Piceance Creek basin, northwestern Colorado; Denver basin, eastern Colorado; Raton basin, southeastern Colorado and northeastern New Mexico; San Juan Mountains and northern San Juan basin, southwestern Colorado; and San Juan basin, northwestern New Mexico.

Reference sections of intertonguing marine and nonmarine rocks of Campanian and Maastrichtian age are present in northwestern and eastern Colorado. Coexisting ammonite and palynomorph zones allow definition of upper Campanian, lower Maastrichtian, and upper Maastrichtian palynomorph intervals on the basis of the ammonite-based stages. Likewise, coexisting vertebrate (dinosaur and mammal) and palynomorph zones allow definition of upper Maastrichtian and lower Paleocene palynomorph intervals on the basis of the vertebrate-based stages.

The most precise Cretaceous-Tertiary boundary definitions are based on vertebrates and palynomorphs in the Denver basin and on palynomorphs associated with an iridium anomaly in the Raton basin. The resulting stage intervals and Cretaceous-Tertiary boundary can be correlated (but only approximately and incompletely) to the southern San Juan basin, New Mexico, where disagreements exist concerning the ages of formations, continuity of deposition, and presence or absence of a boundary unconformity.

This paper confirms old, and offers new, evidence supporting the interpretation of a substantial boundary unconformity, or possibly two unconformities, being present in the San Juan basin, with variable amounts of Maastrichtian (to early Paleocene?) time not represented. Furthermore, some rocks assigned a late Maastrichtian age by paleomagnetic means are no younger than late Campanian and early Maastrichtian, according to biostratigraphic correlations.

INTRODUCTION

Controversy has developed in recent years over geologic ages and stratigraphic relationships of Cretaceous-Tertiary boundary formations in the San Juan basin (see Introduction by Fassett and Rigby, this volume). Because biostratigraphy is a fundamental key to some of the San Juan basin correlation problems, this paper presents biozone and correlation data extending from Colorado to the southern San Juan basin, New Mexico, in order to demonstrate that biozones are present in certain strategic places and that they have relative age and correlation reliability. Some biostratigraphic standards are established in northwestern and eastern Colorado, where definitive zone and stage data are available, and correlations are then made southward into the San Juan basin. The method involves these steps:

1. Compilation of published and unpublished biostratigraphic

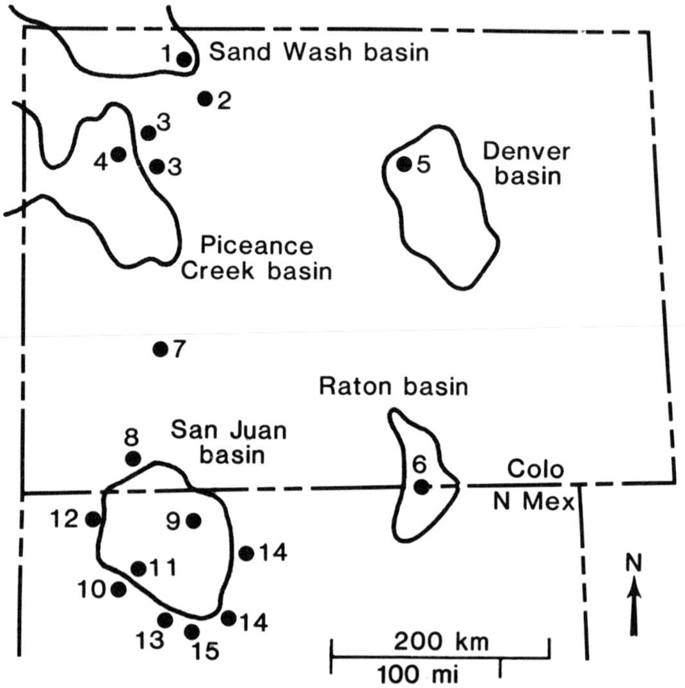

Figure 1. Locations of outcrop and cored sections, Colorado and New Mexico. Basin margins are drawn at base of Tertiary strata. Localities are listed by number in Table 1.

data from outcrop and cored sections in the Sand Wash basin, Williams Fork Mountains, Grand Hogback, and Piceance Creek basin, northwestern Colorado; Denver basin, eastern Colorado; Raton basin, southeastern Colorado and northeastern New Mexico; northern San Juan Mountains and northern San Juan basin, southwestern Colorado; and San Juan basin, northwestern New Mexico. Figure 1 shows the locations of these sites, and Table 1 lists the locality names and sources of data for each.

2. Establishment of criteria for definition of upper Campanian, lower Maastrichtian, upper Maastrichtian, and lower Paleocene strata on the basis of ammonite and vertebrate zonations and determination of palynomorphs characteristic of those stage intervals.
3. Biostratigraphic correlation from north to south to determine relative ages of formations where possible and to assess the magnitude of a boundary unconformity in the San Juan basin.

The San Juan basin is known most of all for its vertebrate fossils, and although some palynologic data have been available for many years (Anderson, 1960; Baltz and others, 1966; Fassett and Hinds, 1971; Tschudy, 1973, 1980), such information has only recently been applied systematically to the current age and unconformity controversy (Newman, 1982, 1983). Earlier efforts were dismissed by Lindsay and others, (1979), who announced that "palynological data are most useful for paleoecological correlations" and that "it would seem extremely risky to assign terrestrial rocks to marine stages by means of palynological data." Nonetheless, a goal of this paper is to show that palynomorph biozones, integrated where possible with other fossil zones, definitely are applicable to relative age determination and time-correlation problems in the San Juan basin, just as has been demonstrated elsewhere for decades.

COLORADO REFERENCE SECTIONS

The composite reference section for northwestern Colorado is shown in Figure 2. Lithostratigraphic and biostratigraphic data are compiled from localities 1 through 4, Table 1. Rock units include the Mancos Shale, Mesaverde Group of northwestern Colorado (divisible into Iles and Williams Fork formations), Lewis Shale of northwestern Colorado, Lance Formation, and Fort Union Formation.

The marine Mancos Shale contains the ammonites *Baculites perplexus* and *Baculites scotti* in its upper parts. Within the Mesaverde Group, the marine shale below the Trout Creek Sandstone has yielded *Exiteloceras jenneyi,* a marine unit in the lower Williams Fork Formation has yielded *Didymoceras cheyennense,* and the marine shale below the Twentymile Sandstone has yielded *Baculites reesidei.* The positions of the zones of *Baculites eliasi* (in the Williams Fork Formation just below the Lewis Shale) and of *Baculites baculus, Baculites grandis,* and *Baculites clinolobatus* (within the Lewis Shale) are projected into the section from eastern Colorado by Izett and others (1971).

Current radiometric age estimates for some of these ammonite zones are shown to range from 75 Ma for *Baculites scotti,* 73.7 Ma for *Exiteloceras jenneyi,* and 73.4 Ma for *Didymoceras cheyennense* to 70.1 Ma for *Baculites baculus* (Obradovich and Cobban, 1975; Fouch and others, 1983). The Campanian-Maastrichtian ammonite boundary in the Western Interior has been placed as high as the base of *Baculites baculus* or as low as the base of *Baculites reesidei.* If the stage boundary age of 72.3 Ma of Lowrie and Alvarez (1981) is accepted, the *Baculites reesidei* boundary is reasonable and is used in this paper for reference section and correlation purposes. If one wishes to argue for a somewhat higher or lower stage boundary, it can be done without affecting the overall conclusions of this paper.

The Lance–Fort Union formational contact in northwest Colorado appears to be unconformable in outcrop sections; therefore, a certain amount of upper Maastrichtian to lower Paleocene section may not be present. It is assumed from regional stratigraphy and from palynomorph correlations that the upper Lance and lower Fort Union formations are equivalent to parts of the *Triceratops* dinosaur and Puercan mammal zones of other basins, although vertebrates of those zones have not been reported from the sections used here. This apparent biostratigraphic control gap is very well filled in by the section at Golden, Colorado, on the west margin of the Denver basin (locality 5). Rock units include the upper Pierre Shale, Fox Hills Sandstone, Laramie Formation, Arapahoe Formation, and Denver Formation (Fig. 3). The upper Pierre Shale contains the *Baculites grandis, Baculites clinoloba-*

TABLE 1. OUTCROP AND CORE LOCALITIES, WITH REFERENCES FOR BIOSTRATIGRAPHIC DATA

Location	Reference(s)
1. Southeastern Sand Wash Basin, Colorado	Ammonite data: Izett, Cobban, and Gill (1971). Palynomorph data: Newman (1961, 1964, 1965).
2. Williams Fork Mountains, Colorado	(same as above)
3. Grand Hogback, Colorado	Ammonite data: Zapp and Cobban (1960), Madden (1983). Palynomorph data: Newman (1961, 1964, 1965), Madden (1983).
4. Piceance Creek basin, Colorado	Vertebrate data: Donnell (1969). Palynomorph data: Newman (1974).
5. Denver basin, Golden, Colorado	Ammonite data: Scott and Cobban (1965). Vertebrate and palynomorph data: Newman (1979).
6. Raton basin, Colorado and New Mexico	Palynomorph data: Tschudy (1973), Pillmore and Flores (this volume), Williams (1984).
7. Northern San Juan Mountains, Colorado	Ammonite data: Cobban (1973). Palynomorph data: Dickinson, Leopold, and Marvin (1968).
8. Northern San Juan basin, Durango, Colorado	Ammonite data: Cobban (1973). Palynomorph data: Newman (1982, 1983), Manfrino (1984).
9. Gasbuggy core, eastern San Juan Basin, New Mexico	Palynomorph data: Tschudy (1973).
10. Southwestern San Juan basin, Fossil Forest core 369, SE1/4 T23N,R12W, San Juan County, New Mexico	Palynomorph data: Newman (this paper).
11. Southwestern San Juan basin, New Mexico	Vertebrate and palynomorph data: Baltz, Ash, and Anderson, 1966.
12. Western San Juan basin, Farmington, SE1/4 T29N,R14W, San Juan County, New Mexico	Palynomorph data: Newman (this paper).
13. Southeastern San Juan basin, Pot Mesa, N1/2 T20N,R6W, McKinley County, New Mexico	Palynomorph data: Newman (this paper).
14. Southeastern San Juan basin, Cuba, New Mexico	Palynomorph data: Anderson (1960).
15. Southeastern San Juan basin, Mesa Portales, New Mexico	Palynomorph data: Tschudy (in Fassett and Hinds, 1971).

tus, and *Sphenodiscus* sp. ammonite zones. In the overlying nonmarine formations, the local range zone of *Triceratops* dinosaurs spans the Laramie, Arapahoe, and lower Denver formations. About 15 m above the highest known remains of *Triceratops* are found mammals of Puercan age, early Paleocene. For the purposes of this paper, *Baculites clinolobatus* is taken to be the boundary between lower and upper Maastrichtian strata, with an age of about 69.5 Ma (Fouch and others, 1983), and the Maastrichtian-Paleocene boundary, lying between the zones of *Triceratops* and Puercan mammals, has an age greater than 66 Ma (Obradovich and Sutter, 1984).

Palynomorphs from the two Colorado reference sections can be divided into several groups of zones within upper Campanian, lower Maastrichtian, upper Maastrichtian, and lower Paleocene strata (Fig. 4). Several very distinctive palynomorphs have range zone tops in upper Campanian strata, within the Mesaverde Group, lower Williams Fork Formation, or equivalents, at about the level of the *Didymoceras cheyennense* ammonite zone. These include *Trudopollis meekeri, Myrtaceoipollenites peritus, Pseudoplicapollis* spp., *Vacuopollis wolfcreekensis,* and *Umbosporites callosus.* The unique marker species for correlation of the upper Campanian in this paper are *Trudopollis meekeri, Myrtaceoipollenites peritus,* and *Pseudoplicapollis* spp., which are known together only from coastal plain and adjacent deposits of the late Campanian regression in the Colorado–New Mexico area.

Three species ranging somewhat higher, about to the level of *Baculites eliasi* in the upper Williams Fork Formation, lowest Maastrichtian, are *Kuylisporites scutatus, Ilexpollenites compactus,* and *Tricolpites interangulus.*

Palynomorphs appearing in lower Maastrichtian strata, with range zone bases in the uppermost Williams Fork Formation and

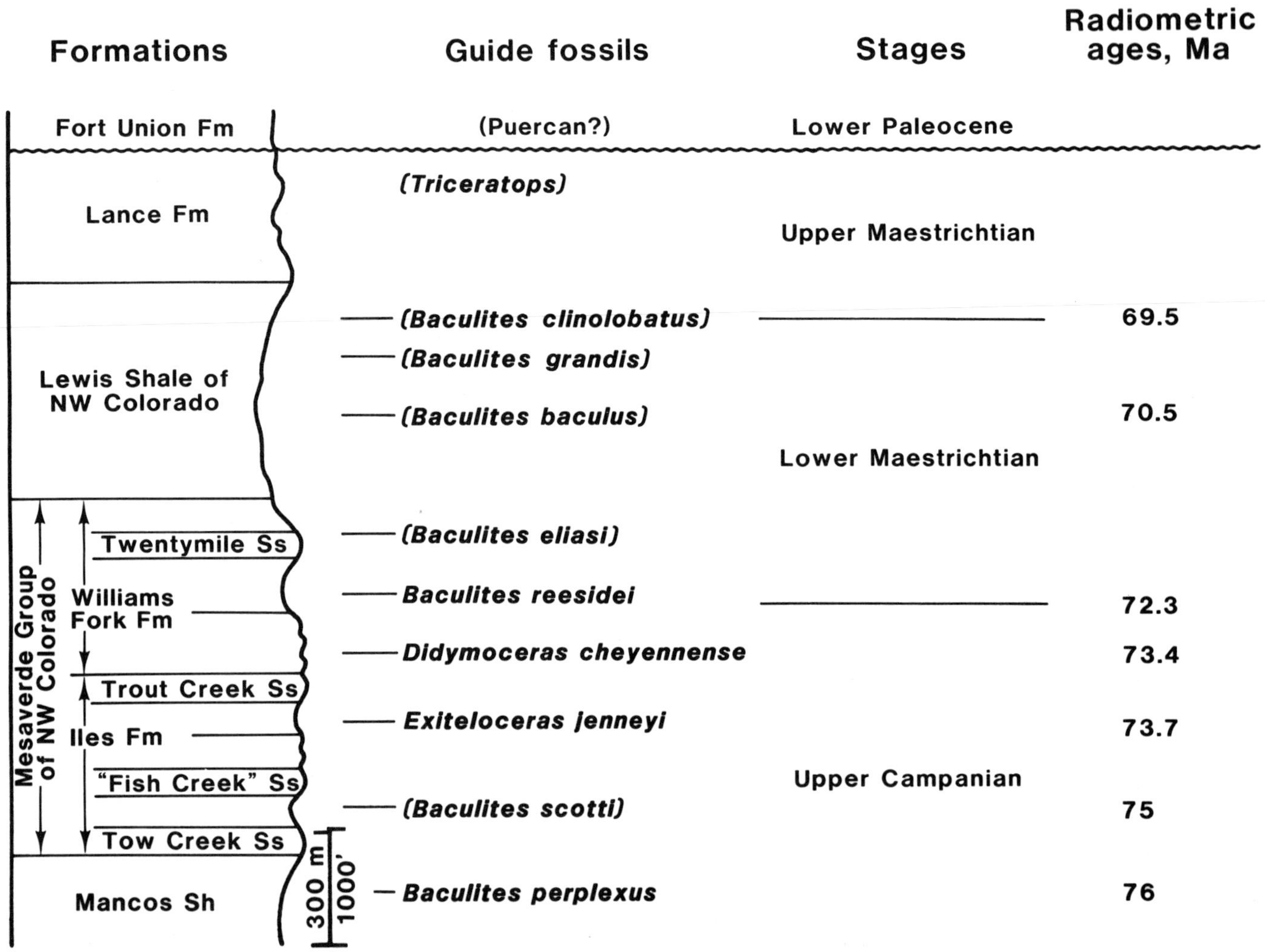

Figure 2. Northwestern Colorado biostratigraphic reference section. Data are composited from localities 1 to 4.

lower Lewis Shale, include *Triporopollenites scabroporus,* Ulmoideipites spp. (three- and four-pored smooth to verrucate forms), *Kurtzipites* spp., *Tricolpites anguloluminosus, Tilia wodehousei, Tricolporopollenites confossus, Gunnera microreticulata,* and *Liliacidites complexus.* The range zone top of *Triporopollenites scabroporus* is near the level of the *Baculites clinolobatus* ammonite zone, mid-Maastrichtian. The rest of the species range to the top of the Cretaceous or into the Paleocene.

A distinctive palynomorph appearing in mid-Maastrichtian strata or above and ranging into the Paleocene is *Thomsonipollis magnificus.*

Palynomorphs having range zones restricted to upper Maastrichtian strata include *Aquilapollenites delicatus* and *Wodehouseia spinata.* These species terminate at or near the top of the Cretaceous along with many species of *Proteacidites, Aquilapollenites, Balmeisporites, Cranwellia, Interpollis,* and *Ulmoideipites.* All of these palynomorphs have been found in the *Triceratops* zone in the Golden section, Denver basin. Also appearing in the upper part of upper Maastrichtian strata, uppermost *Triceratops* zone, are *Triatriopollenites* sp. and *Syncolporites* sp. that continue into lower Paleocene strata.

Lower Paleocene (Puercan) strata at Golden contain a very different pollen assemblage characterized by *Momipites coryloides, Momipites tenuipolis, Ulmoideipites* sp. (triporate, rugulate form), *Triporopollenites rugatus, Triatriopollenites trina,* and *Polyporopollenites* spp.

Further calibration of the upper Maastrichtian and lower Paleocene palynomorph intervals is provided by data from the Raton basin, Colorado and New Mexico (locality 6). Palynomorphs reported by Tschudy (1973) and Williams (1984) confirm the presence of *Thomsonipollis magnificus, Wodehouseia spinata,* and *Triatriopollenites* sp., along with *Proteacidites* spp. and *Aquilapollenites* spp. in the upper Vermejo Formation and lower Raton Formation (Fig. 5). Paleocene species of *Momipites* appear in the Raton Formation above the tops of the range zones of *Proteacidites* and *Aquilapollenites.*

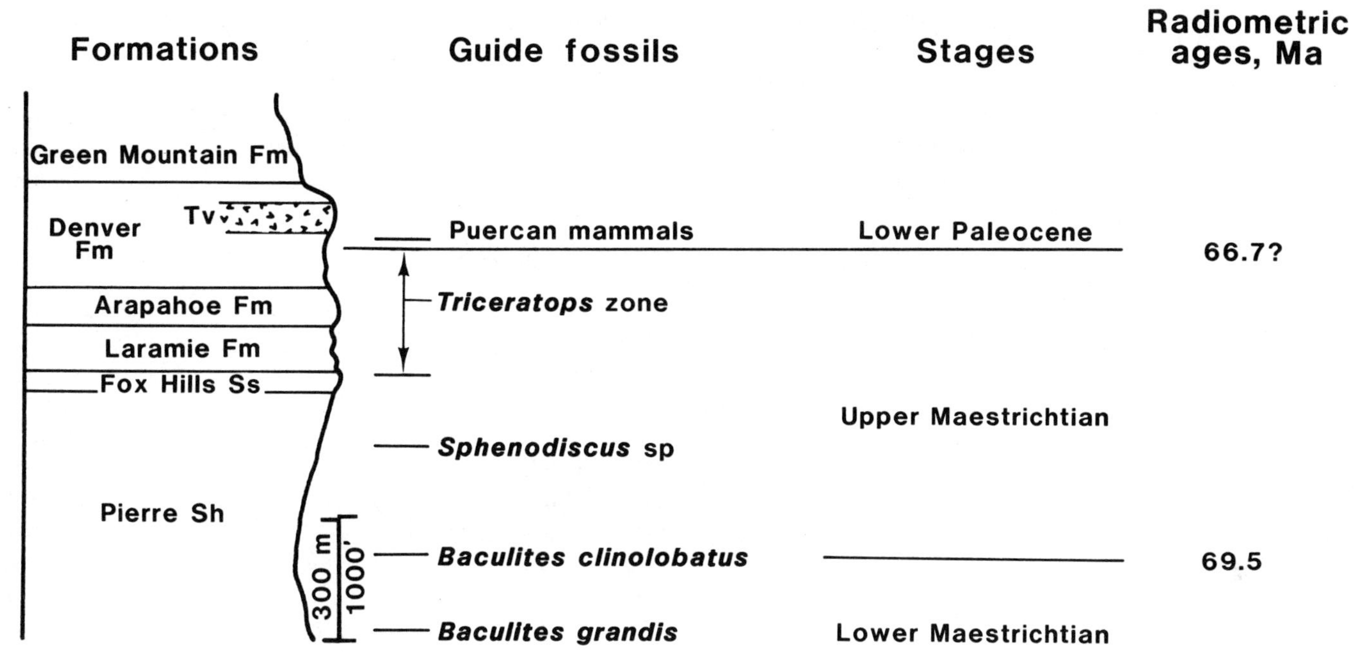

Figure 3. Golden, Colorado, biostratigraphic reference section (locality 5). Tv indicates position of lava flows capping South Table Mountain.

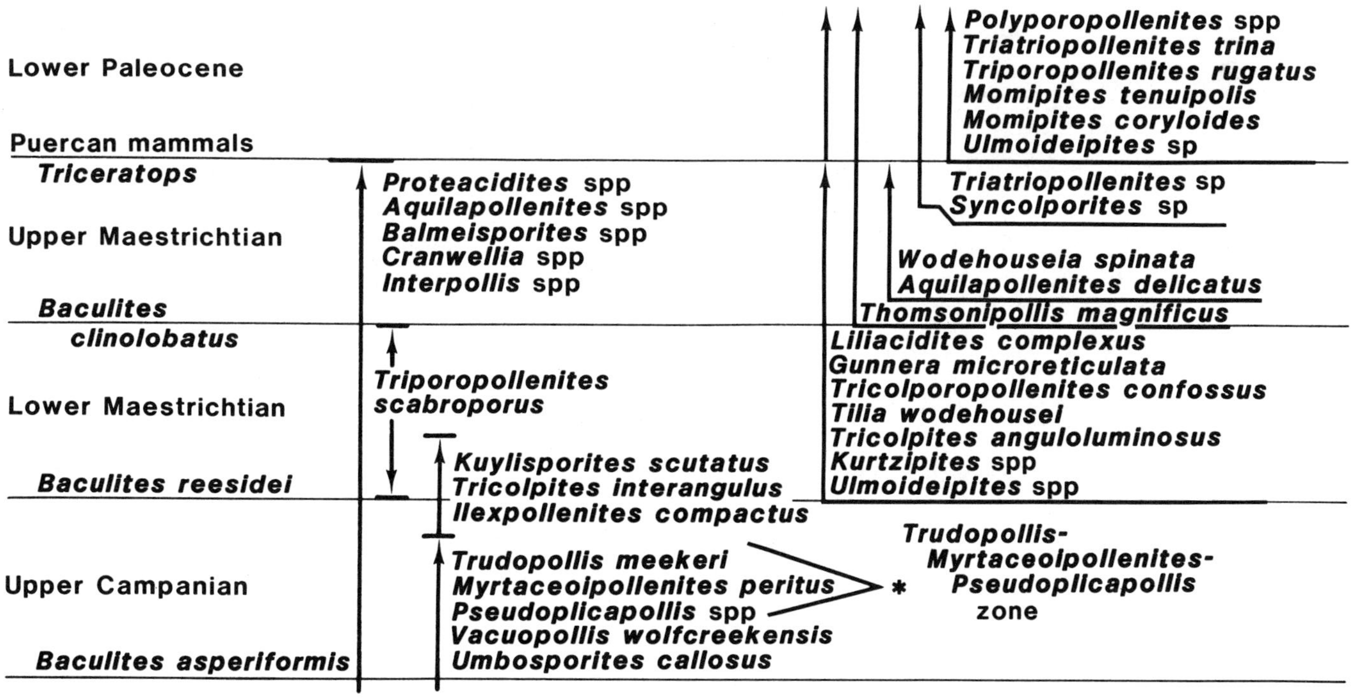

Figure 4. Generalized palynomorph range zones in upper Campanian, lower Maastrichtian, upper Maastrichtian, and lower Paleocene strata, Colorado. Data are composited from localities 1 to 5. Taxon references are listed in Table 2.

TABLE 2. ALPHABETICAL LIST OF PALYNOMORPH TAXONS CITED IN FIGURE 4 AND IN TEXT, WITH REFERENCES TO PUBLISHED ILLUSTRATIONS

Aquilapollenites delicatus Stanley, 1965, plate 49, figs. 1-4.
Aquilapollenites spp.: e.g., Tschudy and Leopold, 1970, plate 1, figs. 1-16.
Balmeisporites spp.: e.g., Stone, 1973, plate 12, figs. 67-69.
Cranwellia spp.: e.g., Leffingwell, 1970, plate 8, fig. 13; Stone, 1973, plate 18, figs. 118-119.
Gunnera microreticulata: Leffingwell, 1970, plate 6, figs. 7-8.
Ilexpollenites compactus Stone, 1973, plate 18, figs. 120-121.
Interpollis spp.: e.g., Tschudy, 1980, figs 3n-o.
Kurtzipites spp.: e.g., Anderson, 1960, plate 2, figs. 15-17; Leffingwell, 1970, plate 10, figs. 1-7.
Kuylisporites scutatus Newman, 1965, plate 1, fig. 1.
Liliacidites complexus: Stanley, 1965, plate 36, figs. 7-14; Leffingwell, 1970, plate 7, fig. 3.
Momipites coryloides: Newman, 1965, plate 1, fig. 9.
Momipites tenuipolis Anderson, 1960, plate 7, fig. 14.
Myrtaceoipollenites peritus Newman, 1965, plate 1, fig. 11.
Polyporopollenites spp.: e.g., Tschudy, 1970, plate 6, fig. 6.
Proteacidites spp.: e.g., Anderson, 1960, plate 2, figs. 1-7.
Pseudoplicapollis (Sporopollis) spp.: e.g., Newman, 1965, plate 1, fig. 13; Tschudy, 1980, figs. 3h-j.
Syncolporites sp.: possibly the same as Cupanieidites inaequalis Leffingwell, 1970, plate 9, fig. 10.
Thomsonipollis magnificus: Tschudy, 1980, figs. 3p-q.
Tilia danei Anderson, 1960, plate 7, figs. 10-11.
Tilia wodehousei Anderson, 1960, plate 2, figs. 11-13.
Triatriopollenites sp.: triangular species with atria about half as deep as in Momipites; possibly the same as Momipites inaequalis Anderson, 1960, plate 6, figs. 7-10.
Triatriopollenites trina: Alnus trina Stanley, 1965, plate 43, figs. 4-5.
Tricolpites anguloluminosus Anderson, 1960, plate 6, figs. 15-17.
Tricolpites interangulus Newman, 1965, plate 1, fig. 3.
Tricolporopollenites confossus Newman, 1965, plate 1, fig. 4.
Triporopollenites rugatus Newman, 1965, plate 1, fig. 7.
Triporopollenites scabroporus Newman, 1965, plate 1, fig. 5.
Trudopollis meekeri Newman, 1965, plate 1, fig. 12.
Ulmoideipites sp. (in lower Paleocene zone): a new species, or a rugulate variety of U. tricostatus Anderson, 1960, plate 4, figs. 9-10.
Ulmoideipites spp. (in Maastrichtian Stage): e.g., Leffingwell, 1970, plate 6, fig. 6; Tschudy, 1970, plate 4, figs. 17-18; Stone, 1973, plate 20, figs. 135, 145, 146.
Umbosporites callosus Newman, 1965, plate 1, fig. 2.
Vacuopollis (Conclavipollis) wolfcreekensis Newman, 1965, plate 1, fig. 10.
Wodehouseia spinata Stanley, 1961, plate 1, figs. 1-12.

Pillmore and Flores (this volume) report the association of the palynomorph Cretaceous-Tertiary boundary with a thin iridium-rich clay layer at a number of localities in the Raton basin. This Raton basin palynomorph boundary presumably is the same as the palynomorph boundary at Golden that is bracketed by the *Triceratops* zone below and Puercan mammals above.

NORTHERN SAN JUAN MOUNTAINS, COLORADO

Data linking parts of the northwestern Colorado reference section with sections in the San Juan basin were published by Dickinson and others (1968) and Cobban (1973) from the Cimarron Ridge and Montrose areas, northern San Juan Mountains (locality 7). Ammonites from the upper Mancos Shale in this area have been interpreted as being from the *Exiteloceras jenneyi* and *Didymoceras cheyennense* zones (Fig. 6). Above the Mancos Shale lie the Pictured Cliffs, Fruitland, Kirtland, and Cimarron Ridge formations. Guide palynomorphs reported from the Fruitland and Kirtland formations include *Kuylisporites* (as *Hemitelia*) and *Tricolpites interangulus.* Those from the overlying Cimarron Ridge Formation include *Liliacidites complexus, Tricolpites anguloluminosus,* and *Tricolporopollenites confossus* (as *Nyssa*). With reference to the northwestern Colorado section, these occurrences indicate a late Campanian to Maastrichtian age range for the Fruitland, Kirtland, and Cimmarron Ridge formations. Dickinson and others (1968) correlated the Fruitland, Kirtland, and Cimarron Ridge formations of the northern San Juan Mountains with the Fruitland, Kirtland, and McDermott formations of the Durango area, northern San Juan basin, on lithologic similarities.

NORTHERN SAN JUAN BASIN, COLORADO

The stratigraphic section along the Animas River at Durango, Colorado, in the northwestern part of the San Juan basin (locality 8), consists of the Lewis Shale of the San Juan basin, Pictured Cliffs Sandstone, Fruitland Formation, Kirtland Formation, McDermott Formation (or McDermott Member of the Animas Formation), and Animas Formation (Fig. 7).

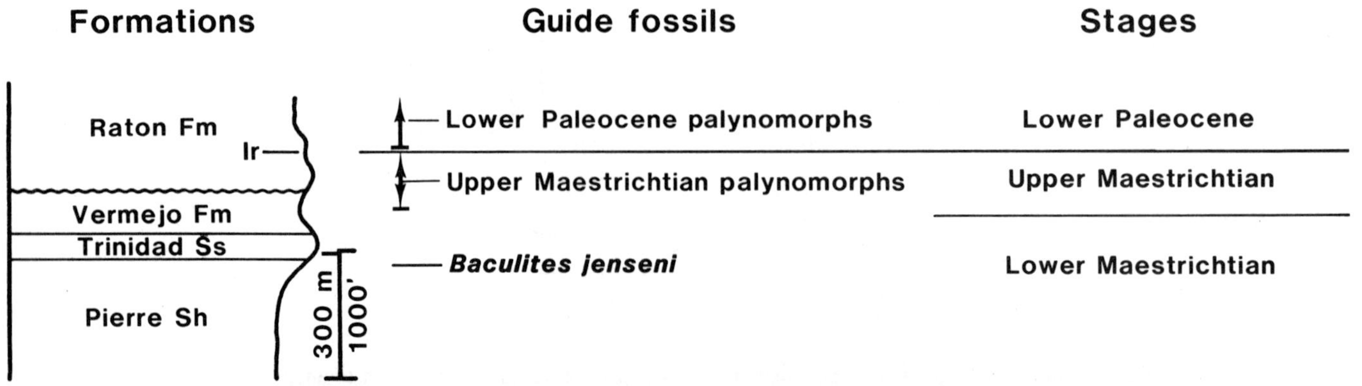

Figure 5. Cretaceous-Tertiary boundary in Raton basin, Colorado and New Mexico (locality 6). Ir indicates position of iridium-rich boundary clay layer.

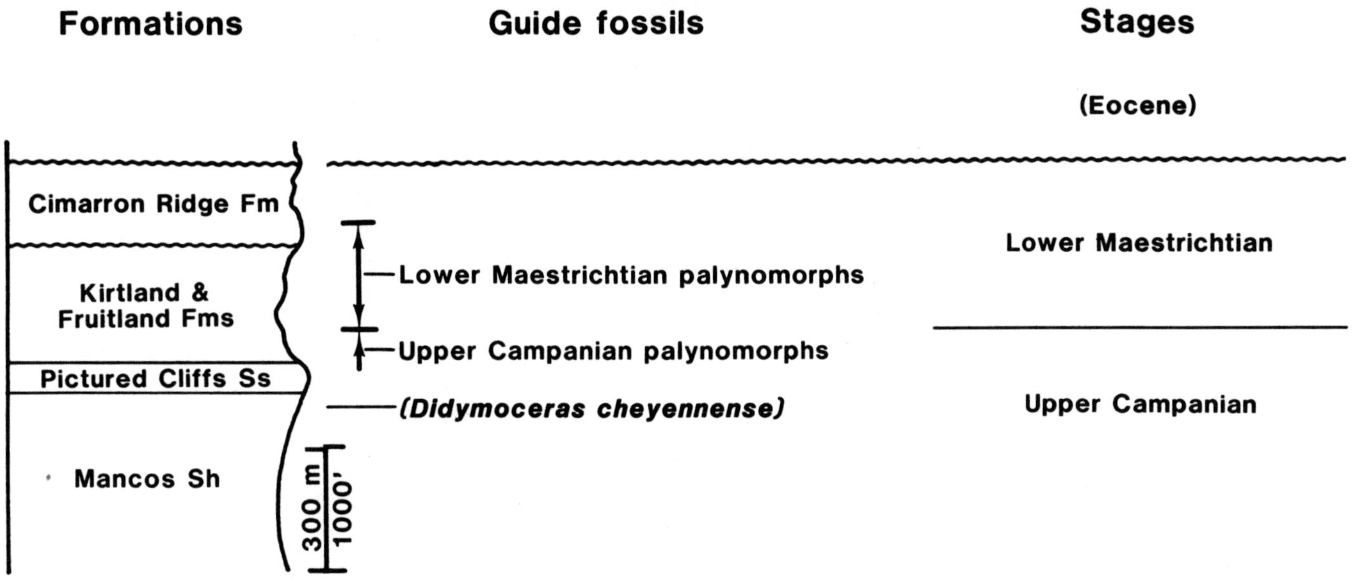

Figure 6. Biostratigraphy of upper Campanian and lower Maastrichtian formations, northern San Juan Mountains, Colorado (locality 7).

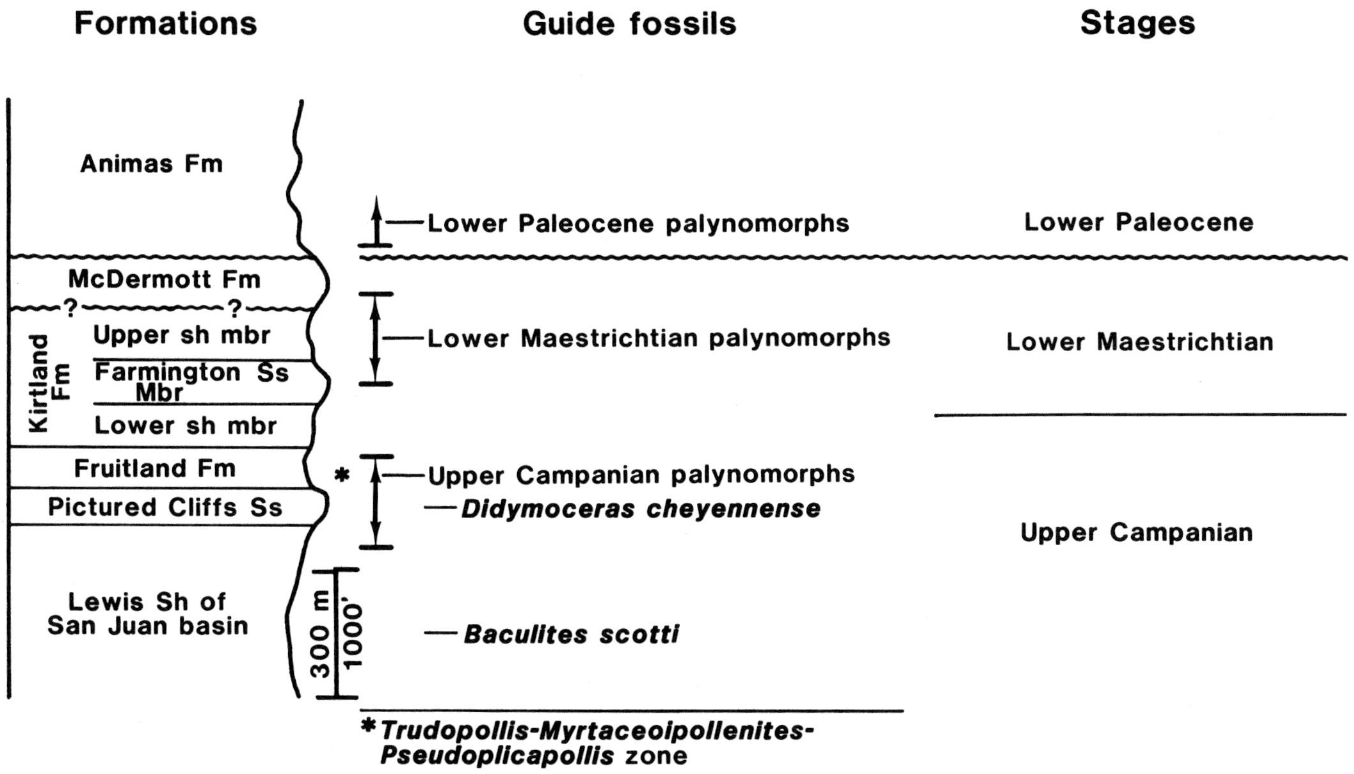

Figure 7. Biostratigraphy of upper Campanian, lower Maastrichtian, and lower Paleocene formations at Durango, northern San Juan basin, Colorado (locality 8).

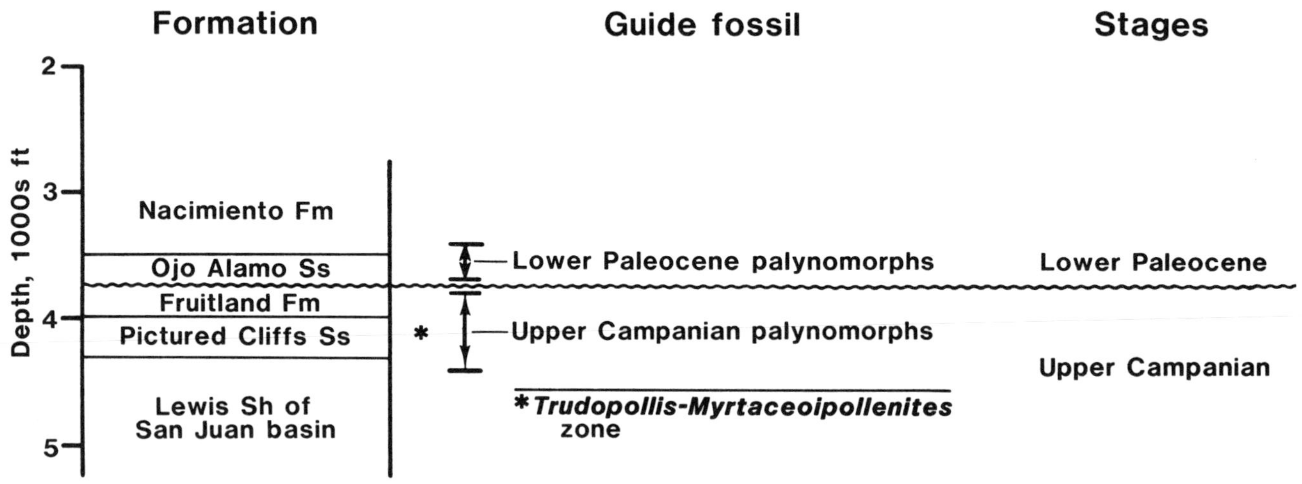

Figure 8. Biostratigraphy of upper Campanian and lower Paleocene formations in Gasbuggy core, eastern San Juan basin, New Mexico (locality 9).

Didymoceras stevensoni has been found in the upper Lewis Shale, and the uppermost Lewis Shale and Pictured Cliffs Sandstone probably lie in the zone of *Didymoceras cheyennense* (in this area) (Cobban, 1973). These ammonites indicate late Campanian age.

Significant palynomorphs occurring in the Pictured Cliffs Sandstone and lower Fruitland Formation include *Trudopollis meekeri, Myrtaceoipollenites peritus, Pseudoplicapollis* sp., *Kuylisporites scutatus, Ilexpollenites compactus,* and *Tricolpites interangulus* (?). With reference to the northwestern Colorado section, this assemblage likewise indicates a late Campanian age.

The lower shale member of the Kirtland Formation has not yet yielded diagnostic palynomorphs in this section, but the Farmington Sandstone and upper shale members have yielded several. Guide palynomorphs found in the Farmington Sandstone Member include *Gunnera microreticulata, Ulmoideipites* sp. (four-pored, verrucate form), and *Tilia wodehousei.* Those from the upper shale member include the same species and possibly *Triporopollenites scabroporus.*

Samples 3 m above the base of the McDermott Formation yield palynomorphs very similar to those of the upper Kirtland, including definite *Triporopollenites scabroporus,* a lower Maastrichtian guide fossil. On the basis of these results, the entire Farmington Sandstone through McDermott Formation interval is interpreted to be early Maastrichtian in age. If there is an unconformity at the base of the McDermott, it is not of great enough magnitude to cause an obvious break in the palynomorph zones.

Palynomorphs from the Animas Formation, about 30 m above the top of the McDermott Formation, include *Momipites coryloides, Momipites tenuipolis,* and *Ulmoideipites* sp. (three-pored, rugulate form), indicating early Paleocene (Puercan) age.

My interpretation of this section is that the youngest Cretaceous rocks present are early Maastrichtian, including the McDermott Formation, which has a highly weathered top. Approximately the upper half of the Maastrichtian Stage is not represented, and perhaps some earliest Paleocene is missing as well between McDermott and Animas strata.

GASBUGGY CORE, EASTERN SAN JUAN BASIN, NEW MEXICO

The Gasbuggy core (locality 9), as described by Fassett and Hinds (1971), includes the upper Lewis Shale, Pictured Cliffs Sandstone, Fruitland Formation, Ojo Alamo Sandstone, and lower Nacimiento Formation (Fig. 8). Tschudy (1973) reported cf. *Myrtaceoipollenites peritus* from the Pictured Cliffs Sandstone and *Trudopollis meekeri* from the upper Lewis Shale to middle Fruitland Formation. Additional guide taxons present in the Fruitland include *Ilexpollenites* sp. and *Kuylisporites* sp. The overlying Ojo Alamo and Nacimiento formations yielded species that I interpret to be *Momipites* spp. and *Ulmoideipites* sp.

This cored section then appears to have late Campanian (to earliest Maastrichtian?) Fruitland Formation overlain unconformably by early Paleocene Ojo Alamo Sandstone and lower Nacimiento Formation. Most of the Maastrichtian and perhaps some earliest Paleocene time are not represented. Tschudy (1973) reached similar conclusions using comparisons with data from Montana, Wyoming, and the Raton basin.

SOUTHWESTERN SAN JUAN BASIN, NEW MEXICO

Figure 9 is a composite section of data from the Fruitland Formation (Fossil Forest core 369, locality 10) and from Kirtland, Ojo Alamo, and Nacimiento formation outcrops at Hunter Wash, Ojo Alamo Arroyo, and Barrel Spring Arroyo (locality 11; Baltz and others, 1966). Approximate formation thicknesses

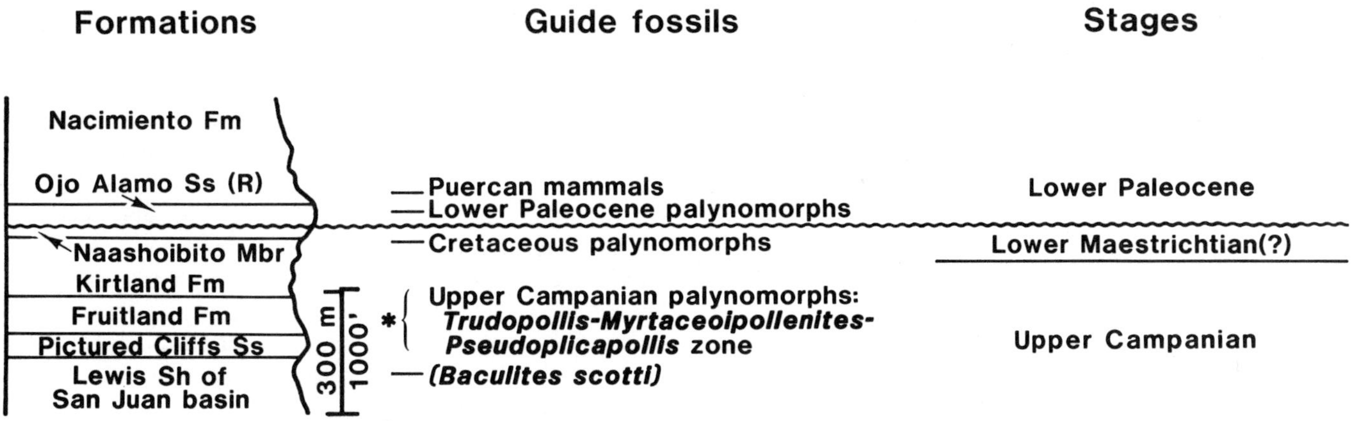

Figure 9. Biostratigraphy of upper Campanian, lower (?) Maastrichtian, and lower Paleocene formations, southwestern San Juan basin, New Mexico (localities 10 and 11).

are taken from the latter authors and from Fassett and Hinds (1971).

Ten samples from 51 m of cored Fruitland Formation have yielded an excellent assemblage of palynomorphs including the guide fossils *Trudopollis meekeri, Myrtaceoipollenites peritus,* and *Pseudoplicapollis* sp. The cored interval lies a few hundred meters above the Huerfanito bentonite of the Lewis Shale, thought to be in the *Baculites scotti* zone (Cobban, 1973). Therefore, the combination of the ammonite and palynomorph zones indicates late Campanian age for the upper Lewis, Pictured Cliffs, and Fruitland formations in this area, just as at Durango.

The Kirtland Formation (undivided) is about 183 m thick in this area and includes the Farmington Sandstone, upper shale, and Naashoibito members in its upper part. Baltz and others (1966) reported dinosaur remains from several levels within the Naashoibito Member and also reported palynomorphs from a sample from the upper shale member of the Kirtland, about 5 m below the base of the Naashoibito Member near Ojo Alamo Arroyo. The taxons reported indicate Late Cretaceous age, but none of the guide fossils used in this paper can be detected in the list. Comparison of the section with that at Durango would lead one to conclude that the upper shale member of the Kirtland is most likely early Maastrichtian in age, based on stratigraphic position alone.

The Naashoibito Member of the Kirtland Formation (or lower Ojo Alamo Sandstone; see Fassett, 1973) in this critical area has not yet yielded palynomorphs, as far as I know, because of the severe weathering of the rocks along the arroyos. Some data that are from distant samples, but that might bear on the Naashoibito or lower Ojo Alamo age, come from samples recently collected near Farmington (locality 12) and at Pot Mesa (locality 13). In each case, mudstone collected above the basal conglomeratic sandstone (= base of Naashoibito Member?, or base of Ojo Alamo Sandstone?) but below the upper conglomeratic sandstone (= upper Ojo Alamo Sandstone?, or Ojo Alamo Sandstone, restricted?) has yielded well-preserved specimens of *Proteacidites, Balmeisporites, Interpollis, Gunnera, Kurtzipites,* and *Ulmoideipites* spp. (three- and four-pored smooth to verrucate forms) that indicate Maastrichtian age. So far, it has not been possible to determine whether the age is early or late Maastrichtian from these assemblages.

The interval of Ojo Alamo Sandstone (restricted) or upper Ojo Alamo Sandstone intertonguing with the lower Nacimiento Formation has yielded palynomorphs in the vicinity of Barrel Springs Arroyo (Baltz and others, 1966). Taxons listed include *Momipites* spp. and *Ulmoideipites tricostatus.* Puercan mammals are known from beds about 9 m higher in the Nacimiento Formation, according to the same authors.

My interpretation of the composite section in this southwestern part of the San Juan basin is that the Fruitland Formation is definitely Campanian (indicated by the *Trudopollis-Myrtaceoipollenites-Pseudoplicapollis* zone); the Kirtland Formation is probably late Campanian and early Maastrichtian (based on stratigraphic position and similarity to the Durango section); the Naashoibito Member of the Kirtland Formation, or lower Ojo Alamo Sandstone, is Maastrichtian and not Paleocene if the Farmington and Pot Mesa palynomorphs are applicable; and the Ojo Alamo Sandstone (restricted), or upper Ojo Alamo Sandstone, and lower Nacimiento Formation are early Paleocene in age (indicated by *Momipites* pollen and Puercan mammals). The Campanian-Maastrichtian boundary, therefore, must lie within the Kirtland Formation, just as at Durango, and the Cretaceous-Tertiary boundary is most likely at the base of the Ojo Alamo Sandstone (restricted). The question still remains whether there might be two unconformities rather than just one, as was suggested long ago by Reeside (1924, p. 48, Fig. 3). The exact age span of the Naashoibito Member of the Kirtland Formation, or lower Ojo Alamo Sandstone, is the critical factor in determining if there is an unconformity at its base as well as at its top. In either case, it appears likely that much of Maastrichtian time and perhaps some earliest Paleocene time as well are not represented in this section.

SOUTHEASTERN SAN JUAN BASIN, NEW MEXICO

Palynologic data have been reported from the southeastern San Juan basin by Anderson (1960) and Fassett and Hinds (1971). Anderson reported palynomorphs from six samples from widely separated localities in the Lewis, Kirtland, Ojo Alamo, and Nacimiento formations, mostly in the vicinity of Cuba, New Mexico (locality 14). I interpret the illustrated palynomorphs to indicate that the upper Lewis Shale is late Campanian in age (indicated by probable *Trudopollis* and *Pseudoplicapollis*); the upper Kirtland Formation is Maastrichtian undifferentiated (indicated by *Tilia wodehousei* and *Kurtzipites trispissatus*); the Ojo Alamo Sandstone is either latest Cretaceous or Paleocene (indicated by possible *Triatriopollenites* sp.); and the lower Nacimiento Formation is Paleocene, but younger than in the southwestern part of the basin (indicated by *Tilia danei* that first appears in rocks of Torrejonian mammal age, according to regional data [Newman, 1980]).

Fassett and Hinds (1971) reported palynomorphs from Mesa Portales (locality 15), southwest of Cuba. Palynomorphs from the undivided Fruitland and Kirtland formations were interpreted by Tschudy, in their report, to indicate Cretaceous age for most of the interval, but Paleocene age for the uppermost part of what is mapped as Fruitland-Kirtland and for the two-sandstone unit mapped as Ojo Alamo.

Overall, the data from the Anderson samples and from the Fassett and Hinds samples are anomalous, considering the basin as a whole, and are conflicting in their age relationships. Because of this, at this time I am unable to compare the data with sections elsewhere in the basin. The Mesa Portales and Cuba area sections are being resampled and restudied independently by the U.S. Geological Survey and me in order to clarify the lithostratigraphic and biostratigraphic relationships.

BIOSTRATIGRAPHIC CORRELATIONS

Biostratigraphic correlations of the northwestern Colorado, Golden (Colorado), Durango (Colorado), and southwestern San Juan basin (New Mexico) sections, based on my interpretations of the data presented, are illustrated in Figure 10. Significant results are the following:

1. Upper Campanian strata (defined by coexisting zones of ammonites *Baculites scotti* to *Baculites reesidei* and of palynomorphs *Trudopollis meekeri, Myrtaceoipollenites peritus,* and *Pseudoplicapollis* spp.) include the Iles and lower Williams Fork formations of northwestern Colorado and the upper Lewis, Pictured Cliffs, and Fruitland formations and part of the Kirtland Formation of the San Juan basin, Colorado and New Mexico.

2. If the base of the *Baculites reesidei* zone is considered to be the Campanian-Maastrichtian boundary, that stage boundary lies below the Twentymile Sandstone Member of the Williams Fork Formation in northwestern Colorado, below the Farmington Sandstone Member of the Kirtland Formation at Durango, Colorado, and somewhere within the Kirtland Formation in the southwestern San Juan basin, New Mexico.

To illustrate and to emphasize this result in another way, a graphic correlation from the northwestern Colorado section to the Durango, Colorado, section is shown in Figure 11. This graph, plotted using the method of Shaw (1964), illustrates a line of time correlation defined by the stratigraphic positions of guide ammonites and palynomorphs present in both areas. Using the line of correlation so defined, the Campanian-Maastrichtian boundary of northwestern Colorado projects to the upper part of the lower shale member of the Kirtland Formation in the Durango section. In this way, a stage boundary reference point is established for the San Juan basin. Unfortunately, however, there are not yet enough common biozone data points available to extend this stage boundary as precisely to the southwestern San Juan basin section. Even so, considering the overall correlations illustrated in Figures 10 and 11, it is difficult to conclude that the Fruitland and Kirtland formations of the southwestern San Juan basin are *late* Maastrichtian in age, as did Lindsay and others (1981) using paleomagnetic methods.

3. Lower Maastrichtian strata (defined by coexisting zones of ammonites *Baculites reesidei* to *Baculites clinolobatus* and of palynomorphs *Triporopollenites scabroporus* and others) include the uppermost Williams Fork Formation and much of the Lewis Shale of northwestern Colorado; the upper part of the lower shale member of the Kirtland Formation, Farmington Sandstone Member and upper shale member of the Kirtland Formation, and the McDermott Formation at Durango; and possibly the upper Kirtland Formation in the southwestern San Juan basin.

If the extended line of correlation in Figure 11 is valid, it indicates that the youngest Cretaceous rocks in the Durango section are equivalent to about the middle of the Lewis Shale in the northwestern Colorado section, within the middle part of the lower Maastrichtian interval. The horizontal segment of the line indicates the thickness, about 700 m, of Maastrichtian strata present in the northwestern Colorado section that has no equivalent in the Durango section.

4. Upper Maastrichtian strata (defined by coexisting zones of ammonites *Baculites clinolobatus* to *Sphenodiscus* sp., through zone of *Triceratops* and of palynomorphs *Thomsonipollis* and others) include the upper Lewis Shale and Lance Formation of northwestern Colorado and the upper Pierre, Fox Hills, Laramie, Arapahoe, and lower Denver formations of the Golden, Colorado, section. This biostratigraphic interval is not detected with certainty in any of the San Juan basin sections. If it is present, it would be in the very thin interval of rocks called Naashoibito Member of the Kirtland Formation or lower Ojo Alamo Sandstone in the western and southern San Juan basin, New Mexico.

5. As shown in Figure 10, preserved strata of Maastrichtian age are approximately 1140 m thick in the northwestern Colorado section, 380 m thick in the Durango, Colorado, section, and 90 m thick in the southwestern San Juan basin, New Mexico, section. It does not appear likely, on sedimentologic grounds

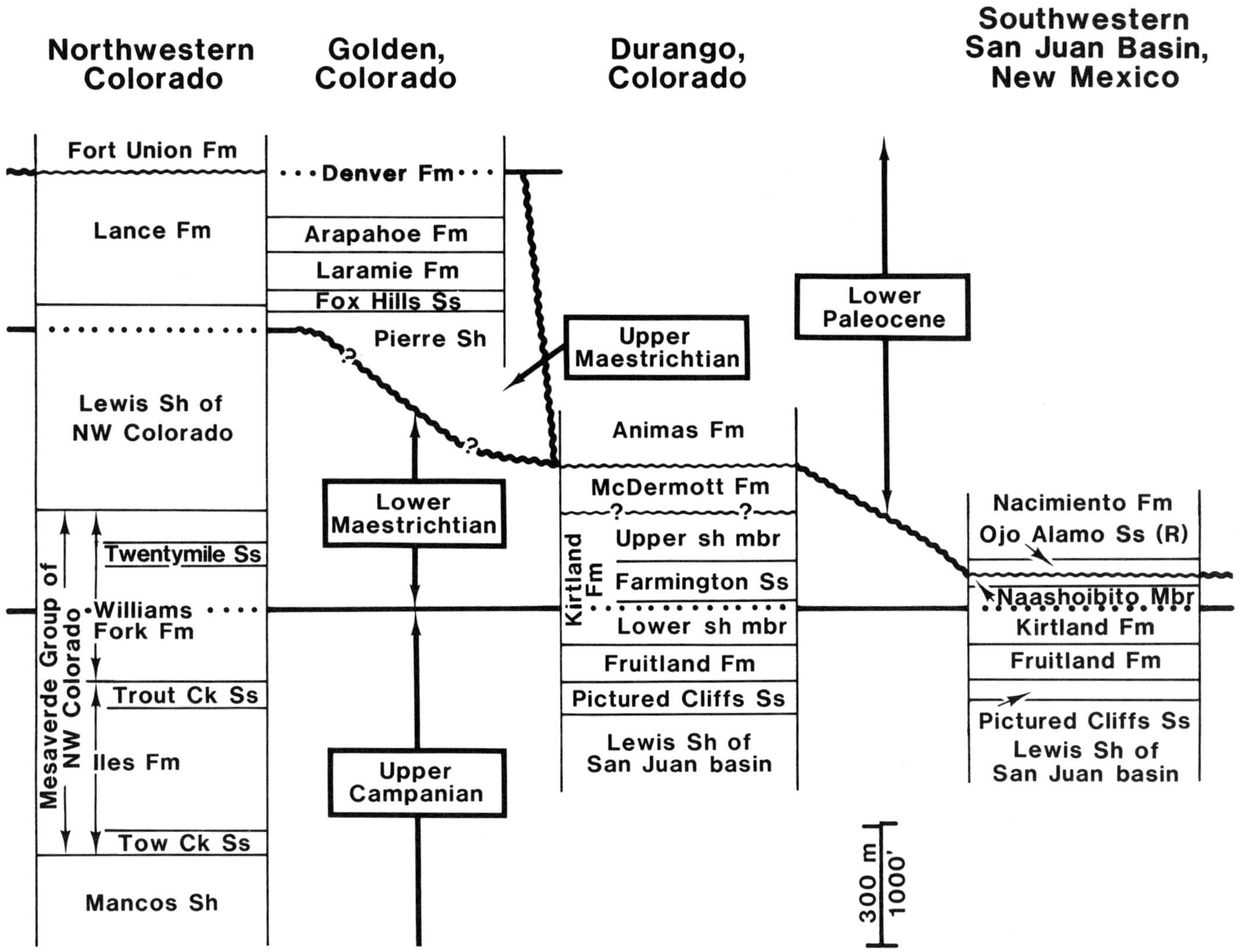

Figure 10. Biostratigraphic correlation of formations from northwestern Colorado and Golden, Colorado, to Durango, northern San Juan basin, Colorado, and to southwestern San Juan basin, New Mexico. Datum is the Campanian-Maastrichtian boundary based on ammonite and palynomorph zones.

alone, that the southwestern San Juan basin section has a complete, continuous record of Maastrichtian time, about 5.6 Ma.

6. Lower Paleocene strata (defined by coexisting zones of Puercan mammals and palynomorphs *Momipites coryloides* and others) include the lower Fort Union Formation in northwestern Colorado; the middle Denver Formation at Golden, Colorado; the lower Animas Formation at Durango, Colorado; and the Ojo Alamo Sandstone (restricted) and lower Nacimiento Formation in the southwestern San Juan basin, New Mexico.

SUMMARY AND CONCLUSIONS

Lithostratigraphic and biostratigraphic data compiled from many outcrops and two cores show that variable combinations of ammonite, vertebrate, and palynomorph zones can be found in rocks assignable to late Campanian, Maastrichtian, and early Paleocene ages in Colorado and that some of the same zones can be found in the San Juan basin of southwestern Colorado and northwestern New Mexico.

The Fruitland Formation and Kirtland shale (excluding the Naashoibito Member) of the San Juan basin, claimed to be of late Maastrichtian age (paleomagnetic chrons 30 and 31) by paleomagnetic methods (Lindsay and others, 1981), are of late Campanian to early Maastrichtian age according to the biostratigraphic ages and correlations of this paper. Therefore, those formations are not a factor in the Cretaceous-Tertiary boundary problem, unless they are truncated by erosion, and their contained dinosaur fossils have no bearing on the problem of the final extinction of dinosaurs.

The Cretaceous-Tertiary boundary, quite well defined in the

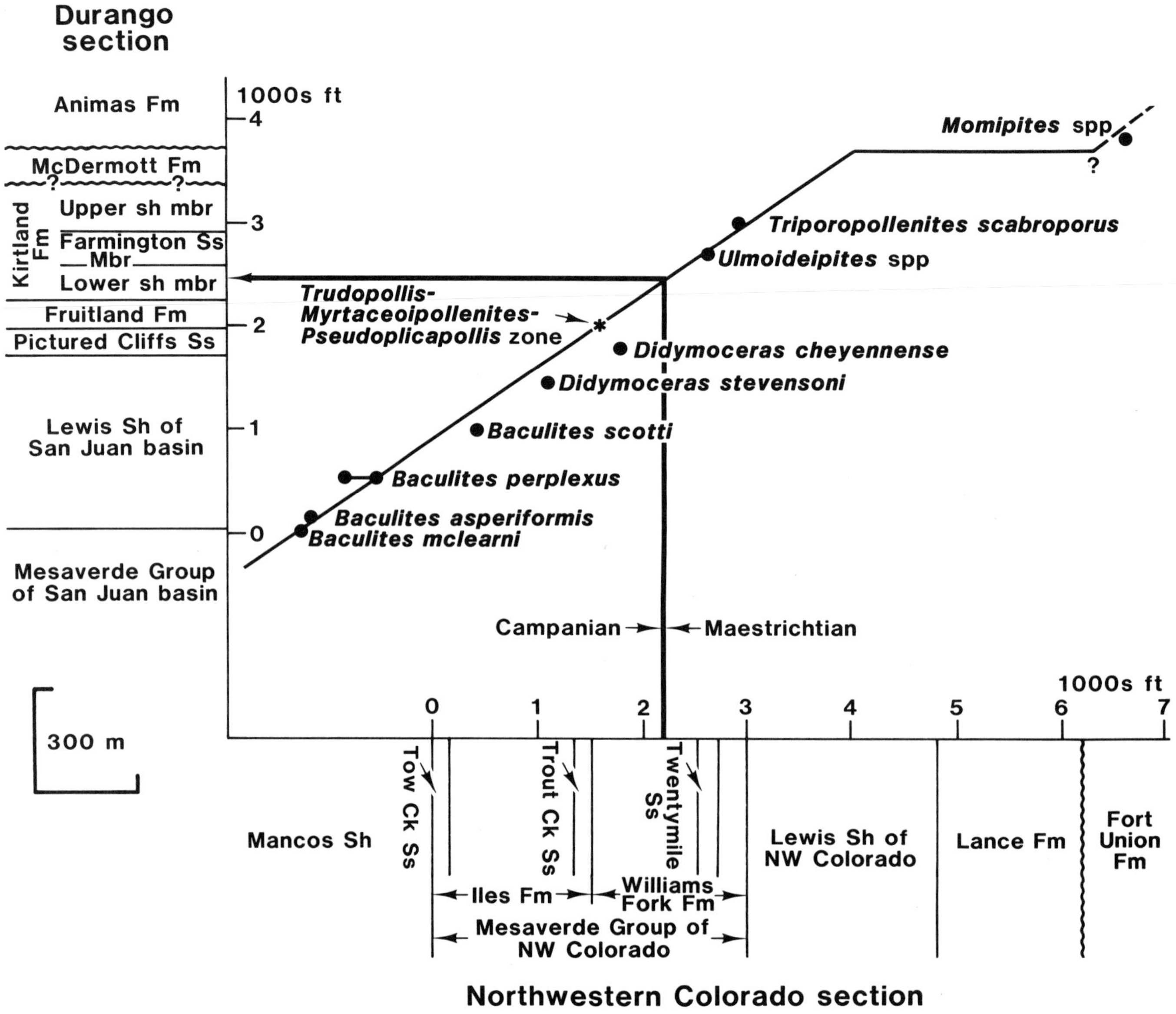

Figure 11. Graphic correlation of formations from northwestern Colorado to Durango, northern San Juan basin, Colorado. *Baculites* and *Didymoceras* data points are the presently known occurrence levels; complete range zones are not known. The *Trudopollis-Myrtaceoipollenites-Pseudoplicapollis* data point is the thin concurrent-range zone of these genera in each section; maximum known thickness is about 91 m. The *Ulmoideipites* spp., *Triporopollenites scabroporus,* and *Momipites* spp. data points are the bases of their range zones, as presently known.

relatively complete Golden and Raton basin sections, is an unconformable boundary throughout the San Juan basin because no complete section relative to the Colorado reference sections can be found. In the San Juan basin sections, the magnitude of the Cretaceous-Tertiary unconformity varies from about half of the Maastrichtian Stage, as at Durango, Colorado, to most of the Maastrichtian Stage, as in the Gasbuggy core, New Mexico. In the western and southern San Juan basin, New Mexico, presently available palynomorph data indicate only that some of the rocks mapped as lower Ojo Alamo Sandstone or Naashoibito Member of the Kirtland Formation are Maastrichtian in age. If they are early Maastrichtian, then at least half of the Maastrichtian Stage is not represented in those areas. If they are upper Maastrichtian in age, then two unconformities are likely, with a thin sheet of upper Maastrichtian lower Ojo Alamo or Naashoibito strata lying unconformably on early Maastrichtian Kirtland Formation and lying unconformably below Paleocene strata. If variable amounts of earliest Paleocene time also are not represented in the San Juan basin sections, that cannot be detected using the palynomorph zones of this paper.

ACKNOWLEDGMENTS

I thank A. T. Cross, J. E. Fassett, and R. H. Tschudy for manuscript reviews; J. E. Fassett, J. K. Rigby, Jr., R. E. Sloan, and R. H. Tschudy for helpful discussions during the past several years; C. R. Robison for providing core samples from the Fossil Forest area, New Mexico; and D. J. Madden and C. Manfrino for their discoveries of guide fossils in the Grand Hogback and Durango areas, Colorado.

REFERENCES CITED

Anderson, R. Y., 1960, Cretaceous-Tertiary palynology, eastern side of the San Juan Basin, New Mexico: New Mexico Bureau of Mines and Mineral Resources Memoir 6, 59 p.

Baltz, E. H., Ash, S. R., and Anderson, R. Y., 1966, History of nomenclature and stratigraphy of rocks adjacent to the Cretaceous-Tertiary boundary, western San Juan Basin, New Mexico: U.S. Geological Survey Professional Paper 524-D, 23 p.

Cobban, W. A., 1973, Significant ammonite finds in uppermost Mancos Shale and overlying formations between Barker Dome, New Mexico, and Grand Junction, Colorado, *in* Fassett, J. E., ed., Cretaceous and Tertiary rocks of the southern Colorado Plateau: Four Corners Geological Society Memoir, p. 148–153.

Dickinson, R. G., Leopold, E. B., and Marvin, R. F., 1968, Late Cretaceous uplift and volcanism on the north flank of the San Juan Mountains, Colorado, *in* Epis, R. C., ed., Cenozoic volcanism in the southern Rocky Mountains: Colorado School of Mines Quarterly, v. 63, no. 3, p. 125–148.

Donnell, J. R., 1969, Paleocene and lower Eocene units in the southern part of the Piceance Creek basin, Colorado: U.S. Geological Survey Bulletin 1274-M, 18 p.

Fassett, J. E., 1973, The saga of the Ojo Alamo Sandstone; or the rock-stratigrapher and the paleontologist should be friends, *in* Fassett, J. E., ed., Cretaceous and Tertiary rocks of the southern Colorado Plateau: Four Corners Geological Society Memoir, p. 123–130.

Fassett, J. E., and Hinds, J. S., 1971, Geology and fuel resources of the Fruitland Formation and Kirtland Shale of the San Juan basin, New Mexico and Colorado: U.S. Geological Survey Professional Paper 676, 76 p.

Fouch, T. D., Lawton, T. F., Nichols, D. J., Cashion, W. B., and Cobban, W. A., 1983, Patterns and timing of synorogenic sedimentation in Upper Cretaceous rocks of central and northeast Utah, *in* Reynolds, M. W., and Dolly, E. D., eds., Mesozoic paleogeography of the west-central United States: Rocky Mountain Secton, Society of Economic Paleontologists and Mineralogists, Rocky Mountain Paleogeography Symposium 2, p. 305–336.

Izett, G. A., Cobban, W. A., and Gill, J. R., 1971, The Pierre Shale near Kremmling, Colorado, and its correlation to the east and west: U.S. Geological Survey Professional Paper 684-A, 19 p.

Leffingwell, H. A., 1970, Palynology of the Lance (Late Cretaceous) and Fort Union (Paleocene) Formations of the type Lance area, Wyoming, *in* Kosanke, R. M., and Cross, A. T., eds., Symposium on palynology of the Late Cretaceous and Early Tertiary: Geological Society of America Special Paper 127, p. 1–64.

Lindsay, E. H., Butler, R. F., and Johnson, N. M., 1979, Reply *to* Comment by Lucas and Rigby: Geology, v. 7, p. 326–327.

—— , 1981, Magnetic polarity zonation and biostratigraphy of late Cretaceous and Paleocene continental deposits, San Juan basin, New Mexico: American Journal of Science, v. 281, p. 390–435.

Lowrie, W., and Alvarez, W., 1981, One hundred million years of geomagnetic polarity history: Geology, v. 9, p. 392–397.

Madden, D. J., 1983, Stratigraphy of Grand Hogback coalfield, Upper Campanian Mesaverde Group, Rifle Gap to Newcastle, Garfield County, Colorado [Ph.D. thesis]: Golden, Colorado School of Mines, 381 p.

Manfrino, C., 1984, Stratigraphy and palynology of the upper Lewis Shale, Pictured Cliffs Sandstone, and lower Fruitland Formation (Upper Cretaceous) near Durango, Colorado [M.S. thesis]: Golden, Colorado School of Mines, 97 p.

Newman, K. R., 1961, Micropaleontology and stratigraphy of late Cretaceous and Paleocene formations, northwestern Colorado [Ph.D. thesis]: Boulder, University of Colorado, 96 p.

—— , 1964, Palynologic correlations of late Cretaceous and Paleocene formations, northwestern Colorado, *in* Cross, A. T., ed., Palynology in oil exploration: Society of Economic Paleontologists and Mineralogists Special Publication 11, p. 169–180.

—— , 1965, Upper Cretaceous–Paleocene guide palynomorphs from northwestern Colorado: University of Colorado Studies, Series in Earth Sciences, No. 2, 21 p.

—— , 1974, Palynomorph zones in early Tertiary formations of the Piceance Creek and Uinta basins, Colorado and Utah, *in* Murray, D. K., ed., Guidebook to the energy resources of the Piceance Creek basin, Colorado: 25th Field Conference, Rocky Mountain Association of Geologists, p. 47–55.

—— , 1979, Cretaceous/Paleocene boundary in the Denver Formation at Golden, Colorado, USA, *in* Christensen, W. K., and Birkelund, T., eds., Proceedings, Cretaceous-Tertiary boundary events symposium, II: University of Copenhagen, p. 246–248.

—— , 1980, Paleocene time scale for Rocky Mountain region [abs.]: American Association of Petroleum Geologists Bulletin, v. 64, p. 757.

—— , 1982, Biostratigraphy of Fruitland, Kirtland, and Animas Formations (Cretaceous and Paleocene), northern San Juan basin, Colorado: Geological Society of America Abstracts with Programs, v. 14, p. 344.

—— , 1983, Biostratigraphic correlation of K-T rocks from northwestern Colorado to San Juan basin, Colorado and New Mexico: Geological Society of America Abstracts with Programs, v. 15, p. 307.

Obradovich, J. D., and Cobban, W. A., 1975, A time-scale for the Late Cretaceous of the Western Interior of North America: Geological Association of Canada Special Paper 13, p. 31–54.

Obradovich, J. D., and Sutter, J. F., 1984, What is the age of the Cretaceous-Tertiary boundary as recognized in continental deposits?: Geological Society of America Abstracts with Programs, v. 16, p. 612.

Reeside, J. B., Jr., 1924, Upper Cretaceous and Tertiary formations of the western part of the San Juan basin of Colorado and New Mexico: U.S. Geological Survey Professional Paper 134, p. 1–70.

Scott, G. R., and Cobban, W. A., 1965, Geologic and biostratigraphic map of the Pierre Shale between Jarre Creek and Loveland, Colorado: U.S. Geological Survey Miscellaneous Geological Investigation Map I-439.

Shaw, A. B., 1964, Time in stratigraphy: New York, McGraw-Hill, 365 p.

Stanley, E. A., 1961, A new sporomorph genus from South Dakota: Pollen et Spores, v. 3, p. 155–162.

—— , 1965, Upper Cretaceous and Paleocene plant microfossils and Paleocene dinoflagellates and hystrichosphaerids from northwestern South Dakota: Bulletins of American Paleontology, v. 49, no. 222, p. 175–384.

Stone, J. F., 1973, Palynology of the Almond Formation (Upper Cretaceous), Rock Springs uplift, Wyoming: Bulletins of American Paleontology, v. 64, no. 278, p. 1–135.

Tschudy, B. D., and Leopold, E. B., 1970, Aquilapollenites (Rouse) Funkhouser; Selected Rocky Mountain taxa and their stratigraphic ranges, *in* Kosanke, R. M., and Cross, A. T., eds., Symposium on palynology of the Late Cretaceous and Early Tertiary: Geological Society of America Special Paper 127, p. 113–167.

Tschudy, R. H., 1970, Palynology of the Cretaceous-Tertiary boundary in the northern Rocky Mountain and Mississippi embayment regions, *in* Kosanke, R. M., and Cross, A. T., eds., Symposium on palynology of the Late Cretaceous and Early Tertiary: Geological Society of America Special Paper 127, p. 65–111.

—— , 1973, The Gasbuggy core; A palynological appraisal, *in* Fassett, J. E., ed., Cretaceous and Tertiary rocks of the southern Colorado Plateau: Four Corners Geological Society Memoir, p. 131–143.

—— 1980, Normapolles pollen from *Aquilapollenites* province, western United States: New Mexico Bureau of Mines and Mineral Resources Circular 170, 14 p.

Williams, C. A., 1984, Palynology of late Cretaceous and early Tertiary strata from the northern Raton basin, south-central Colorado [Ph.D. thesis]: Boulder, University of Colorado, 313 p.

Zapp, A. D., and Cobban, W. A., 1960, Some late Cretaceous strand lines in northwestern Colorado and northeastern Utah: U.S. Geological Survey Professional Paper 400-B, p. 246–249.

MANUSCRIPT ACCEPTED BY THE SOCIETY MARCH 24, 1986

Printed in U.S.A.

Geological Society of America
Special Paper 209
1987

Paleocene and latest Cretaceous mammal ages, biozones, magnetozones, rates of sedimentation, and evolution

Robert E. Sloan
Department of Geology and Geophysics
University of Minnesota
310 Pillsbury Drive
Minneapolis, Minnesota 55455

ABSTRACT

North American Paleocene land mammal ages are the Mantuan, Puercan, Torrejonian, Tiffanian, and Clarkforkian. These ages (and associate stages) are subdivided into 16 zones or subzones, varying in duration from 0.1 to 2.9 m.y., defined by widespread species. Although gross evolutionary changes during the first four of these ages are about equal, their durations are very unequal. As defined by magnetostratigraphy and fossil occurrence, the Mantuan is about 0.2 m.y., the Puercan about 1.1 m.y., the Torrejonian about 3.1 m.y., the Tiffanian about 6.1 m.y., and the Paleocene part of the Clarkforkian about 1.3 m.y. in duration. Puercan encompasses normal magnetozone 29, Torrejonian zones 28 and 27, and the Tiffanian–Clarkforkian boundary falls in zone 25. The type Rio Chico Formation of Patagonia is of mid-Tiffanian to Clarkforkian age.

Problems in the identification of magnetozones in the San Juan Basin have arisen because an unconformity is present between the Kirtland Shale and the Ojo Alamo Sandstone, and for some years an extra normal chron was falsely identified. When this hiatus is taken into account, marine and terrestrial fossil correlations agree with magnetozone correlations throughout Upper Cretaceous and Paleocene rocks.

The Danian stage in marine rocks in North Dakota is equivalent to Mantuan through early Tiffanian; the Thanetian is exactly equivalent to mid-Tiffanian to early Clarkforkian. The rate of Paleocene sedimentation in the major basins of North America does not depart from linearity much more than the contemporary rate of seafloor spreading. Terrestrial rates of sedimentation vary from a peak of 568 bubnoffs (b) (meters per million years) for the Hoback Formation at the Rocky Mountain front to 99 b in the San Juan Basin and 15 b in the Black Peaks Formation in Texas, compared to 2.7 b at Gubbio, Italy. Sedimentation rates along a transect through the Bighorn, Powder River, and Williston basins follow the equation $Y = 200X^{-0.25}$, where Y is the sedimentation rate in bubnoffs and X is the radial distance in kilometers from the Absaroka thrust.

Absolute taxonomic and morphologic rates of evolution of the most rapidly evolving mammals during the Bugcreekian–Mantuan interval across the Cretaceous–Paleocene boundary peak at 5 genera per m.y. and 3.85 darwins (a rate of measurement defined in the text), the fastest rates known in the fossil record, and decline exponentially to more normal rates of 1 species per m.y. and 0.5 darwins by the Tiffanian. Range charts of 299 species of ungulates, primates, and multituberculates permit ready identification of zones. Seven new species of multituberculates are described, and shape and metrical properties of latest Cretaceous–Paleocene neoplagiaulacid multituberculates are summarized for ease in identification.

INTRODUCTION

Four of the land mammal ages established by the Wood Committee (Wood and others, 1941) for the Paleocene are still valid and usable. In order of decreasing age, these are the Puercan, Torrejonian, Tiffanian, and Clarkforkian. The fifth age of the Wood Committee, the Dragonian, is no longer considered to be of age status since it is not as well differentiated at the generic level as the others. The durations of these ages were unknown at the time of the committee report. It was generally (but incorrectly) presumed that since roughly equal amounts of generic change were involved in each age, the ages were roughly equal in duration as well. Although little progress in the study of the terrestrial Paleocene was made during and shortly after World War II, since 1960 there has been more than a fivefold increase in number of local faunas and localities known. This increase in knowledge was brought about primarily by the development of the screen-washing technique.

An additional age, the Mantuan, was established by Van Valen (1978); although shorter than any other age, it is very distinct at the generic level. All the age definitions of the Wood Committee are now obsolete because of the increase in knowledge since 1941; stratigraphic range extensions and extensive taxonomic review consequently require alteration of these definitions. Recent general summaries of Paleocene litho- and biostratigraphy include Van Valen and Sloan (1965), D. E. Russell (1967), Sloan (1970), Gingerich (1975b, 1976, and 1980a), and Savage and Russell (1983).

Most of the problems of Paleocene biostratigraphy that remain involve discrimination of time intervals that are an order of magnitude smaller than the usual Cenozoic land mammal ages. The five conventional land mammal ages included wholly or in part within the Paleocene Epoch are here subdivided into sixteen mammalian zones or subzones, spanning the approximately 12 million years (Harland and others, 1982) of the epoch (Fig. 1; in pocket inside back cover). It likely will be possible, through careful attention to stratigraphic position of collections and metrical properties of species, to further subdivide these zones into usable subzones. Increasing the resolution of Paleocene biostratigraphic correlation to about 0.1 million years is a goal likely to be attained in the next decade. The principal reason for this happy state of affairs is the rapid rate of evolution during the initial stages of the adaptive radiation of mammals during and after the extinction of dinosaurs. This rapid rate of evolution means that the use of stratigraphic or chronospecies that are statistically distinct, and overlapping ranges of such species, can lead to the establishment of vertically separable stratigraphic zones or subzones as thin as 8 m, even in sections within 50 km of major sources of sediment. Farther from the major sources of sediment, zones and subzones thin. The observed zones in any one vertical section thicken with the passage of time in the Paleocene as a result of the observed exponential decline in rates of speciation and phylogenetic transformation (Sloan, 1970).

The Paleocene Epoch represents approximately the first 12 to 9 million years of the Tertiary Period. Schimper in 1874 defined the Paleocene to include Thanetian and Sparnacian rocks of the Paris Basin in France and Belgium (Gingerich, 1975a; Schorn, 1971, 1975). More recently the Sparnacian has generally been referred to the Eocene, and the rocks of Danian and Montian age are now generally added to the Paleocene as post-Cretaceous in age.

Little rock of Paleocene age accumulated in Europe, where Paleocene rocks are some 30 to 50 m thick in the Paris Basin and the Mediterranean area. In North America, in contrast, the marine Paleocene of the Gulf Coast region is about 150 m thick, and the terrestrial sediments of the Rocky Mountain region reach a maximum thickness of 3500 m in western Wyoming. These nonmarine sedimentary rocks were deposited by rivers and streams in channels, on flood plains, and occasionally in lakes and swamps, east and downstream from the rapidly rising Rocky Mountains. The areas of outcrop of these continental sediments are customarily referred to as basins. These basins are structural depressions, for the most part produced by late Paleocene or Eocene uplift of the intervening mountain ranges. The remaining Paleocene rocks in the basins are erosional remnants of a former continuous sheet of sediments that stretched from the heart of the Rocky Mountains to the Gulf coastal plain.

It is often convenient to refer to the rocks in terms of the Eocene to Recent structural basins in which they occur. Figure 2 is an index map showing the names of the major basins containing Paleocene sediments and the locations of the major Paleocene mammal localities within the basins. The only major mammal localities of Paleocene age in North America not included in Figure 2 are the Goler Formation localities of southern California and the famous oil well in Louisiana that produced the type and only specimen of *Anisonchus fortunatus.*

The climate of North America during the Paleocene was considerably warmer than at present, and the vegetation consisted of forests and marshes. Wolfe and Hopkins (1967) state that "No known Paleocene flora in North America indicates a climate cooler than warm-temperate, and sub-tropical climate existed at least as far north as latitude 62N in Alaska during part, if not all of the Paleocene." Sloan (1970) analyzed the Rocky Mountain paleobotanical data and suggested a trend from an equable subtropical climate during the late Cretaceous to a less equable, warm temperate climate during the Paleocene, followed by a return to an equable subtropical climate during the early Eocene. This climatic nadir of this cycle was during the middle of the late Paleocene, or Tiffanian (Sloan, 1970; Hickey, 1980).

For two reasons, the methodology used in this paper and in Paleocene studies in general differs from that used in mammalian biostratigraphy in later Tertiary rocks. First, the Paleocene is unique among Tertiary epochs in that it is the time of the placental mammal radiation. During the early Paleocene, there occurred rates of evolution as great as ten times faster than at most other times in the Cenozoic. Second, it has become apparent that a more refined methodology is needed to overcome the significant errors of geologic correlation and interpretation of structural, geo-

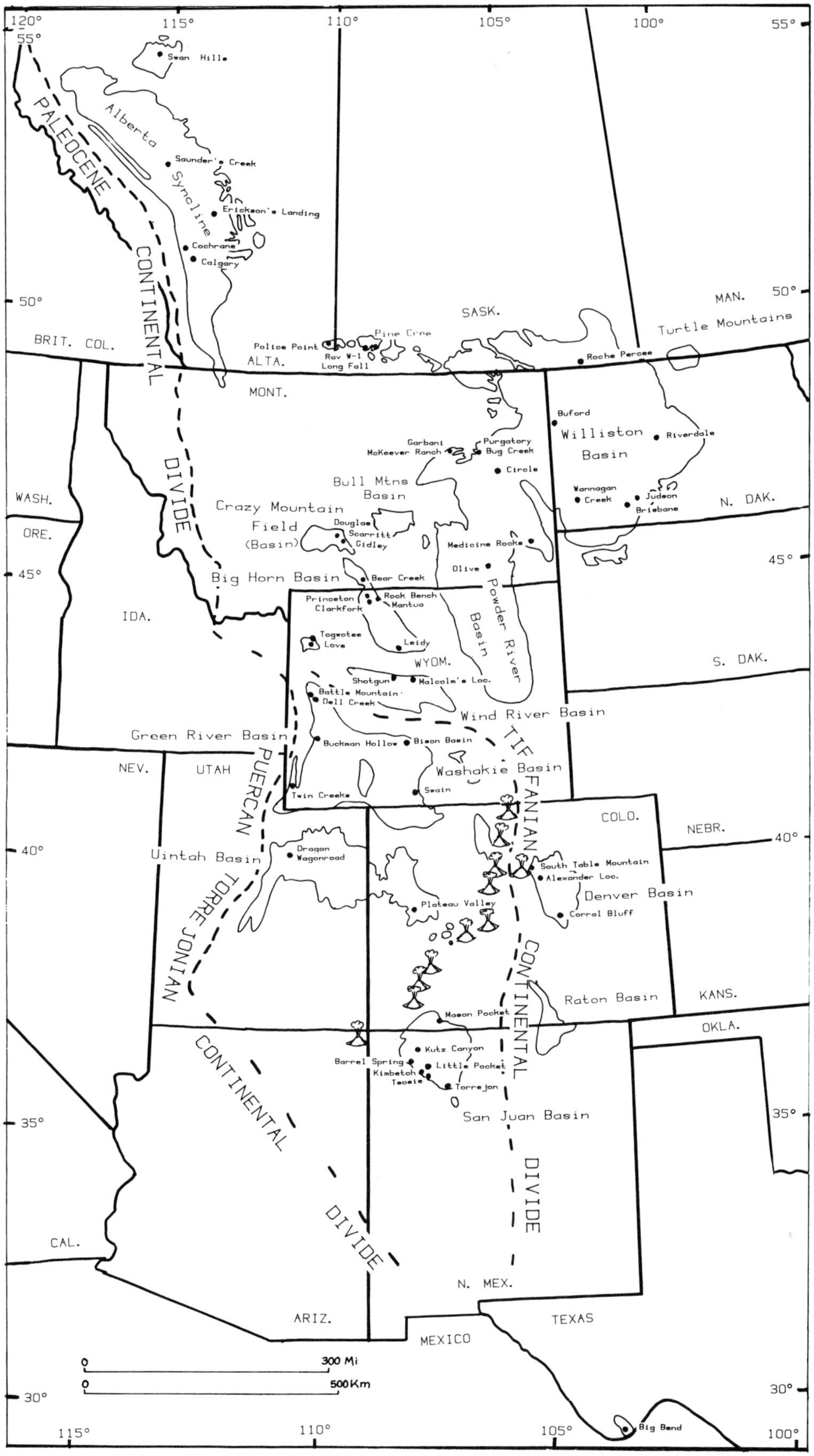

Figure 2. Location map of the Rocky Mountain and Great Plains region of North America showing major basins and Paleocene mammal localities. Volcano symbols locate latest Cretaceous and early Paleocene volcanic sources (Steven and others, 1972).

logic, and paleontologic history routinely being made by geologists not aware of the logical consequences of this rapid rate of evolution.

The five Paleocene mammal ages average 2 million years in duration as compared to averages of 5 million years for each of the Eocene ages, 4 million years for the Oligocene mammal ages, 4 million years for the Miocene mammal ages, 3 million years for the Pliocene mammal ages, and 1 million years for those of the Pleistocene. These mammal ages are rather coarsely defined and are based mainly on the distribution of certain genera. In both the Paleocene and the Pleistocene the practical unit of correlation is significantly shorter than a mammal age. In both cases the faunas were changing rapidly, and documentation of the rapid changes greatly improves the precision of useful correlation.

Lack of recognition of this rapid rate of evolution and its consequences has led many geologists to postulate major unconformities in Paleocene rocks on the grounds that late Paleocene faunas occurred stratigraphically too close to the Cretaceous sediments for early and middle Paleocene rocks to be present. Such unconformities rarely exist; in most parts of North America deposition was continuous. The original Paleocene land mammal ages were established for roughly equal amounts of morphological change, but because of the initial rapid rates of evolution the durations of the early and middle Paleocene ages as defined by mammals are significantly shorter than those of the late Paleocene.

The present Paleocene zonation is an approximation limited by the available number of fossil samples. Except in the early Paleocene (Mantuan and Puercan), the number of faunas required to subdivide a zone probably will be three or four times the present number of faunas in that zone. In the early Paleocene there are still sufficient gaps in species lineages so that new subzones can be established. I originated the zonal abbreviations used here in 1976 in charts freely distributed; they are based on the initial letter or letters of the age and numbered sequentially in an age. They have been widely used.

Some vertebrate paleontologists have been reluctant to accept the plausibility of zonation as opposed to biofacial interpretations of faunal differences on the grounds that some species are present in more than one zone. Not all species lineages evolved at the same rate; those that evolved more rapidly are more useful for precise correlation. The ultimate test of zonation is repeatability and predictability. For example, the differences between *Deltatherium* and *Pantolambda* zones in the San Juan Basin are duplicated in the Crazy Mountain Field of Montana. Species in evolving species lineages appear in the same stratigraphic order in different basins and thus are predictable.

PHYSICAL STRATIGRAPHY OF THE NORTH AMERICAN PALEOCENE

As is usually the case, the rock-stratigraphic nomenclature of the Paleocene developed independently in the various basins of North America. Formation names change abruptly at state and national boundaries or from basin to basin. Adequate descriptions of type sections are the exception, and some widely used names have never had type sections designated. Adequate stratigraphic descriptions and names by vertebrate paleontologists have often been ignored or, even worse, altered out of recognition by U.S. Geological Survey geologists, creating a confused tangle of names. The misguided attempt to make formation boundaries coincide with period and epoch boundaries rather than with significant lithologic changes is usual in the literature on North American Paleocene strata. Convenient summaries of Paleocene strata are Taylor and others (1964), Robinson (1972), and McDonald (1972). A more detailed volume by Flores and Kaplan (1985) appeared while this paper was in preparation; half of the included papers concern the North American Paleocene.

The Fort Union Formation, the most widely recognized Paleocene rock unit in North America, is no longer given formation rank at its type locality, Buford, North Dakota (originally Fort Union), at the junction of the Missouri and Yellowstone rivers. It now has group status with five included formations in the type area and surrounding parts of the Williston Basin (100 km to the north across the Canadian boundary in Saskatchewan the same rocks in the same basin are known as the Ravenscrag Formation).

Because of various nomenclatorial problems, I provide a basin-by-basin discussion of the rock units. As previously noted, Figure 2 is an index map of the Paleocene basins of North America, and the major mammal localities, local durations of these formations in terms of mammal ages, zones, and subzones are shown in Figure 1.

In the southern Alberta Syncline the Willow Creek Formation includes latest Cretaceous and early Paleocene rocks. The overlying and partial lateral equivalent Porcupine Hills Formation grades upward and northerly into the late Paleocene Paskapoo Formation. The base of the Paskapoo Formation includes synorogenic cobbles derived from eastern British Columbia, a source cut off during the Eocene (Taylor and others, 1964). Deposition of these sediments was originally continuous with those of the Williston Basin to the east, but Eocene to mid-Miocene erosion separated the outcrop belts.

The Canadian portion of the Williston Basin extends from southeastern Alberta to southwestern Manitoba. In the Alberta and Saskatchewan portions of the basin, Paleocene sediments are referred to the Ravenscrag Formation. Known fossil localities within the Ravenscrag Formation show its age to be from Mantuan at the base to late Tiffanian at least. The absence of Torrejonian mammal localities appears to be entirely the result of limited collecting. In southeastern Saskatchewan the marine Cannonball Formation replaces part of the lower Ravenscrag Formation.

On the eastern border of the Williston Basin, straddling the international boundary between North Dakota and Manitoba an isolated block of Paleocene sediments forms the Turtle Mountains. This low set of hills rises some 300 m above the surrounding plain. The marine Cannonball Formation occurs at the base of the Paleocene, and the overlying Turtle Mountain

Formation includes the balance of the Paleocene. The rocks are poorly exposed and have not been searched for fossil vertebrates.

On the U.S. side of the border the Paleocene rocks of the Williston Basin in eastern Montana and the Dakotas are subdivided into a number of formations. The basal unit, the Tullock Formation, is coal bearing, in contrast to the underlying Cretaceous Hell Creek Formation. Several fossil localities date the Tullock Formation as Mantuan and Puercan in age. The overlying non-coaly Lebo Shale includes most but not all of Torrejonian age-equivalent strata. The Tongue River Formation, largely sandstone, includes rocks of late Torrejonian to late Tiffanian age. The Sentinel Butte Formation is also rich in coal and is of late Tiffanian to Clarkforkian age. All of the rocks within a 20-km radius of old Fort Union (now Buford)—the type locality of the Fort Union Formation—belong to the Sentinel Butte Formation. These formations collectively form the Fort Union Group. The latest Paleocene is included within the dominantly early Eocene Golden Valley Formation of North Dakota. In northwestern South Dakota the Tullock and Lebo formations are collectively referred to the Ludlow Formation. In North and South Dakota the Tullock, Lebo, and lower Tongue River formations intertongue with the marine Cannonball Formation, which was deposited in a seaway extending from the Gulf of Mexico through central Texas north into Canada. Foraminifera from the Cannonball Formation show that it is equivalent in age to the marine planktonic foraminiferal zones P1 and P2 (Fox and Olsson, 1969), widely recognizable over most of the world. The intertonguing between marine and nonmarine beds in North Dakota permits direct correlation between land mammal ages and zones and the marine Paleocene sequence. The Danian age (stage) of Berggren and others (1982) is directly correlatable with the Mantuan to mid-Tiffanian land mammal ages.

The Crazy Mountain (fossil) Field of west central Montana is an isolated outcrop area of Paleocene sediments included within the Crazy Mountains Basin. The western portion of the outcrop area includes a 2-km-thick pile of volcaniclastic rocks known as the Livingston Formation, ranging in age from Late Cretaceous (Campanian) to late Paleocene (Tiffanian). The western part of the area is heavily forested, exposures are sparse, and no one has yet attempted to subdivide this unit there. The eastern half of the outcrop area, however, is ranch land, exposures are better, and there the Livingston is subdivided into four formations. The lowest is the Hell Creek Formation of latest Cretaceous age, overlain by the Bear Formation of Simpson (1937b), essentially the same as the Mantuan and Puercan Tullock Formation of eastern Montana except that coarser gravels are present. On top of the Bear Formation are the Torrejonian Lebo Shale and the latest Torrejonian and Tiffanian Melville Sandstone. Through the collecting efforts of Albert Silberling, the paleontology of the Lebo and Melville is well known (Simpson 1936, 1937a, 1937b).

Paleocene sediments of the Powder River Basin of southeastern Montana and northeastern Wyoming are generally referred to the Fort Union Formation despite the fact that Tullock, Lebo, Tongue River, and Sentinel Butte formations can be recognized. The Powder River Basin was separated from the Williston Basin by the uplift of the Black Hills and the Cedar Creek Anticline. This uplift took place during the Uintan late Eocene and the early Oligocene as is shown by Uintan radiometric dates on the intrusives of the northern Black Hills (Shapiro, 1971) and by the gravels and sands in the early Oligocene Chadron Formation of South Dakota eroded from a Black Hills source.

The Big Horn Mountains of north central Wyoming and adjacent Montana separate the Bighorn Basin from the Powder River Basin. The first sign of uplift of the Big Horn Mountains is the earliest Eocene Kingsbury Conglomerate near Sheridan, Wyoming, on the western side of the Powder River Basin. This synorogenic conglomerate contains *Hyracotherium*. Thus, during the Paleocene the Bighorn, Powder River, and Williston basins were all confluent. Jepsen (1940) defined the Paleocene Polecat Bench Formation to include three members—the Mantua Lentil, the Rock Bench Quarry Beds, and the Silver Coulee Beds—in the Bighorn Basin. Subsequent U.S. Geological Survey publications have generally referred this sequence of rocks to the "Fort Union Formation" despite the general lack of coal in all but the uppermost part of the Polecat Bench Formation and despite coarser sediments than those in the type area of the Fort Union.

In Jepsen's 1930 and 1940 papers and references therein, he provided all that was necessary to define a formation at that time. Although more information would be required today to define a new formation, Jepsen's designation was valid at the time, and since real differences exist between the upstream and downstream portions of the sedimentary record, Jepsen's name is still valid. In fact, Jepsen's three members are mappable and could be raised to formation rank. This has been borne out by Gingerich's work in the Bighorn Basin (as summarized in Gingerich, 1980a), where the Silver Coulee beds and the Polecat Bench Formation are mapped. Gingerich (1980a) is notable for the fact that U.S.G.S. geologists call these rocks by the inappropriate name Fort Union Formation, whereas non-U.S.G.S. geologists term them the Polecat Bench Formation. Unfortunately, this dual nomenclature will probably persist for years.

The tectonic uplift of the Bighorn Basin that resulted in the erosion of the Puercan to early–late Torrejonian (To3a of Fig. 1) sediments and deposition of the Rock Bench Quarry Beds had an effect downstream toward North Dakota. The coarsening of sediments at the transition from the Lebo Shale to the Melville Sandstone in the Crazy Mountain Field and from the Lebo Shale to the Tongue River Formation in eastern Montana is the same age as the Rock Creek Quarry Beds (To3b) and is the result of the same increase in volume of sediments due to the steepening of the grade of the depositing rivers.

To the south of the Bighorn Basin, separated from it by the Owl Creek Mountains, lies the Wind River Basin. Terrestrial Paleocene sediments of this basin are referred to the usual catchall, the "Fort Union Formation," and divided into two members. The lower unit is the marine or estuarine Waltman Shale deposited in a body of water apparently connected to the Cannonball

Sea to the east. The base of the upper terrestrial Shotgun Member is dated by the Shotgun local fauna of earliest Tiffanian age.

The uplift of the Owl Creek Mountains began during the late Tiffanian, about zone Ti4 (Fig. 1). The first sign of unroofing of the rising mountains is the pebbles of Cretaceous marine shale that occur in late Paleocene sediments along the north edge of the basin. Successively higher beds of early Eocene sediments of the Wind River Formation contain pebbles and cobbles of Mesozoic, Paleozoic, and Precambrian formations that appeared in inverse stratigraphic order as the Owl Creek Mountains were eroded.

McKenna (1980) summarized the geology and paleontology of the Togwotee Pass area, a region of structural complexity and extensive lateral facies change in short distances. Paleocene faunas of late Middle Tiffanian and Clarkforkian age are known from the synorogenic Pinyon Conglomerate and the overlying "lower variegated beds" of Love (1973). These small faunas are of considerable interest from the viewpoint of providing a detailed chronology of tectono-stratigraphic events in northwestern Wyoming and a look at higher altitude faunas.

South and west of the Wind River Basin southwestern Wyoming is separated by the Rock Springs Uplift into the western Green River Basin and the eastern Washakie Basin. The Gros Ventre and Wind River ranges, which separate the Wind River and Green River basins, were clearly elevated during the late Paleocene. The extreme northwestern corner of the Green River Basin has been called the Hoback Basin. In the Hoback Basin Paleocene sediments of late Paleocene age (Ti3 to Cf1; Fig. 1) are known as the Hoback Formation and reach the maximum thickness known for the Paleocene of 3500 m.

Throughout the Green River–Washakie Basin the main source of sediment was the Thrust Belt at the western edge of Wyoming. The Paleocene rocks of this basin are known as the "Fort Union Formation" although the coal beds so prominent in the type Fort Union are sparse and deposition throughout the Paleocene was in a basin separate from that of the type Fort Union. The Rock Springs Uplift that now separates the Green River and Washakie basins did not develop until the Eocene. The Washakie Basin extends into northwestern Colorado where it is called the Sand Wash Basin.

In the extreme southwestern corner of Wyoming and adjacent Utah the Paleocene rocks consist of synorogenic conglomerates derived from the Wasatch Range some 60 km to the west. These rocks are referred to the Henefer, Pulpit, and Sawmill conglomerates. Grain size rapidly decreases to the east, and in Fossil Syncline, still on the west side of the Green River Basin, these conglomerates grade into the Evanston Formation and overlying Clarkforkian Chappo Member of the Wasatch Formation. Evanston Formation is probably the most suitable name to apply to Paleocene rocks throughout the Green River and Washakie basins.

South of the Green River–Washakie Basin in northeastern Utah and northwestern Colorado lie the Uintah Basin and its eastern extension, the Piceance Basin. The western edge of the Uintah Basin is the north-south trending Wasatch Range in central Utah, the major western sediment source for this basin. The terrestrial North Horn Formation ranges in age from Campanian to Torrejonian and probably to early Tiffanian. Gazin (1941) separated the Paleocene portion of the North Horn Formation as the Joes Valley Member. Overlying the Joes Valley Member is the Flagstaff Limestone, a lacustrine unit resulting from the first total blockage of easterly drainage. The Flagstaff Limestone grades upward into the overlying lacustrine Wasatchian and Bridgerian Green River Formation. In the eastern end of the Uintah Basin the Flagstaff Limestone grades into the lower part of the terrestrial Colton Formation.

The Piceance Basin did not become separated from the Uintah Basin until the late Paleocene. The basal unit of the Paleocene in the Piceance Basin is the Ohio Creek Conglomerate, which is coarser to the east and finer grained to the west. Fossils are absent from the Ohio Creek Conglomerate, but the conformably overlying Atwell Gulch Member of the Wasatch Formation contains the latest Tiffanian and Clarkforkian Plateau Valley local fauna and extends upward into the early Wasatchian. It is quite apparent that there was a late Tiffanian uplift in central Colorado that produced these sediments and quite probable that this uplift produced the topographic barrier that closed the basin in which Flagstaff Lake, the earliest of the Green River lakes, developed.

North Park, a small intermontane basin in north central Colorado, contains the Paleocene and Eocene Coalmont Formation. No Paleocene mammals have yet been collected from these rocks, and their precise age is not yet established. The southern part of this basin is sometimes separated as Middle Park and contains the Paleocene Middle Park Formation. Again no fossil mammals have as yet been collected.

East of the Front Range in Colorado lies the Denver Basin in which the latest Cretaceous to early Paleocene (Puercan) Denver Formation occurs. In the southern part of the Denver Basin the term Dawson Arkose has been used for the same rock unit. The Denver Formation is a volcaniclastic unit. The volcanic source for this pile of sediments appears to be associated with the Colorado Mineral Belt (Steven and others, 1972). All that is left of the source volcanoes is a series of granodiorite and quartz monzonite stocks, plugs, dikes, and sills intruded into Cretaceous and earlier rocks along a northeast-trending lineament from the Carrizo Mountains of northeastern Arizona to Boulder, Colorado. These intrusions are the sources for many of the precious metal ore deposits of Colorado. K/Ar age determinations on this group of intrusions range from 72 to 65 Ma. The northeastern part of this series forms the source for the Paleocene rocks of Colorado's Middle Park and Denver Basin and includes the intrusives near Boulder, Blackhawk, Central City, Georgetown, Breckenridge, Climax, Leadville, and the West Elk Mountains in Colorado. To the southwest, this series of intrusives continues at Ouray, Rico, and LaPlata, Colorado, and the Carrizo Mountains of Arizona and forms a source for the Paleocene volcaniclastic sediments of the San Juan Basin of southern Colorado and New Mexico. These now eroded mountains clearly formed a major drainage divide and sediment source during the latest Cretaceous through at least the Torrejonian middle Paleocene. Early and middle Paleocene sediments in all basins thicken toward this source.

South of the Denver Basin on the east side of the Sangre de Cristo Range in northeastern New Mexico and adjacent Colorado lies the Raton Basin. In the southern part of the Raton Basin the Raton Formation includes latest Cretaceous and early Paleocene sediments. In the central part of the basin, the Raton Formation is overlain unconformably by the late Paleocene Poison Canyon Formation, and in the northwestern part of the basin the Poison Canyon Formation rests unconformably on Cretaceous sediments. Clearly the Poison Canyon Formation source was the same central Colorado Tiffanian uplift yielding the Ohio Creek Conglomerate and Atwell Gulch Member of the Wasatch Formation. Raton Basin rocks have yet to produce fossil mammals.

West of the Raton Basin in northwestern New Mexico and adjacent Colorado lies the San Juan Basin. This is the type area for the Puercan, Torrejonian, and Tiffanian mammal ages. The Paleocene sediments in the northern part of the basin are referred to the volcaniclastic Animas Formation of latest Cretaceous to middle Paleocene (Torrejonian) age and to the Tiffanian "Tiffany Member" (named but undefined) of the otherwise Eocene San Jose Formation. In the southern part of the basin in New Mexico, Puercan and Torrejonian sediments are referred to the Nacimiento Formation (Fig. 3, in pocket inside back cover). Mantuan sediments are probably included in the upper part of the dinosaur-bearing Ojo Alamo Sandstone, and late Paleocene sediments are missing by erosion. Early and middle Paleocene sediments clearly thicken and coarsen to the northwest, and the degree of maturity of the sandstone also decreases in that direction. Drainage was to the south, and the region was not yet a basin. During the late Paleocene the eastern side of the area was uplifted, forming the closure of the basin, and the major sediment source shifted to the east.

The last basin to be mentioned is a small one in the Big Bend National Park of Texas in which fluvial Late Cretaceous to early Eocene rocks, including the Maastrichtian to Torrejonian Javelina Formation and the Tiffanian Black Peaks Formation, were deposited.

In summary, early and middle Paleocene sediments were deposited on a coastal plain some 1000 km broad between the continental divide on the west—in the thrust belt mountains of eastern British Columbia, western Montana, westernmost Wyoming, and central Utah—and the Cannonball Sea to the east. A major bay of the Cannonball Sea in which the Waltman Shale was deposited extended 500 km into the middle of Wyoming, separating the coastal plain into a northern province composed of Alberta, Saskatchewan, Manitoba, Montana, the Dakotas, and northern Wyoming and a southern region. The southern region in turn was separated by the volcanic mountains of the Colorado Mineral Belt into the southern Wyoming, Utah, and northwestern Colorado province and the New Mexico and southeastern Colorado province.

During the Tiffanian and Clarkforkian many but not all of the central ranges of the Rocky Mountains in Wyoming, Colorado, and New Mexico were uplifted, creating a new continental divide and blocking easterly drainage to produce Flagstaff Lake.

CHRONOLOGY OF THE PALEOCENE

Although the original type sections of the Tertiary epochs were established in marine rocks in northwestern Europe, in many ways the most important reference section for the late Cretaceous to Eocene is that at Gubbio, Italy (Alvarez and others, 1977; Napoleone and others, 1983). This section is important because it is relatively complete, with no major unconformities between the early Cretaceous and late Eocene. It is richly fossiliferous with both calcareous nannoplankton and planktonic foraminiferal zones well represented. In addition it has preserved an excellent stratigraphic record of magnetic anomalies. Last, but by no means least, it preserves the thin iridium-rich clay layer at the Cretaceous-Tertiary boundary that provides the reference spike in many other sections in the world that can serve to locate this boundary precisely (Alvarez and others, 1980). The only features this section of pelagic sediments lack that would be useful for global correlation are terrestrial mammals and radiometrically datable minerals. Here, then, planktonic zones, magnetozones, and the iridium spike all occur together and can easily be correlated with the less complete, or terrestrial, type and reference sections elsewhere in the world.

The Cretaceous-Paleocene boundary is, of course, defined in marine rocks on the basis of marine extinctions. Recently precise redefinitions of the marine boundary were made for Gubbio, Italy (Alvarez and others, 1977), and for Zumaya, Spain (Percival and Fischer, 1977), in terms of micropaleontology and magnetostratigraphy. In terrestrial rocks, the boundary is not precisely definable at present except by correlation and the identification of the iridium anomaly. The present approximation in North America is the Roland Brown formula (Brown, 1962), placing the boundary at the next highest coal bed above the highest dinosaur bone found in place. This is approximately correct in eastern Montana and western North Dakota on the basis of physical correlations between the marine Fox Hills and Cannonball Formation on the one hand and the nearby terrestrial rocks on the other. This definition serves adequately for geologic mapping purposes but makes the untestable assumption that dinosaurs became extinct simultaneously everywhere at the same time as the widespread marine extinctions. Dinosaurs in the Hell Creek Formation appear to have become extinct before the Danian since the overlying Cannonball Formation is of early Danian age (Fox and Olsson, 1969). Similarly, basal Danian foraminifera immediately overly Maastrichtian dinosaurs in southern France and northern Spain (Plaziat, 1970a and b).

There are four adequate dates on the Cretaceous-Tertiary boundary. The one that is marine—in the late Maastrichtian Peedee Formation of North Carolina—was done on glauconite by Harris (1982) and measured as 66.7 Ma. (ICC constants, Steiger and Jäger, 1978). The other three are in terrestrial sediments of the western interior. The most southerly is at South Table Mountain, Golden, Colorado, in sediments containing late Mantuan Fossils (Everndon and others, 1964) originally dated as 64.8 Ma, now recalculated (ICC constants) as 66.4 Ma. Another is at Brownie Butte, Garfield County, Montana, in the Z coal bed (upper Z of Archibald, 1982). Originally dated as 64.0 ± 1.2 Ma. (Baadsgard and others, 1964), it has now been recalculated to 65.7 ± 1.3 Ma (ICC). This locality is closely bracketed by dinosaurs and Mantuan mammals. The third is an ash bed in the Nevis coal bed in the Scollard Formation of the Edmonton Group, Red Deer River, Alberta, a few kilometers west of the town of Scollard. This coal bed is 1 m below the major pollen change, 4.5 m above the highest dinosaurs (Lerbekmo and others, 1979), and a few meters above the late Lancian Trochu mammal fauna from the same section (Lillegraven, 1969). This sample was dated as 63.3 ± 2 Ma (ICC, Lerbekmo and others, 1979). The grand mean of these four dates—65.5 Ma—represents the present best estimate of the age of the Cretaceous-Tertiary boundary.

There are two prime dates for the Paleocene-Eocene boundary. One is from the late Thanetian of Reniul, France, dated by Bonhomme and others (1968) as 53.6 ± 2.5 Ma by Rb/Sr methods. This is the source of the usually quoted 53.5 Ma date of Hardenbol and Berggren (1978) for the Paleocene-Eocene boundary. Owens and Sohl (1973) reported a date (recalculated to ICC constants) of 54.9 ± 1.8 Ma for the middle part of the Manasquan Formation of New Jersey. This is dated as planktonic foraminiferal zone P6b or earliest Eocene. Other dates exist for both late Paleocene and early Eocene, but these are the most critical. The grand mean of these ages is 54.2 Ma.

Several current time scales exist for the early Tertiary. Curry and Odin (1982) have proposed on the basis of the above data that the Cretaceous-Tertiary boundary is 65 + 1/−1.5 Ma, writing off the variations listed above as "this discrepancy probably results from correlation problems which can be serious in continental sequences. . . ." From the standpoint of both the magnetic and vertebrate correlations, these four age samples ought not to be more different in age than about 0.1 m.y., so the variations in measured ages would seem to be a function of the method, of diagenesis, or of interlaboratory error.

Curry and Odin have also proposed an estimate of the Paleocene-Eocene boundary of 53 ± 1 Ma, based mainly on the Reniul date and other northwestern European dates. This gives an estimate of 12 m.y. for the duration of the Paleocene.

Another time scale, presented by Lowrie and Alvarez (1981), is a composite that includes the dates of Hardenbol and Berggren (1978) updated to the new ICC constants and a revision of the magnetic polarity time scale based on Ness, Levi, and Couch (1980). The latter is in turn based on a major revision of seafloor spreading rates, also based on Hardenbol and Berggren. Lowrie and Alvarez set the Cretaceous-Paleocene boundary at 66.7 Ma and the Paleocene-Eocene boundary at 54.9 Ma. On this scale the duration of the Paleocene is 11.8 m.y.

The Harland and others (1982) time scale differs from that of Lowrie and Alvarez (1981) in the assumption of a smoothed seafloor spreading rate and a date of 65 m.y. for the Cretaceous-Tertiary boundary; the date of 54.9 m.y. for the Paleocene-Eocene boundary remains unchanged. The total duration of the Paleocene is 10.1 m.y. in this time scale.

TABLE 1.

Mammal Zone	Lowrie and Alvarez '81		Harland and others '82		DNAG '83	
	End (Ma)	Length (m.y.)	End (Ma)	Length (m.y.)	End (Ma)	Length (m.y.)
Cf1	54.9	1.3	54.9	1.1	57.8	.9
Ti5	56.2	1.5	56.0	1.2	58.7	1.1
Ti4	57.7	0.6	57.2	0.5	59.8	0.4
Ti3	58.3	2.9	57.8	2.4	60.2	2.0
Ti2	61.2	0.6	60.1	0.5	62.2	0.4
(Danian)	61.1		60.2		62.3	
Ti1	61.8	0.5	60.6	0.4	62.6	0.4
To3	62.3	0.9	61.0	0.8	63.0	0.7
To2	63.2	0.7	61.8	0.6	63.7	0.5
To1	63.9	1.5	62.4	1.4	64.2	1.2
P2	65.4	0.7	63.8	0.6	65.4	0.5
P1	66.1	0.4	64.4	0.4	65.9	0.3
M	66.5	0.2	64.8	0.2	66.2	0.2
BC	66.7	0.3	65.0	0.3	66.4	0.3

The Decade of North American Geology (DNAG) time scale (Palmer, 1983; Berggren and others, 1985a, 1985b) has a shorter duration of the Paleocene of 8.5 m.y., between 66.4 and 57.8 Ma. This is based on smoothed seafloor spreading rates between chrons 21 and 34, although Berggren and others (1985a) report that the biochronology of radiometric dates near the Paleocene-Eocene boundary suggests that boundary might be as young as 56.5 Ma. This would make the duration of the Paleocene about 9.9 m.y.

The Paleocene is then between 11.8 and 8.6 m.y. in duration; further refinement of the time scale is likely with more data. The assumption of uniform rates of seafloor spreading seems unlikely in view of the known major changes in sea level (Vail and others, 1977) and their suspected relationship to spreading rates (Hays and Pitman, 1973; Pitman, 1978). The DNAG time scale is likely to be too short.

In any event, the Paleocene mammal zones are all well dated in terms of the magnetochrons, as shown later in this paper, and well dated with respect to the marine boundary between the Danian and the Thanetian or Selandian. The Danian equivalent is composed of zones M through Ti2; the Thanetian or Selandian equivalent is composed of zones Ti3 through Cf1. Table 1 shows the ages and durations of the mammal zones calculated from the ages of the magnetochrons of three of the time scales. Revisions of these numbers to fit future time scales will then be easy. Numerical dates in this paper are those of the Lowrie and Alvarez time scale, with DNAG dates in parentheses.

The various mammal zones of the Paleocene can be dated on the basis of joint magnetic polarity studies and fossil collections reported in Lerbekmo and others (1979) for Alberta; Archibald and others (1982) for Montana; Butler and others (1981) for Wyoming; Rapp and others (1983) for Texas; Tomida and Butler (1980) for Utah; and Lindsay and others (1981) for New Mexico. In the last two works, paleomagnetic chron numbers have been increased by one, for reasons to be discussed below. An independent check is available in North Dakota where the marine Cannonball Formation interfingers with terrestrial mammal-bearing sediments, making a direct correlation possible between planktonic foraminiferal zones and mammal zones and where Lund (1976) demonstrated that the extinction of dinosaurs occurs in reversed magnetic-polarity rocks.

The late Maastrichtian Trochu mammal fauna of Lillegraven (1969) comes from the Scollard Formation of Alberta at the same localities from which Lerbekmo and others (1979) reported on the K/Ar dates, pollen, and magnetic polarity. The mammal fauna immediately precedes the Bug Creek Anthills Fauna; *Cimolestes cerberoides* from the Trochu fauna comes from an interval of 0 to 27 m, with most specimens from a level of 20 m above the Knee Hills Tuff. Lerbekmo and others report this interval to be in the upper part of chron 30 and 1/5th of the way through the reversed interval after chron 30. They also report the highest dinosaurs to be 43.5 m above the Knee Hills tuff and 6 m below the pollen change that is the closest local approximation of the Cretaceous-Tertiary boundary, all within that reversed interval. Lerbekmo and others term this chron 30 and the reversed interval below chron 29.

Archibald and others (1982) report that "the latest dinosaurs in the Hell Creek Formation in Garfield and McCone Counties, Montana, come from the middle of a reversed interval which also contains the Bug Creek and Mantuan mammal faunas, the highest dinosaurs, and the pollen boundary." They inferred the reversed interval is the same one encountered in the Scollard Formation, but quibbled about its identification—mainly because the highest dinosaurs in the San Juan Basin were reported to occur in a normal interval, since shown to be false (Butler, 1984). Lund (1976) had earlier shown the highest dinosaurs in North Dakota occur in a reversed zone at the top of the Hell Creek Formation. In view of the proximity of Lund's study to the immediately overlying marine Danian Cannonball Formation, it seems likely that this reversed interval is indeed the one between chrons 29 and 30 and that Lerbekmo and others (1979) were correct in their identification.

Butler and others (1977) and Lindsay and others (1978, 1979a and 1979b, 1981) have strongly taken the position that there is no significant unconformity between the Ojo Alamo Sandstone and the underlying Naashoibito Member of the Kirtland Shale containing the Alamo Wash fauna on the one hand and the rest of the Kirtland Shale on the other. This position and their initial identification of the magnetochrons in the Fruitland and Kirtland formations automatically placed the classical Puerco faunas of the Nacimiento Formation in chron 28. On the other hand, Alvarez and Vann (1979), Fassett (1979), Lucas and Rigby (1979), and Lucas and Schoch (1982) have argued as strongly that the unconformity is present at that horizon and that the numbering of anomalies by Lindsay and others (1978) is wrong.

Lehman (1983) concludes that the Alamo Wash fauna from the Naashoibito Member is late Maastrichtian or Lancian in age, distinctly younger than the fauna of the rest of the Kirtland Shale. Fassett and others (this volume) describe four localities in the overlying Ojo Alamo Sandstone from which unreworked dinosaur bones have been collected. Fassett also lists several pollen samples of Paleocene age collected from the Ojo Alamo Sandstone. Rigby (1983) describes an important specimen of *Anisonchus (Gillisonchus) gillianus,* a mammal of early Puercan (P1) age, from the upper part of the Ojo Alamo Sandstone. Several alternatives present themselves. It would seem likely that the Cretaceous-Tertiary boundary and Mantuan age strata fall within the rock unit Ojo Alamo Sandstone as presently restricted by the Baltz and others (1966) redefinition. It is possible that dinosaurs survived into the Tertiary in the San Juan Basin, although that cannot be proved with present data. Physical evidence of an unconformity at the base of the Naashoibito Member, the lower conglomerate of Brown (1910), and the presence of distinctly younger dinosaurs in the Naashoibito Member than in the balance of the Kirtland Shale suggest that Brown (1910) was correct in associating the rocks now named the Naashoibito and the Ojo Alamo Sandstone (restricted) as a depositional unit. It would be useful to reexpand the Ojo Alamo Sandstone to its original limits despite its widespread use in its restricted sense. Less useful would be to separate the Naashoibito from the rest of the Kirtland Shale and raise it to separate formation rank.

Rigby (this volume) discusses the reasons for identifying chrons 30 and 31 in the Hunter Wash–Barrel Spring Arroyo section of Lindsay and others (1981) as chron 33. In Montana, Puercan faunas are clearly included in chron 29 as shown by data in Lerbekmo and others (1979) and Archibald and others (1982). This suggests that the duration of the unconformity between the upper Shale Member of the Kirtland and the Naashoibito Member is on the order of 7 million years or about two-thirds of the Maastrichtian.

Butler and others (1981) report the magnetic polarity sequence of the late Paleocene and early Eocene strata of the Big Horn Basin. They find that chron 25 straddles the Tiffanian-Clarkforkian boundary and that chron 26 straddles the Ti3-Ti4 zonal boundary. They noted without sufficient comment the problem of chron 26 of the Bighorn Basin being late Tiffanian and chron "26" of the San Juan Basin of Lindsay and others (1978) being late Torrejonian. Gingerich was not comfortable with the implied equivalence, and Lindsay and Butler suggested the difference might be due to discordance in faunal correlation. They again dismissed the possibility of error in the San Juan Basin. However, Rapp and others (1983) discuss the magnetic polarity sequence of the Big Bend region, Texas, and find the magnetic polarity sequence there agrees with the Bighorn Basin rather than with the San Juan Basin. Again, the problems of magnetic polarity in the San Juan Basin disappear if the Paleocene chron numbers of Lindsay, Jacobs, and Butler (1981) are increased by one. This alteration brings all mammal correlations and magnetic zones into full agreement and is followed in the balance of this paper.

Finally, in 1984, Butler showed that the numbering of chrons in the Paleocene of the San Juan Basin of Lindsay and others (1978) was in error. He demonstrated that the chron labeled 29n in the earlier reports, located in the Ojo Alamo Sandstone, was an artifact of remagnetization. Butler and Lindsay (1985) reported in detail on the revised magnetic chronology of the San Juan Basin, providing consistent and sequential correlations between the magnetic time scale and the mammal ages. These now agree with those of this paper, which were first written in 1981. One problem remaining is their continued insistence that there is no break in deposition between the upper Kirtland shales and the Naashoibito Member and that a northwest source exists for the Fruitland Formation and the Kirtland Shale.

Nonetheless, the original reference of Lindsay and others (1978) was so widely cited that it has had a great impact on other correlations and time scales. Such important primary and secondary sources as Savage and Russell (1983), the DNAG time scale (Palmer, 1983; Berggren and others, 1985a and 1985b), and Marshall and others (1981) will be widely but mistakenly cited as supporting correlations in error by one chron or durations of the Puercan different by 1 m.y. from the true value.

Figures 4, 5, and 6 (in pocket inside back cover) represent the zonal distribution of ungulates, primates, and multituberculate species during the Bug Creekian to Clarkforkian and can be used to identify the age of unknown samples.

THE PALEOCENE LAND MAMMAL AGES

Bugcreekian Age (Stage)

Although earlier than the Paleocene, the Bugcreekian stage (*Protungulatum* zone) must be discussed at this point. The Bugcreekian age (stage) (new) is principally known from three successive faunas in superposed stream channels of the upper 25 m of the Hell Creek Formation in McCone County, Montana, and from additional University of California localities of comparable age in adjacent Garfield County (Archibald, 1982). The faunas are described in Sloan and Van Valen (1965), Van Valen and Sloan (1966, 1977a), Sloan (1970, 1976), and Van Valen (1978); and Sloan and others (1986). This newly defined age (stage) is typified by *Procerberus formicarum, Protungulatum donnae, P. gorgun, Mimatuta morgoth, Oxyprimus erikseni, Purgatorius ceratops, Stygimys kuszmauli,* and *Catopsalis joyneri,* all of which are restricted to this age (stage) so far as is presently known. This age (stage) is contemporary with the latest part of the *Triceratops* community and zone and is, as closely as can be determined, latest Maastrichtian in age.

The three main superposed faunas in McCone County, Montana—Bug Creek Anthills, Bug Creek West, and Harbicht Hill— are distinct and show progressive evolutionary changes of increase in the number of mammal species, increase in the size of many placental mammal species, diversification of dental morphology (Van Valen, 1978), and progressive decrease in numbers of dinosaur species (Sloan, 1970, 1976; Archibald and Clemens, 1982; Brandenburg, 1983; Sloan and others, 1986). Other faunas in this zone are Rick's Place in Garfield County, Montana (Archibald, 1982); Black Butte in Sweetwater County, Wyoming (Breithaupt, 1982); and a Saskatchewan locality, Frenchman 1 from the upper Frenchman Formation near Ravenscrag (Johnston, 1980; Johnston and Fox, 1984). Localities V-5920 and V-5921 of Clemens (1963, 1965, and 1973) and Estes (1964) from the Lance Formation of Niobrara County, Wyoming, are sufficiently close to the Roland Brown formula boundary so that they may be of this zone, but no diagnostic fossils have yet been found. On the basis of paleomagnetics and the presence of *Cimolestes cerberoides,* the immediate ancestor of *Procerberus formicarum,* Lillegraven's (1969) Trochu fauna from the Scollard Formation of Alberta, appears to be just older than Bugcreekian.

On the basis of Archibald and others (1982) and Lerbekmo and others (1979), the Bugcreekian falls in the lower part of the reversed interval between magnetochrons 29 and 30 or from 66.7 to 65.4 Ma., with a maximum duration of 0.3 million years. Our best evidence suggests the asteroid impact layer of Alvarez and others (1980), the precise boundary of the Cretaceous and Paleocene, falls at the boundary between the Bugcreekian and the Mantuan.

Mantuan Age (Stage)

The Mantuan age (stage) defined by Van Valen (1978), is typified by the fauna of Mantua Quarry in the Mantua Lentil of the Polecat Bench Formation of Jepsen (1930, 1940) in Park County, Wyoming. Other localities of Mantuan age are Leidy Quarry in the basal "Fort Union" Formation of the southern Big Horn Basin in Hot Springs County, Wyoming (Van Valen, 1978); Worm Coulee, Hell Hollow, and McKeever Ranch No. 1 in the basal Tullock Formation, Garfield County, Montana (Archibald, 1977 and 1982); the Long Fall Locality of southwestern Saskatchewan (Johnston and Fox, 1984); and the South Table Mountain and Alexander localities in the Denver Formation near Denver, Colorado (M. Middleton, pers. comm., 1980). The Mantuan age (*Catopsalis alexanderi* zone) is of short duration, entirely within the latter half of the reversed interval between chrons 29 and 30, or from 66.7 (66.4) to 66.5 (66.2) Ma, about 0.2 million years in duration (Archibald and others, 1982). Nonetheless at the generic level it is as distinct from the Puercan as that age is from the Torrejonian.

The Long Fall locality of Saskatchewan; the Garfield County, Montana, localities of Archibald (1977, 1982); and the Mantua Quarry and Leidy Quarry of Wyoming all represent the earlier subzone of the Mantuan, Ma. I do not expect Ma to be found south of Thermopolis, Wyoming. Two localities near Denver, Colorado—the South Table Mountain locality and the Alexander locality—represent a later subzone of the Mantuan, Mb. This subzone differs from Ma in having the more advanced arctocyonid *Baiaoconodon denverensis* and the earliest species of

Anisonchus, but is earlier than P1 in having the typical Mantuan multituberculates *Mesodma ambigua, Cimexomys hausoi,* and *Catopsalis alexanderi* (Middleton, 1982 and pers. comm.). The latter is replaced in P1 by its larger descendant *Catopsalis foliatus* (Sloan, 1981).

The Mantuan is typified by *Eoconodon copanus, Ragnarok nordicum* (last occurrence), *Earendil undomiel, Mimatuta minual, M. makpialutae, Oxyacodon josephi, Maiorana noctiluca, Oxyprimus galadrieli, O. putorius,* an undescribed species of *Procerberus* ancestral to the taeniodonts, and the first occurrences of *Stygimys gratus* and *Catopsalis alexanderi.* I suspect this stage will not be found south of central Colorado.

Puercan Age (Stage)

Wood and others (1941) based the Puercan age on the "Puerco formation" of the San Juan Basin of New Mexico, an ill-defined rock unit now considered to be the lower 50 ± m of the Nacimiento Formation. In practice, the Puercan stage of the type area has been interpreted to be the combined *"Ectoconus"* and the *Taeniolabis* zones of the area between Ojo Alamo and Barrel Spring Arroyo, Kimbeto Arroyo, and "Arroyo Eduardo," now better known as Betonnie Tsosie Arroyo. The Mantua Lentil was regarded as the principal correlative but has now been demonstrated to be older.

Wood and others (1941) listed *Carsioptychus, Conacodon, Ectoconus, Eoconodon, Loxolophus, Onychodectes, Oxyacodon, Taeniolabis,* and *Wortmania* as index fossils. Taxa listed as appearing for the first time were *Anisonchus,* Condylarthra (excluding the Arctocyonidae now generally referred to that order), Creodonta (including only Acreodi = Arctocyonidae+Mesonychidae), *Eucosmodon, Oxyclaenus,* Taeniodonta, and Taligrada (Periptychidae only).

It is now possible to define the two zones of the Puercan, informally recognized in the past, at least as clearly as the Puercan was defined by Wood and others (1941). The earlier, lower zone—P1 of Figure 1—was named the *"Ectoconus"* zone for the abundance of that genus (Sinclair and Granger, 1914), although it is also present in the next zone, a fact that disturbed many biostratigraphers. It is not necessary that the name bearer of a zone be restricted to a single zone, but there are ready substitutes available. The genus and species *Hemithlaeus kowalevskianus* will be used as the name bearer of zone P1. This zone is also known from central Utah (Gastank fauna) and from the Denver Basin of Colorado (Corral Bluff). P1 is marked by the first occurrence of the genera *Loxolophus, Desmatoclaenus, Platymastus, Mimotricentes, Oxyclaenus, Promioclaenus, Bomburia, Protoselene, Choeroclaenus, Periptychus, Hemithlaeus, Eucosmodon, Ptilodus, Neoplagiaulax,* and *Parectypodus.* The first and last occurrence of *Catopsalis foliatus* is in this zone. The precise species involved are indicated on Figures 4, 5, and 6 (in pocket inside back cover). I anticipate that future discoveries will lead to the subdivision of this zone.

P2, the later of the two Puercan zones, is the classic *Taeniolabis* zone of Sinclair and Granger. It is known from the classic localities of the San Juan Basin, from the Wagonroad localities of central Utah, from Garbani and Purgatory Hill localities of Montana, and from West Bijou Creek in the Denver Basin, Colorado. Two minor localities in Canada appear to be of this age. The base of this zone is also marked by the first occurrence of *Goniacodon, Ellipsodon, Haploconus, Tinuviel, Chriacus, Litomylus,* and such species as are shown on Figures 3 and 4.

The *Taeniolabis* zone can be subdivided into two subzones. One is the *Taeniolabis* n. sp. subzone (P2a), known from Garbani Quarry in Montana and Rav W-1 in southwestern Saskatchewan (Johnston and Fox, 1984). The later *Taeniolabis taoensis* subzone (P2b) is known from Barrel Spring Arroyo in the San Juan Basin, from Wagonroad Quarry in Utah, from West Bijou Creek in the Denver Basin, Colorado (M. Middleton, pers. comm., 1980), and from Purgatory Hill, Montana.

The Puercan is marked by continued radiation of both allotherian and placental mammals. Based on published and unpublished records, no placental mammal lineages became extinct until the end of P1 and few until the end of P2. The base of P1 is at the base of magnetochron 29 as is shown by Lindsay and others (1981), as modified by Butler and Lindsay (1985). The Wagonroad local fauna is the youngest Puercan known and is in the lower part of the reversed interval between chrons 29 and 28 (Tomida and Butler, 1980, where it termed to be between chrons 28 and 27). The estimated duration of the Puercan is from 66.5 (66.2) to 65.4 (65.4) Ma. or about 1.1 (0.8) million years.

Torrejonian Age (Stage)

Wood and others (1941) based the Torrejonian age on the "Torrejon formation" of the San Juan Basin of New Mexico, another ill-defined and misused rock unit that is now regarded as the upper 250 ± m of the Nacimiento Formation. The Wood committee also defined a Dragonian age based on the Dragon local fauna (Gazin, 1941) occurring in a 100 ± m interval within the North Horn Formation of Utah. A similar 100-m-thick interval generally lacking mammalian fossils occurs within the Nacimiento Formation below the fossiliferous horizons of the Torrejonian and above the top of the *Taeniolabis* zone of the Puercan.

Every stated criterion for the definition of the Dragonian age as well as some of those for the Torrejonian has become obsolete. It is clear that the Dragonian is not of age (stage) dimensions; rather it appears to be more or less comparable to the other zonal subdivisions of age (stage) and is best regarded as the earliest zone of the Torrejonian, To1 (Tomida and Butler, 1980). The base of the Torrejonian is at present defined by the first occurrence of *Conoryctella, Mimotricentes subtrigonus, Promioclaenus mantiensis, Protoselene griphus, Litaletes shepherdi* and *Periptychus gilmorei.* Further collecting in Wyoming and Montana may indicate a need for redefinition of the base, since many of these genera and species appear to be confined to the area south and west of the Paleocene continental divide (Sloan, 1970). For the present, the lowest zone of the Torrejonian may be called the *Anisonchus dracus* zone in an allusion to the old Dragonian age. Fossils from this zone are presently known mainly from the Dragon local fauna. The only fossil localities of this age known from the northeastern zoogeographic region are localities 65, 78, and 9 of the lowest 100 m of the Lebo Formation of the Crazy Mountain Field, Montana (Simpson, 1937b). Even though they have not been extensively collected, they have produced a small neoplagiaulacid, a *Mimotricentes* species, *Oxyclaenus pugnax,* and from the highest of these localities, *Tetraclaenodon* cf. *symbolicus* (=*puercensis*). Of these species, only the *Tetraclaenodon* is known from higher strata. *Oxyclaenus pugnax* is also known from zone P2.

Tomida (1981) suggested that this zone be termed the *Loxolophus-Periptychus* zone, but this terminology depends on the maintenance of a separate genus for *Carsioptychus,* a practice abandoned by most authors. Tomida and Butler (1980) and Tomida (1981) have shown that the zone occurs in the upper part of chron 28 (27 in their terminology) in both Utah and New Mexico. The zone thus has a duration from 65.4 (65.4) to 63.9 (64.2) Ma or 1.5 (1.2) million years. The most interesting features of this zone are the first extinctions of ungulate species and the development of regionally distinctive faunas.

The second zone of the Torrejonian, To2, is the classic *Deltatherium* zone of Sinclair and Granger (1914) found at many localities in the San Juan Basin and best typified in Wilson (1956) and Taylor (1981). The *Deltatherium* zone represents approximately the middle third of the Torrejonian stage and is about 150 m thick in the San Juan Basin. The genus *Triisodon* appears to be restricted to this zone as are the species *Deltatherium fundaminis, Deuterogonodon noletili, Haploconus angustidens,* and *H. corniculum.* The *Deltatherium* zone is further marked by the first occurrence of *Conoryctes, Tetraclaenodon puercensis, Mioclaenus, Chriacus baldwini, Chriacus pelvidens, Palaecthon, Paromomys, Parectypodus clemensi,* and *P. trovessartianus* in the San Juan Basin. In northern localities this interval is poorly sampled; the best localities so far known are 73, 51, 24, 25, and 81 of the middle third of the Lebo Formation of the Crazy Mountain Field, about 150 m thick (Simpson, 1937b). *Deuterogonodon montanus* appears to be restricted to this interval, and this genus may serve as an additional zonal name bearer. It is further marked by the first appearance of *Protictis haydenianus, Promioclaenus aquilonius,* and *Anisonchus sectorius.* This fauna appears to be the northeastern equivalent of the *Deltatherium* zone.

The *Deltatherium* zone straddles chron 28 as shown by Tomida and Butler (1980); Tomida (1981); Lindsay and others (1981); and Butler and Lindsay (1985). Lucas and O'Neill (1981) have found the lowest *Pantolambda* in Kutz Canyon in the San Juan Basin to be at a level that places the top of To2 low in the reversed interval between chrons 28 and 27. This makes the duration of the zone from 63.9 (64.2) to 63.2 (63.7) Ma, or 0.7 (0.5) million years. *Deltatherium fundaminis,* the name bearer of the zone, does not appear to be descended from any known North American ancestor and more likely can be interpreted as the first pantodont to migrate into North America from Asia. It is the first placental mammal in North America since *Protungulatum* appeared that seems to be an immigrant.

The last of the major zones of the Torrejonian is the *Pantolambda* zone (Sinclair and Granger, 1914), To3. It is some 50 m thick in the central and northwestern San Juan Basin but missing by erosion on the east side of the basin. It is widely recognizable in Wyoming, Montana, and New Mexico. In the Crazy Mountain Field it has a thickness of about 150 m of the upper Lebo and lower Melville formations and includes the well-known Gidley and Silberling quarries. In the Bighorn Basin, it encompasses the Rock Bench Quarry beds of the Polecat Bench Formation (the late Puercan and early and middle Torrejonian are missing by erosion). The taxon for which this zone is named, *Pantolambda,* is restricted to this zone. It apparently is ancestral to the later North American Paleocene pantodonts. *Pantolambda* and *Deltatherium* most likely immigrated from Asia. The base of the *Pantolambda* zone is also marked by the lowest occurrence of *Arctocyonides ("Claenodon") ferox, Arctocyonides ("Neoclaenodon") procyonoides, Pronothodectes, Elphidotarsius, Torrejonia, Plesiolestes, Elpidophorus,* and *Jepsenella.* The most rapidly evolving mammals during the *Pantolambda* zone time were primates and pantodonts. Gingerich (1976) and Rose (1977) have been able to divide this zone into subzones: the earlier *Pronothodectes matthewi* subzone (To3a) of Gidley Quarry and the later *Pronothodectes jepi* subzone of Rock Bench Quarry (To3b). In many taxa the Rock Bench Quarry specimens are notably more advanced and usually larger than the comparable taxa from Gidley Quarry. The differences are sometimes but not always significant at the species level. Gingerich (1980a) now refers to these zones as To3 and To4, but they are not recognizable outside the Bighorn Basin as yet.

The *Pantolambda* zone occurs mostly in chron 27 based on data in Lindsay and others (1981), where it is termed 27 by Butler and Lindsay (1985), but the base is low in the reversed interval between chrons 28 and 27 (Lucas and O'Neill, 1981). Data of less than desirable quality in Butler and others (1981, Fig. 4) suggest that the top of To3b is at the top of anomaly 27. Tsentas (1981) describes major additions to the fauna of the type *Pantolambda* zone. The duration of To3 is 63.2 (63.7) to 62.3 (63.0) Ma or about 0.9 (0.7) million years.

A major uplift in western Wyoming resulted in the transition between the Lebo Shale and the overlying Melville Formation in the Crazy Mountain Field and between the Lebo Shale and the overlying Tongue River Formation farther east in Montana. The base of the Rock Bench Quarry beds of the Polecat Bench Formation appears to be the same coarsening event in sedimentation.

Tiffanian Age (Stage)

Wood and others (1941) defined the Tiffanian age based on the Tiffany local fauna of La Plata and Archuelata counties,

Colorado, described by Granger (1917), Matthew and Granger (1921), and Simpson (1935a). Most of the specimens came from the Mason Pocket, a carnivore den(?) of small animal remains; the others came from eight surface localities of large animal remains scattered over a distance of 80 km and a vertical interval of at least 300 m. The specimens occur in the middle of the Animas Formation of the northern San Juan Basin, a lateral equivalent of the Nacimiento Formation of New Mexico. The formation is a stream-deposited volcaniclastic sequence of conglomerate, sand, and shale derived from the contemporaneous eruptions of the volcanoes and intrusive centers of La Plata, Rico, and Ouray, Colorado. The most thorough description of the stratigraphy of these rocks is in the texts accompanying two geologic maps, Barnes (1953) and Zapp (1949). Later Simpson (in Simons, 1960) referred the strata enclosing the type of *Ignaciolambda barnesi* to the "Tiffany formation" (*sic*), an undescribed rock unit. *Ignaciolambda* occurs about 180 m above the base of the Animas Formation; Mason Pocket occurs about 280 m higher, about 120 m below the base of the Lysitian(?) San José Formation. The type Tiffanian stage thus extends over about 500 m of rock, whereas the Puercan and Torrejonian stages at about the same distance from the sediment source extend over about 300 m. Since the entire group of localities of the type Tiffanian is located in the Southern Ute Reservation, it appears unlikely that extensive recollecting can be done in the near future.

The faunal definition of the Tiffanian was based only in part on the Tiffany local fauna. The fauna of the Melville Formation of the Crazy Mountain Field and Bear Creek of Montana, the Silver Coulee Beds of Wyoming, and the Plateau Valley of Colorado also were used in defining the stage. The awkward definition of the Tiffanian eked out from many largely disparate faunas has made it the most troublesome of all the Paleocene land mammal ages. Neither the top nor the bottom of the Tiffanian was well defined. A majority of the Paleocene local faunas discovered since Wood and others (1941) are Tiffanian in age, and most of the thickness of Paleocene strata is Tiffanian in age.

In essence anything later than the Torrejonian and earlier than the Clarkforkian has become Tiffanian by default. There are few problems with the base of the Tiffanian other than to describe the known but as yet undescribed taxa involved. The boundary between the Tiffanian and the Clarkforkian was redefined by Gingerich (1976) and Rose (1980). As a result of the small collections from the type Tiffany local fauna and the lack of opportunity to make new collections with good stratigraphic control, the operational stratotype by default has become the Silver Coulee Beds of the Polecat Bench Formation of the northern Bighorn Basin. The well-documented work by Jepsen from 1929 to 1970 and the more recent work by Gingerich and his students (Gingerich, 1980a; Gingerich and others, 1980) have defined both the paleontology and stratigraphy better than any other Tiffanian sequence. The magnetostratigraphy of this sequence was defined by Butler and others (1981) and clearly defines the Tiffanian as from the top of chron 27 to the middle of chron 25, or from 62.3 to 56.2 Ma., a duration of 6.1 million years. In particular, the Polecat Bench Formation is the proper name for these deposits rather than the more commonly used "Fort Union" Formation. The rocks of the type Fort Union region have not been adequately described since Weed (1896) and Wilder (1904) and are only similar to those of the Polecat Bench Formation to the extent that they are both Paleocene and both fluviatile.

The most useful taxa for establishing Tiffanian zones are pantodonts, plesiadapid and carpolestid primates, and neoplagiaulacid multituberculates. Gingerich (1975b) formally established these zones, which were elaborated by Rose (1977).

The lowest zone of the Tiffanian, Ti1, is the *Plesiadapis praecursor* zone (Gingerich, 1975b, 1976) typified by that species, the earliest and smallest of that genus. Also limited to this zone are *Nannodectes intermedius* and the carpolestids *Elphidiotarsius shotgunensis* and *Carpodaptes* cf. *hazelae.* The zone is presently known from the Douglass Quarry in the Crazy Mountain Field (a Princeton locality, 65 of Simpson's 1937b list) and from Shotgun or "Keefer Hill" in central Wyoming. The base of the range of *Ectocion ("Gidleyina")* appears to be in this zone. The neoplagiaulacid multituberculates of this zone are surviving late Torrejonian species, *Mimetodon silberlingi, Neoplagiaulax nelsoni, Ectypodus aphronorus, Parectypodus sinclairi, P. clemensi,* and *Mesodma pygmaea.* The pantodont is apparently transitional between *Pantolambda cavirictus* and *Caenolambda* and may be referrable to the latter genus. In addition, the Bangtail locality (Houde and Olson, 1982; Gingerich and others, 1983) in the western part of the Crazy Mountain Field is of this age. The Ti1 zone fails entirely in the base of the reversed zone between chrons 27 and 26 and by interpolation lasted from 62.3 (63.0) to 61.8 (62.6) Ma., a duration of 0.5 (0.4) million years. This short duration is due entirely to the rapid evolution of plesiadapid primates in the early Tiffanian.

The second zone of the Tiffanian, Ti2, is the *Plesiadapis anceps* or *Caenolambda pattersoni* zone. The principal localities of this zone are the Saddle locality of Bison Basin and Highway Blowout, both of Wyoming; Scarritt Quarry of the Crazy Mountain Field, Montana (Simpson, 1936, 1937a and b); and White Site or Medicine Rocks II in eastern Montana. In addition to the nominal taxa, Zone Ti2 is typified by the lowest occurrence of *Nannodetes gazini, Carpodaptes hazelae, Litolestes notissimus, Thryptacodon, Neoplagialax hunteri,* and *Ectypodus powelli.* In Ti2, there was a major extinction of surviving Torrejonian ungulate species, coupled with continued evolution of primates. By interpolation from the data of Butler and others (1981), Ti2 lasted from 61.8 (62.6) to 61.2 (62.2) Ma. or a duration of 0.6 million years. It falls entirely within the reversed interval between chrons 26 and 27.

The third and longest Tiffanian zone is the *Plesiadapis rex* range zone, Ti3, which is also typified by *Nannodectes simpsoni* and *Chiromyoides minor.* The zone can be subdivided by the transition from *Neoplagiaulax hunteri* (Ti3a) to *N. mckennai* (Ti3b). Localities from the *Plesiadapis rex–Neoplagiaulax hunteri* composite range zone are Brisbane and Judson in North Dakota, Circle in Montana, and Cedar Point in Wyoming.

Neither species of *Neoplagiaulax* is presently known from the following localities of the *Plesiadapis rex* zone: Battle Mountain, Chappo 17, Ledge Locality, Melville 13, Saddle Annex, West End, Erickson's Landing, and Twin Creeks. The last, due to the presence of *Litomylus ishami,* may well belong to the *P. rex–N. hunteri* composite range zone Ti3a. Love Quarry presently represents the *P. rex–N. mckennai* composite range zone (Tf3b). The lowest occurrence of *Thryptacodon australis, Litomylus ishami, Phenacodus primaevus, Carpodaptes cygneus, C. hobackensis, Titanoides primaevus, T. gidleyi,* and *Neoplagiaulax jepi* is in the *P. rex–N. hunteri* composite range zone, as are the highest occurrences of *Carpodaptes hazelae* and *Neoplagiaulax jepi.* Ti3 is the longest of the Tiffanian zones, extending from 61.2 (62.2) to 58.3 (60.2) Ma. for a duration of 2.9 (2.0) million years. The zone occupies the later part of the reversed interval between chron 27 and 26 and the lower half of chron 26. Primate evolution had clearly slowed compared to that earlier in the Tiffanian. The global Danian transgression ended early in, or just before, this zone, as shown by the close vertical proximity of the Judson and Brisbane faunas to the last deposits of the Cannonball sea in North Dakota (Holtzman, 1978).

The fourth Tiffanian zone is the *Plesiadapis churchilli* zone, also typified among the other primates by *Nannodectes gidleyi, Chiromyoides caesor, Carpodaptes aulacodon,* and *C. hobackensis.* Among the multituberculates, *Ectypodus musculus* is known only south of central Wyoming in this zone, whereas northeast of the divide *Mimetodon silberlingi* and *Neoplagiaulax mckennai* have their highest occurrences. The lowest occurrences of *Champsosaurus gigas, Neoplagiaulax hazeni, Phenacolemur pagei, Litolestes ignotus, Leptolambda schmidti,* and *Haplolambda quinni* are represented in this zone. The principal localities of this age are Mason Pocket and Joe's Bonebed among the southern localities and Olive (Wolberg, 1978), Malcolm's Locality, Long Draw Quarry, Airport Locality, and Divide Quarry among Montana and Wyoming localities. Also in this zone are Roche Percée in Saskatchewan (Krause, 1977) and Riverdale and Wannagan Creek Quarry (Erickson, 1972, 1976) of North Dakota. It is during this time that the earliest known invasion of the southern localities by plesiadapid and carpolestid primates took place, providing a useful tie to the better-known northern sequence. Ti4 occurs in the upper part of chron 26 and the lower half of the reversed interval between chrons 26 and 25. Its duration is from 58.3 to 57.7 Ma. or a total of 0.6 million years, based on both Bighorn Basin data (Butler and others, 1981) and that of the Big Bend region of Texas (Rapp and others, 1983).

The fifth and youngest Tiffanian zone is the *Plesiadapis simonsi* zone, Ti5, which is based primarily on the large collections from Princeton Quarry (Jepsen 1930, 1940; Jepsen and Woodburne, 1969). Other localities referred to this zone include Bayfield, Chappo 1, Dell Creek, Fossil Hollow, Middle Sand Draw, Schaff Quarry, and the *Titanoides* locality. Other primates present in this zone include *Plesiadapis fodinatus, Nannodectes* aff. *gidleyi, Chiromyoides potior, Carpolestes dubius,* and *Phenacolemur pagei. Arctostylops steini* (clearly an immigrant from South America) has its lowest occurrence at this horizon as do *Phenacodaptes sabulosus, Probathyopsis praecursor, P. newbilli, Bathyopsoides* cf. *harrisorum,* and *Lampadophorus.* The base of the Sentinel Butte Formation of eastern Montana and North Dakota is of this zonal age. In addition, a much discussed single specimen of *Hyracotherium* cf. *angustidens* apparently represents a stray specimen from the Graybullian outcrops 14.5-km to the southwest. Typical multituberculates include *Neoplagiaulax hazeni,* the lowest northern occurrence of *Neoliotomus* (the latter genus was earlier present in the Torrejonian Laudate fauna of California and the earliest Tiffanian Shotgun fauna of central Wyoming), and the lowest occurrences of *Mimetodon churchilli* and *Parectypodus laytoni.* Ti5 extends from the middle of chron 26 to the middle of chron 25, a duration of 57.7 (59.8) to 56.2 (58.7) Ma. or 1.5 (1.1) million years.

Clarkforkian Land Mammal Age (Stage)

Wood and others (1941) defined the Clarkforkian age based on the fauna of the Clark Fork beds of the northern Bighorn Basin of Wyoming. The age was defined with *Plesiadapis cookei* as the index fossil, and *Didymictis, Ectypodus, Phenacodus, Probathyopsis,* and *Thryptacodon* as characteristic fossils. It was further marked by the occurrence of *Coryphodon, Ectocion, Esthonyx,* and *Oxyaena* and by the last appearance of *Carpolestes* and *Plesiadapis.* Wood (1967) suggested that the evidence was insufficient to support the existence of Clarkforkian. Rose (1981a) has not only thoroughly demonstrated the existence of the Clarkforkian but has defined it in such a precise way that it is the best defined of all of the Paleocene ages. Three zones are recognized in the Clarkforkian: the *Plesiadapis gingerichi* zone, the *Plesiadapis cookei* zone, and the *Phenacodus-Ectocion* acme zone. Rose refers the first zone to the Paleocene and the other two to the early Eocene. The first zone then extends from the middle of chron 25 into the reversed interval between chrons 25 and 24, for an estimated duration of 56.2 (58.7) to 54.9 (57.8) Ma. or a total of 1.3 (0.9) million years.

Traditionally the Clarkforkian has been referred to the Paleocene in its entirety. After some considerable discussion in the literature (D. E. Russell, 1967; Pomerol, 1969; Schorn, 191, 1975; Gingerich, 1975a), Gingerich (1976) formally defined the Eocene-Paleocene boundary as between the Thanetian and Sparnacian stages of the Paris Basin, in effect placing the boundary between the faunas of Cernay and Berru (Thanetian) and that of Meudon (Sparnacian). This placement did the least violence to the classic correlations of age (stage) and epoch boundaries. It does have the slight disadvantage of causing the Clarkforkian to straddle the Paleocene-Eocene boundary, since the *Plesiadapis cookei* zone correlates with the Meudon fauna, and the *Plesiadapis gingerichi* zone correlates with the Cernay and Berru faunas.

Above the level of Princeton Quarry, but below the *Plesiadapis cookei* zone, Rose (1980) has found seven localities containing a paramyid rodent close to *Paramys atavus* and the species *Plesiadapis gingerichi* which is intermediate between *P. simonsi* and *P. cookei.* This species closely resembles *P. tricuspidens* from the Cernay fauna of the Paris Basin both in size and structure. These localities serve as the basis for a sixth late Paleocene zone, Cf1. *Didymictis* and *Viverravus* have evolved from *Protictis* and *Bryanictis* respectively by this zone, and *Phenacodaptes* has evolved into the earliest *Apheliscus nitidus.* The lowest occurrence of *Ectocion osbornianum* is at this level. Bear Creek in Montana and some of the Plateau Valley, Colorado, localities are part of this zone. This zone is correlative with the well-known Cernay fauna of France of late Thanetian age.

The second Clarkforkian zone, Cf2, was shown by Gingerich (1976) to be correlative with the earliest Sparnacian Meudon locality of France and therefore is earliest Eocene by definition. This zone, the *Plesiadapis cookei* zone, includes Paint Creek, *Franimys* Hill, Upper Sand Draw, and Little Sand Coulee, all in the Bighorn Basin. Buckman Hollow and the lower variegated beds of Togwotee Pass are other localities of this zone.

There is a 119-m zone in the lower Willwood Formation in which *P. cookei* is missing, *Ectocion* and *Phenacodus* are particularly abundant, and *Probathyopsis* has its highest occurrence, but which is below the first occurrence of the typical Graybullian *Microsyops-Pelycodus-Hyopsodus-Diacodexus-Hyracotherium-Homogalax* fauna. This represents the top of the Clarkforkian, zone Cf3.

It now appears that the Clarkforkian age (stage) is the only Paleocene North American land mammal age (stage) that can be recognized readily and precisely in other continents. For the first time since the beginning of the Bugcreekian and the *Deltatherium* zone, the evolving North American Paleocene chronofauna received significant additions by immigration from other continents. These immigrants include *Arctostylops steini,* which came from South America in Ti5. Its descendants *Palaeostylops* and *Sinostylops* appear shortly thereafter in Asia. Appearing at Cf1 are the primitive dinoceratans *Probathyopsis* (including *Bathyopsoides*), surely immigrant to North America and derivable from *Prodinoceras* (including *Mongolotherium,* per Dashzeveg, 1982) in Asia or from the Riochican xenungulate *Carodnia* of South America. This sudden opening of a filter bridge between three continents previously isolated for some 7 million years is surely more significant at the age (stage) level than the transformation of one species to another by anagenesis. Such intercontinental range extensions make the best possible age (stage) boundaries.

ZOOGEOGRAPHIC PROVINCES

A minor problem in precise correlation in the Paleocene of North America is the existence of at least two zoogeographic provinces, particularly during the Torrejonian and Tiffanian. The two major provinces were first discussed by Sloan (1970) as northern and southern provinces: Colorado, southern Wyoming, Utah, New Mexico, Texas, Louisiana, California, and Baja California localities represent the southern province; and northern Wyoming, Montana, North Dakota, and Canada represent the northern province. Schiebout (1974) pointed out that the Texas localities, which range in age from Ti1 to Ti5 or Cf1, contain many "northern" elements. The Texas localities were clearly on the coastal plain of the Gulf of Mexico, while all of the other "southern" localities were clearly part of the Pacific drainage. There thus seem to be several components to provinciality in the Torrejonian and Tiffanian. A latitudinal component clearly exists, for the Texas localities do have many species otherwise restricted to the south. Equally clearly there is an altitudinal and drainage basin component, since the provincial boundary in Wyoming is fuzzy but appears to coincide with the contemporary continental divide. The continental divide in Wyoming and northern Colorado as far south as Gunnison appears to be in the same place in the Tiffanian as at present. It then continues south, east of the San Juan Basin during the Tiffanian. Earlier, during the Puercan and Torrejonian, the divide was on the west side of the San Juan Basin, which was not a basin at the time but rather the inland limit of the coastal plain of the gulf. There was a time when it was fashionable to worry about the upland faunas and how they might differ from the lowland faunas we knew. It now appears that we had upland faunas all along and that such faunal localities as Dell Creek and Plateau Valley originated right at the foothills, perhaps at altitudes of 2,000 m and within tens of kilometers of the continental divide.

RELATIVE RATES OF SEDIMENTATION

For many years, most stratigraphers of the nonmarine Paleocene, as well as others, have been concerned with variability of sedimentation rates in time and space and about the difficulties of projecting data from one place to another. In general, most have been reluctant to use relative stratigraphic position as a guide to geologic age, and some have been reluctant to use superposition in a single measured section as a guide to significant differences in age. Starting with the preliminary attempts to refine stage (age) terminology (Van Valen and Sloan, 1966; D. E. Russell, 1967; Sloan, 1970) and continuing into later versions (Gingerich, 1975b; Rose, 1977), it has become progressively apparent that for the Paleocene of North America, such concerns are not of primary importance. Instead, as Shaw (1964) showed for the Late Cambrian, relative rates of sedimentation were approximately constant for a given location, though they vary greatly from place to place.

As in the case of Shaw's (1964) Cambrian data, the relative stratigraphic position of zoned faunas agrees when all of these sections are put on a common vertical scale. The following rates of sedimentation were calculated by assigning 11.8 million years to the interval Mantuan (M) through Clarkforkian I (Cf1) (Fig. 1) and constructing a composite standard section from the Polecat Bench section (M), Torrejonian 3b or To3B to Clarkforkian 1 Cf1, P1 to To3a missing) and the Crazy Mountain Field section (Simpson, 1937b, M to Ti4). The composite standard section was constructed according to Shaw's (1964) methods. Other sections with dated intervals were then compared to the composite standard to determine the relative and absolute rates of sedimentation. The technique is crude but illustrative and can be refined and revised in the future. It will permit estimates of the amount of erosion at an unconformity and will permit calculations of rate of uplift.

Assuming a duration of 11.8 million years for zones Mantuan through Clarkforkian 1, the most rapid rate of sedimentation found in North America was 568 b (bubnoff = 1 m per million years) at the western edge of Hoback basin for an interval of Tiffanian 3 to Clarkforkian 1, within 10 km of the thrust belt that was the continental divide. The rate in the Crazy Mountain Field of Montana is 225 b for the interval Mantuan to Tiffanian 3. The rate of sedimentation and thickness of the Polecat Bench Formation and Fort Union Group along an east-northeast line from the Absaroka fault to Judson (near Bismarck, North Dakota) follows a simple power curve of the form $Y = aX^k$ where Y is the rate of sedimentation in b (or thickness in meters), X is the radial distance from the Absaroka fault in kilometers, a is 200 b (or 2360 m), and the exponent K is –0.25. Rates of sedimentation in the Bighorn Basin from west to east are 200 to 90 b, 51 b at Olive (M to Ti5), 38 b at Dickinson (Mantuan to Graybullian), and 27 b at Judson (Mantuan to Tiffanian 3). It is reasonable to group all of these sections because the Bighorn Basin, Powder River Basin, and Williston Basin were all a single depositional basin during the Paleocene. The earliest synorogenic Kingsbury Conglomerate contains *Hyracotherium angustidens,* an indication that the Big Horn Mountains did not begin to uplift until the Graybullian and not until then were the Bighorn Basin and the eastern basins separated. Radiometric ages of 40 to 45 Ma are known for the synorogenic intrusives of the northern Black Hills (Shapiro, 1971); thus the Cedar Creek anticline, the northern extension of the Black Hills, did not become elevated until the Uintan. The Powder River and Williston basins were separated at that time. In the northern San Juan Basin near Tiffany, Colorado, the rate of sedimentation is 132 b for an interval P1 to Ti4, whereas in the southwestern San Juan Basin near the stratotype for the Puercan, the rate is 99 b for Puercan 1 to Torrejonian 3. The lowest sedimentation rate found was that of the Black Peaks Formation of Big Bend, Texas, at 15 b for an interval Tiffanian 1 to Clarkforkian 1.

CORRELATION OF NORTH AMERICAN LAND MAMMAL ZONES WITH PLANKTONIC FORAMINIFERAL ZONES

Fox and Olsson (1969) have shown that the entire Cannonball Formation of North Dakota is early Danian in age. They showed that the entire 120 m of Cannonball Formation at the Garrison Dam on the Missouri River in North Dakota contains foraminifera characteristic of Berggren's (1972) marine planktonic zones P1b and P1c. At the same dam site, the Riverdale locality and fauna (Holtzman, 1978) of late Tiffanian (Ti4) age occurs 75 m above the Cannonball Formation. The Cannonball Formation is laterally equivalent to at least Mantuan to Tiffanian (M to Ti2) in age and probably extends to at least Ti3 in view of the abundance of shark teeth at the Brisbane and Judson Localities, 7 and 24 m respectively above the local top of the Cannonball Formation. On this basis, the mammal zones Ti3 to Cf1 would appear to correlate with the Thanetian, marine zones P3 to P6a. Berggren and others (1978) suggest an age of 57 to 58 Ma (58 to 60 Ma ICC constants) for the middle Thanetian planktonic foraminiferal zone P4, which would appear to correlate approximately with mammal zone Ti4. Figure 7 demonstrates the correlations between the section at Gubbio, Italy, and that in the Williston Basin.

It is interesting to note the similar rapid rates of diversification and short zones in mammals and foraminifera following the terminal Cretaceous extinctions. In each case this would appear to be the result of rapid evolutionary diversification due to reduced competition.

DURATIONS OF MAMMAL ZONES AND THE RATES OF EVOLUTION OF MAMMALS

It should be obvious from examining Figures 1 and 4–7 that the mammal zones are not of equal duration; some are exceedingly short. The duration of the zones is inversely proportional to the rates of evolution of the mammals themselves. A distinguishing feature of the Paleocene is the number of observable zones and the sustained high rate of evolution compared to rates of other epochs. Zones are particularly short in duration during times of rapid evolution: the Mantuan-Puercan zones due to rapid evolution of ungulates and the late Torrejonian-early Tiffanian zones due to rapid evolution of primates. Many Paleocene rates of evolution are close to the maximum rates shown by Gingerich (1983) for the time intervals involved. At other times, as in the mid-Tiffanian zone Ti3, there was relative stasis—at least for the Paleocene.

Haldane (1949) proposed a measure of the rate of evolution—appropriately called the darwin—in which 1 darwin represents a change of a factor of e in 1 million years, where e represents the base of the natural logarithms or approximately 2.718. . . . Most paleontological rates of evolution are measured in tens of millidarwins; Haldane noted several examples of 0.04 darwins. Downs (1961) found horse tooth characters had rates of 12 to 124 millidarwins. Gingerich (1980 a and b) found a brief rate of 1 darwin in the change from *Esthonyx xenicus* to *Esthonyx grangeri* during the early Clarkforkian, immediately after the genus immigrated from Asia. During the primary adaptive radiation of ungulates in North America, there were brief periods of even faster rates. The obvious place to look for a very fast rate of evolution is in lineages including the largest late Puercan and mid-Torrejonian ungulates, in particular the sequence *Protungulatum donnae–Ragnarok harbichti–Eoconodon coryphaeus–Triisodon quiverianus.* Length of M_2 is a satisfactory measure available for all the species. The results are shown in Table 2.

The mean rate of evolution over this sequence during an interval of 3.1 m.y. is 0.57 darwins for a size increase of 482 percent. The rate during the Mantuan is 3.85 darwins for an

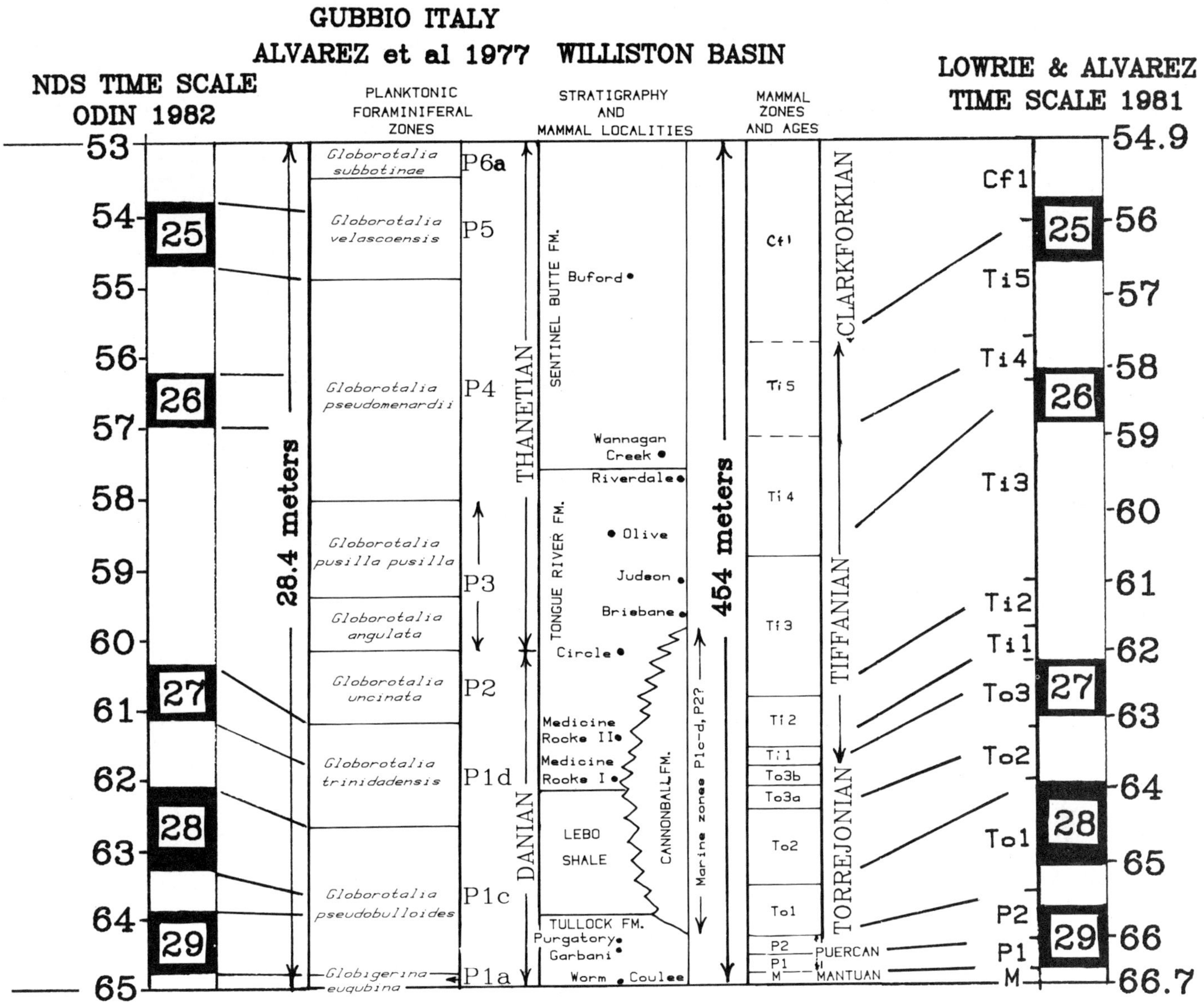

Figure 7. Comparison of Paleocene fossil zones and magnetozones from Gubbio, Italy, and the Williston Basin.

increase of 209 percent in 0.3 million years. This is clearly one of the fastest rates of evolution in the fossil record (cf. Gingerich, 1980b, 1983).

Among multituberculates the most rapid rate of evolution known is from *Catopsalis joyneri* to *Taeniolabis taoensis*—an increase of 360 percent in size of M_1 in 1 million years or a rate of 1.92 darwins.

The classic study of quantum or very rapid evolution in the mammalian fossil record is that of Patterson (1949) on taeniodonts. These rare and strange beasts were hog-sized herbivores whose dentitions combined the best food preparation features of parrots, rodents, and ground sloths; they had huge digging claws on all four feet. With little in the way of metric time to go on

TABLE 2.

Species	Length M_2	Age	Rate in Darwins
Protungulatum donnae	3.81 mm	67.0 (66.8)	
			2.85 (2.85)
Ragnarok harbichti	5.90 mm	66.8 (66.5)	
			3.85 (3.85)
Eoconodon coryphaeus	12.34 mm	66.5 (66.2)	
			0.21 (0.27)
Triisodon quiverianus	18.35 mm	63.9 (64.2)	
		OVERALL AVERAGE	0.57 (0.63)

TABLE 3.

		Length P_4-M_1 (mm)	Age (Ma)	Rate (Darwins)
Cimolestes cerberoides	(L)	12.6	67.5	
				-0.78
Procerberus formicarum	(BCa)	11.6	67.0 (66.7)	
				1.75 (1.75)
Procerberus sp.	(Mantuan)	20	66.7 (66.4)	
				0.51 (0.51)
Onychodectes tisonensis	(Pl)	25.6	66.5 (66.2)	
				0.09 (0.11)
Conoryctes comma	(To2)	41	63.9 (64.2)	
Procerberus sp.	(Mantuan)	20	66.7 (66.4)	
				1.33 (1.33)
Wortmania otariidens	(Pl)	34.5	66.5 (66.2)	
				0.01 (0.01)
Psittacotherium multifragmum	(To2)	36.1	63.9 (64.2)	

(Patterson greatly overestimated the duration of the Puercan at 5 million years) and with the discovery of the Cretaceous ancestry still thirteen years in the future, he nevertheless came to the reasonable and correct conclusion that these beasts evolved very rapidly and that the most bizarre stylinodontines did so the most rapidly of all then known mammals.

Subsequently Sloan and Van Valen (1965) found *Procerberus formicarum* at Bug Creek, and Lillegraven (1969) found its ancestor *Cimolestes cerberoides* at Scollard. Sloan (1970) and Lillegraven independently suggested that a still larger species of *Procerberus* at Mantua Quarry gave rise to *Onychodectes tisonensis* and *Wortmania otariidens,* the oldest taeniodonts from the type Puercan. The larger species of *Procerberus* has also been found at the Mantuan South Table Mountain fauna (M. Middleton, pers. comm., 1980). We now know the intermediate species that Patterson correctly inferred in his Figure 7.

It is instructive to calculate the rate in darwins for the very obvious size changes. As could be expected, the rates of evolution of teeth show the same features as in the case of the triisodontines, i.e.: relative stasis in the late Cretaceous, followed by very rapid evolution during the Bugcreekian and Mantuan, still rapid but declining rates of evolution during the Puercan, and near stasis in size during and after the Torrejonan (Table 3).

Rates would be at least as high, if not higher, if canine or ungual evolution were to be considered, but data are not presently available to make precise calculations on these more interesting features. Nevertheless I estimate the length of the ungual phalanx of *Procerberus formicarum* to have been about 3 mm and that of *Wortmania otariidans* about 33 mm. The estimated rate of evolution between the two is 2.75 darwins, which is exceedingly high, as Patterson predicted.

In addition to morphologic rates, taxonomic rates of evolution can be measured. These include numbers of contemporary species in a larger taxon and duration of species. The rapid speciation during and after the process of dinosaur extinction led to almost five doublings of ungulate species during the Bugcreekian, Mantuan, and the *Hemithlaeus* zone, an interval of only about 0.8 million years. Even by the end of that early time in the ungulate radiation, these species—all derived from the first species of ungulate, *Protungulatum donnae*—varied in length by a factor of 6 and in mass by a factor of about 200. Variation in dental and limb proportions indicates specific habitats varied at least as much. Few paleontologists who have not examined these fossils in detail can really appreciate the rapidity of this primary radiation. Figure 8 shows the number of North American ungulate species by zone. The rapid increase in diversity is obvious. Pronounced dips in the curve below the general trend during To1, Ti1, and Ti2 are most likely due to the very restricted number of faunas of those ages, either very small or, in the case of Shotgun (Ti1), mostly unpublished. Similarly species durations during the radiation are very short at first, on the order of 0.3 million years, but by the middle of the Tiffanian, mean ungulate species durations are much longer by a factor of 10. The mean duration of species has a direct impact on the duration of zones. The early Tiffanian zones are as short as they are mainly because of rapid evolution in primates and pantodonts; Bugcreekian to Puercan zones are short because of the rapid evolution of the ungulates. When most ungulate niches were filled by the Torrejonian and primate niches filled by the mid-Tiffanian, species and zone durations approached the norm for the rest of the Tertiary.

CORRELATION OF THE EXTRA-NORTH AMERICAN TERRESTRIAL PALEOCENE

Due to intercontinental migration of mammals, it is relatively easy to date extra-North American terrestrial Paleocene faunas in terms of the North American mammal zones. Figure 9 shows the location of known Paleocene mammal faunas of the world.

South America

The oldest probably Paleocene fauna in South America is

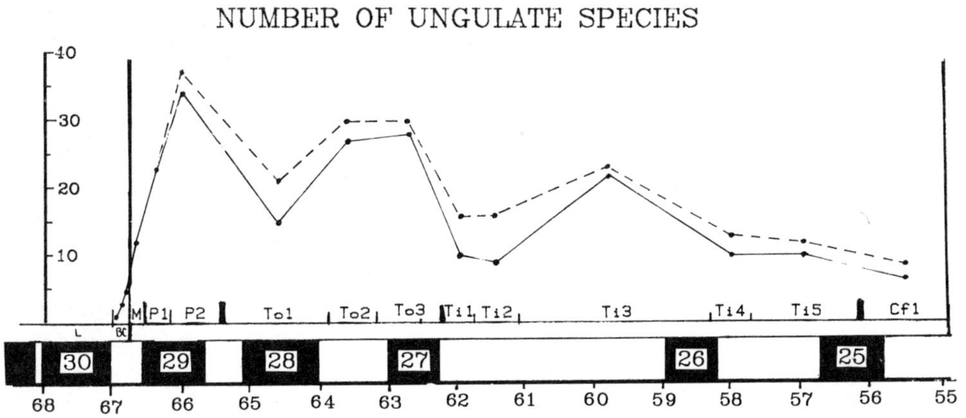

Figure 8. Numbers of ungulate species per zone in the latest Cretaceous and Paleocene of North America. Standard zonal abbreviations, magnetochrons, and age in m.y. (Lowrie and Alvarez [1981] time scale on horizontal axis).

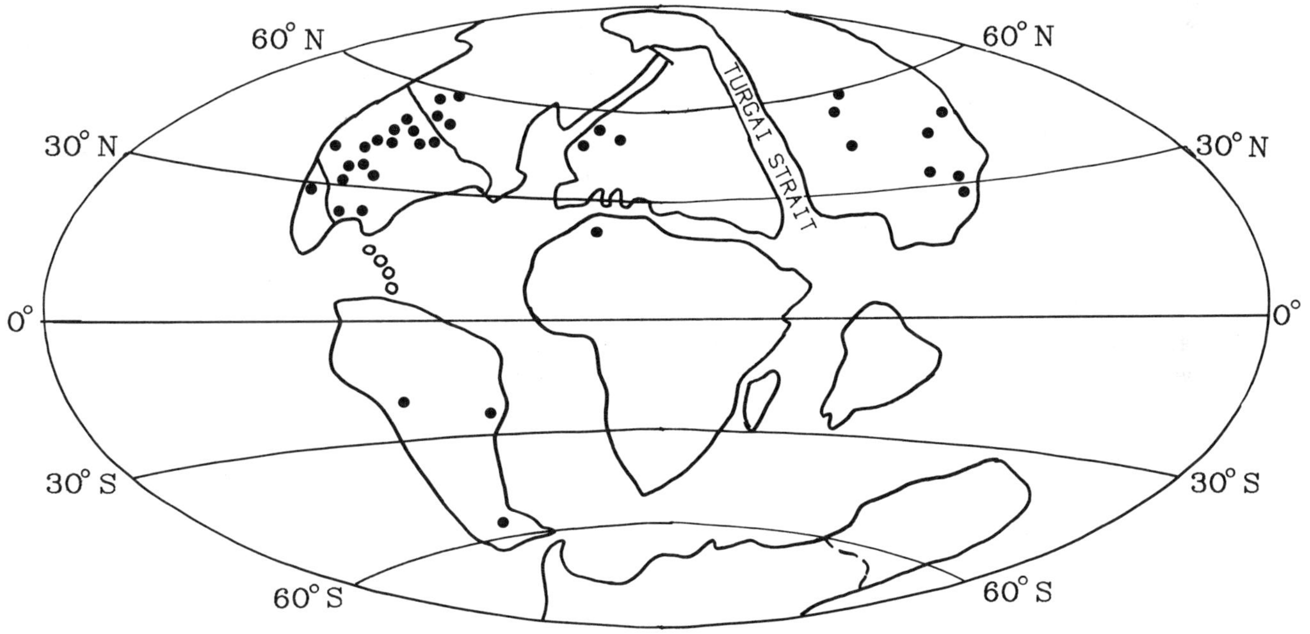

Figure 9. Location of known Paleocene mammal faunas of the world.

that of Laguna Umayo near Lake Titicaca in equatorial Peru (Grambast and others, 1967; Sigé, 1972; Marshall and others, 1983). The fauna was considered to be latest Cretaceous in age by its authors, mainly on the basis of oogonia of charophytes and some probably dinosaurian egg shells. The mammalian part of the fauna includes the marsupial *Alphadon austrinum,* the genus of which has a range of Aquillan to P2; a possible member of the Pediomyidae, otherwise known only from the Aquillan to Lancian of North America; and the condylarth *Perutherium altiplanense.* Simpson (1978) notes the resemblance of the lower molars of *Perutherium* to those of the Phenacodontidae and Didolodontidae, but Van Valen (1978) instead refers the genus to the Periptychidae as a separate subfamily. The inferred earliest possible age for the Laguna Umayo fauna is M, and a more likely age in view of the very low trigonids is P1 to To1. More recently, Marshall and others (1983) demonstrated convincingly that *Perutherium* is in fact the earliest notoungulate, and that the early notoungulate radiation was entirely South American until the late Tiffanian.

The Riochican age (stage) of South America is known from four faunules from Argentinan Patagonia and a fauna from Itaborai in Brazil. These faunas consist of marsupials, xenarthrans, and ungulates (Simpson, 1948, 1967, 1978; Paula Couto, 1952,

1970). The marsupials are all derivable from the Aquillan to P2 genus *Alphadon.* The xenarthrans are not directly derivable from anything yet known in the Northern Hemisphere except possibly from palaeanodonts (Rose, 1978a). The oldest known palaeanodont is *Amelotabes simpsoni* from Ti4; the most likely but definitely unproved ancestry for that species is the Pantolestidae, which has a known range of To1 or To2 to Bridger, in particular from the genus *Coriphagus,* known only from To3. If (and that is a very large if!) this does represent the origin of xenarthrans, the earliest possible age for the Riochican would be Ti1, and an age of Ti3 or Ti4 is more likely.

The ungulates present more useful information on the age of the Riochican. Riochican notoungulates include many forms conspicuously closer to didolodontid condylarths (as well as to loxolophine arctocyonids) than the known Northern Hemisphere notoungulates *Arctostylops, Palaeostylops,* and *Sinostylops.* The oldest well-dated occurrence of *Arctostylops* is from Ti5, the Princeton Quarry level of the Polecat Bench Formation of the Bighorn Basin (Jepsen and Woodburne, 1969); the other known occurrence is from Cf2, the *Plesiadapis cookei* zone. On faunal grounds it would seem apparent that the Riochican at least includes the equivalent of the Clarkforkian and possibly late Tiffanian strata as well.

Marshall and others (1981) measured K/Ar ages and magnetostratigraphy of the type Rio Chico Formation of Chubut Province, Patagonia, Argentina. The terrestrial Rio Chico Formation rests conformably on the foraminiferally dated marine Danian Salamanca Formation. They unambiguously measured magnetochrons 25 and 26 near the top of the Rio Chico and showed that chron 27 had to be within the Salamanca Formation. Dates of 61 ± 5 Ma near the top and 62.8 ± 0.8 Ma and 64.0 ± 0.8 Ma near the base of the Salamanca Formation confirm the Danian age.

Marshall and others (1981) assigned the Rio Chican to the late Torrejonian, Tiffanian, and Clarkforkian on the basis of magnetochron identification in the San Juan and Bighorn Basin sections. It has been shown above and in Butler and Lindsay (1985) that the San Juan section was misnumbered. This implies that no part of the Rio Chican correlates with any part of the Torrejonian.

The age of the Salamancan is further suggested to be Danian (= M to Ti2) by the data in Vail and others (1977), which state that a global drop in sea level of about 140 m took place at the end of the Danian and that this retreat was very rapid. This implies the age of the base of the immediately overlying Rio Chican is Ti1 to Ti3, exactly comparable to the situation in Wyoming and North Dakota. Considering the measured ages in the Salamanca Formation, Ti3 seems more likely, as in North Dakota.

From the stratigraphy of Simpson (1935b) and the magnetochronology of Marshall and others (1981), the rate of deposition of the Rio Chican in the type area would appear to be about 22 b. Since the oldest date of Marshall, 64 ± 0.8, is about 55 m above the highest known dinosaurs in Chubut (Marshall and others, 1981), dinosaur extinction at about 45 degrees south would appear to be at about 66 Ma or, as closely as can be measured, contemporary with that at 45 degrees north.

Simpson arranged the local faunas from the Rio Chico Formation into three zones: the basal *Carodnia* zone, the medial *Kibenikhoria* zone, and the upper *Ernestokokenia* zone. The *Carodnia* zone and the *Kibenikhoria* zone occur in the upper part of the Las Violetas Member, and the *Ernestokokenia* zone occurs in the upper part of the Visser Member. By interpolation from Marshall and others (1981) and Simpson (1935b), the *Carodnia* and *Kibenikhoria* zones occur in the reversed interval between chrons 27 and 26; the *Ernestokokenia* zone is about at chron 25. This is a direct correlation of the *Ernestokokenia* zone with late Ti5 and early Cf1 and implies a correlation of the *Carodnia* zone and the *Kibenikhoria* zone with Ti3.

The oldest xenarthran edentates occur in the Itaborai fauna and beds probably correlatable with the *Carodnia* zone at Las Violetas at Chubut. Rio Chican edentates are rare even in a large fauna. Four specimens are known from Argentina, scutes of three individuals from three localities ranging from the base to the top of the formation. Itaborae specimens include only four dasypod scutes and a pair of astragali, one each from a dasypod armadillo and from the oldest known glyptodont (Cifelli, 1983a). Cifelli specifically resurrects the old Matthew and Granger hypothesis of a close relationship between Paleanodonta, Xenarthra, and Pholidota, explicitly rejected by Emry (1970).

Cifelli (1983a, 1983b) made a careful revision of the Itaborai ungulates on the basis of dentitions and isolated tarsals, coming to the conclusion that didolodont condylarths, litopterns, and notoungulates shared a recent common ancestry. His best estimate of that ancestry is a mioclaenine hyopsodont such as the To1 to To3b *Litaletes.* An earlier ancestry is possible but would require great parallel evolution to essentially duplicate *Litaletes.* Cifelli was less certain about the same origin for *Carodnia* and *Tetragonostylops,* the first genera of the Pyrotheria and Astrapotheria respectively. On size change alone the ancestry of *Tetragonostylops* would had to have evolved at an average rate of 0.77 darwins over the 1.1 m.y. from To3b to Ti3, and *Carodnia* would have had to evolve at an average rate of 1.8 darwins over the same interval. These rates seem unreasonable in view of the peak rates of ungulate evolution during a greater ecological vacuum in the North American BC to P interval. Pushing the immigration of *Litaletes* farther back into the Torrejonian would make the network of similarities and crossing specializations of the Itaborai and Rio Chico didolodonts, litopterns, and notoungulates so well documented by Cifelli less obvious than they are. Cifelli suggested that the astrapotheres and pyrotheres must have come from an earlier ungulate immigration, perhaps *Perutherium.* In any event, McKenna's (1975) mirorder Meridungulata appears to be of dual origin, with a Maastrichtian immigration providing the ancestry of the astrapotheres and pyrotheres and another distantly related one during the Torrejonian providing the ancestry of didolodonts, litopterns, and notougulates.

Cifelli (1983a) suggested that the ankle of *Carodnia* resembled proboscideans more than it did the Dinocerata, in opposition to the well-known dental resemblance.

Asia

Until 1973 the main Paleocene faunas known in Asia were the classic Gashato fauna (44°5′, 103°40′) and its nearby correlative, the Naran Bulak fauna (43°30′, 101°), both of the southern Peoples Republic of Mongolia (Szalay and McKenna, 1971). Two brief notes (Chow, 1960; Young and Chow, 1963) described a fossil from Turfan, Sinkiang Province, China, and mentioned some fossils from Nan-Xiong (=Nanyung, Nanhsiung) County, north Guangdong (=Kwangtung) Province, that clearly were of earlier Paleocene age.

Since 1973, the amount of published information on Chinese Paleocene faunas has grown so rapidly that the Chinese record is better known than any but the North American record. The two most useful summaries of the Chinese Paleocene in English are South China "Redbeds" Research Group, 1977, and Li and Ting, 1983. The second updates but does not replace the first. I here update these with later literature, and my own observations of January to March 1986 at the Institute of Vertebrate Paleontology and Paleoanthropology, Beijing. In the discussion that follows, I will take into account the recent change in Roman letter orthography of Chinese names. I will usually present names both ways, with the current Pin-yin spelling first and the older Wade-Giles spelling in parentheses.

The Paleocene strata of China can be divided into two regional parts on geographic and tectonic grounds. The two regions are the Sino-Korean craton or North China Block and the Yangtse craton or South China Block. The line of division is the east to west striking Qinling mountain belt (Mattauer and others, 1985). Within the North China Block (which includes the Peoples Republic of Mongolia), the only Paleocene faunas known are those strictly correlative with the Gashato fauna, the very latest Paleocene. A major unconformity separates late but not latest Cretaceous strata from the latest Paleocene strata everywhere the Paleocene rocks are known in this block. Four areas of the North China Block are known to have Gashatan mammals. In contrast, throughout the eastern part of the South China Block, a continuous sequence of strata, thickening rapidly to the southeast, was deposited during the full duration of the Paleocene, only to be broken up into many small basins (actually grabens) by Eocene northeast striking block faulting. Seven of these grabens are known to contain Paleocene mammals. Others probably do also, but the outcrops are under intensive cultivation in South China, and discoveries are rare. Presumably the source of these Late Cretaceous and Paleocene sediments of South China was a coastal mountain range produced by subduction of the Pacific plate crust along the southeast coast of China. Gravels in these sediments indicate a source composed mainly of continental crust, not andesite volcanoes. The uniform red color of these sediments implies heavy laterite weathering in the source area.

Asian Paleocene faunas can be divided into two major units, a lower unit known only from the South China Block and an upper unit known from all parts of Asia. The lower unit ("Lower Assemblage" or the "Early-Middle Palaeocene faunal complex" of South China "Redbeds" Research Group, 1977) is composed almost exclusively of the pantodont family Bemalambdidae as large herbivores, varied members of the order Anagalida as small herbivores, and mesonychid condylarths as large carnivores. Two small mioclaenid condylarths migrated in early from North America but gave rise to no descendants that we know. Notable absences from this record include multituberculates, miacines, creodonts, and any ancestry for the dinoceratans. The record is quite good, including 16 families, 28 genera, 42 species, and hundreds of fossils from 7 basins. Li and Ting (1983) assigned the name Shanghuan as a provincial age term to this unit. By crossing migrations this provincial age is exactly correlative with North American zones To1 to Ti3 or basal Ti4 inclusive. It does indeed represent the first half of the Paleocene, less the Puercan-Mantuan, but continued confusion will result from the use of early, middle, and late as adjectives for subdivision of Paleocene time because of the historic use of the mammal ages in North America. Those who are not familiar with this subject will automatically be misled into very wrong correlations.

The upper unit of the Paleocene is known from all over Asia. It was termed the "Upper Assemblage" or the "Late Palaeocene Faunal Complex" by the South China "Redbeds" Research Group (1977) and "Nongshanian" by Li and Ting (1983), but there is a prior synonym in the form of "Gashatan Asiatic Age" of Romer (1966). Szalay and McKenna (1971) defined the Gashatan as characterized by the joint overlapping time ranges of *Palaeostylops, Pseudictops, Prionessus,* and *Eurymylus,* a definition that is very precise but restricts Gashatan to the duration of a short zone and is not of stage (age) dimensions. In view of the fact that the Gashato was the first Paleocene fauna to be found in Asia, its priority as a stage (age) name, and the restriction of most of the families represented in the Gashato to the "Upper Assemblage," I think it appropriate to use the term *Gashatan* in the broader sense. Two stages are sufficient for the Paleocene; further refinement should be in the form of detailed zonation. This age subdivision has the further advantage of making the Gashatan approximately equivalent to the Thanetian of Europe and to the Riochican of South America as demonstrated in this paper.

The most widely distributed faunas are those of the Gashatan, correlative with the Late Tiffanian and Clarkforkian stages (Ti5 to Cf3) of North America. Such faunas include the classic Gashato and Naran Bulak faunas, the Turpan (=Turfan) fauna from the Taizicun Formation of Xinjiang (=Sinkiang) province (42°, 89°; Tong, 1973; Zhai, 1973); the important Naomugen (=Nomogen) fauna (42°90°) from the Naomugen Formation of Haliut and Gonghydong, Naomugen Commune, Nei Mongol (Inner Mongolia) Province (Chow and others, 1976; Chow and Qi, 1978); and three faunas from the south: the faunas of the Doumu Formation and upper Wanghudun Formation of the Qianshan (=Chienshan) basin, Anhui (=Anhwei) Province (30°25′,116°20′); the faunas of the Nongshang Formation in the

Nanxiong Basin, Guangdong Province (24°48′,113°43′; Li, 1977; Zhang, 1978; Chow and Wang, 1978); and those of the Chijiang and Pinghu formations of the Chijiang Basin 20 km immediately to the north (Li and Ting, 1983).

This wide-ranging fauna extends over an area of 2100 km north and south and 2600 km east and west. The local faunas are composed of the classic mixture of advanced taeniolabidoid multituberculates (northern four localities only), the pantodont *Archaeolambda,* anagalids, mimotonids, eurymylids, the successive Asian notoungulates *Asiostylops, Sinostylops,* and *Palaeostylops,* and—in the latest part—the primitive dinoceratan *Prodinoceras* (=*Mongolotherium*; Flerov, 1957; Dashzeveg, 1982). The total absence of any ancestry for *Prodinoceras* in the earlier Asian Paleocene suggests strongly that both it and the North American *Probathyopsis* (=*Bathyopsoides*; Patterson, 1939; Wheeler, 1961) are derived from the Riochican *Carodnia.* The Asian notoungulates are clearly derived from the North American *Arctostylops* and ultimately from some Riochican notoungulate. One additional mammal of South American origin is to be found in the lower part of this unit. *Ernanodon* (Ting, 1979; Radinsky and Ting, 1984) is an edentate—the only pre-Miocene edentate known outside of South America. As seen in the section on South America, armadillo scutes are known in the lowest Riochican, so this is not the oldest edentate. A possible primate, *Petrolemur,* is also known from this level. The Gashatan record is very good: Some 68 species are known, belonging to 41 genera and 18 families. With the appearance of *Homogalax* cf. *protapirinus* in the Wutu Formation of the Changlo district, Shandong Province (36°40′, 118°51′), the Gashatan, the late Tiffanian-Clarkforkian equivalent of Asia, came to a close.

The Shanghuan faunas of Asia are typified by the genus *Bemalambda,* which does not appear to extend into any of the Gashatan localities or horizons. Four zones are easily recognized on the basis of *Bemalambda* species and on genera of the Mesonychidae and Anagalida (Chow and others, 1977). The Shanghuan is presently known from six localities in South China: the Shanghu Formation of the Nanxiong Basin of northern Guangdong Province (24°48′, 113°43′) and the very similar Shizikou Formation that occurs 20 km away in the next basin north (the Chijiang Basin of Jiangxi Province), the lower and middle Wanghudun Formation of the Qianshan Basin (Chienshan) of Anhui (Anhwei) Province (30°25′, 116°20′), the Zaoshi (Zhaoshih) Formation of the Chaling Basin of Hunan Province (26°40′, 113°30′), the Gaoyugou Formation of the Tantou Basin of Henan (Honan) Province located in a graben within the southern Qinling Mountains (34°, 111°46′), and the Fangou Formation of the Shimen graben well within the Qinling Mountains in Shaanxi (Shensi) Province (34°06′, 110°10′; Xue, 1982). Complete sequences of the Shanghuan are preserved only in the Nanxiong and Qianshan basins.

A distal end of a femur and an axis vertebra of a large mammal estimated to be over 1 m in length and over twice the size of any known Cretaceous mammal were found between two beds containing dinosaur eggshell fragments near Changde (28°, 110°40′) in Hunan province (Ting [Ding], demonstration and oral communication, March 1986).

In the company of Mrs. Ting, I examined four outcrops of the lower contact of the Shanghu Formation with the dinosaur-bearing Nanxiong Formation spread over a distance of 25 km along strike. I found no unconformity between the formations at all; the differences are very subtle except for the disappearance of dinosaur eggshell fragments and the appearance of Paleocene mammals. The lowest zone of the Shanghuan (Shanghu 1 of Fig. 1) is about 200 m thick in the Nanxiong Basin and is typified by *Bemalambda pachyoesteus,* the only Asian pantodont primitive enough to provide an ancestry for *Pantolambda or Deltatherium* from North America. This zone contains two mioclaenid genera; *Youdon,* derived from some North American species of *Protoselene* (P1 to Ti3), and *Palasiodon,* derived from some North American species of *Litaletes* (To1 to To3; see Van Valen, 1978). The top of this zone contains the first mesonychid in Asia, *Dissacusium shanghoensis,* which is very similar to the oldest mesonychids in North America, *Dissacus navajovius* and *Microclaenodon,* both first known from To2. Since the entire ancestry of mesonychids is known from North America, this suggests strongly that by crossing migrations, the base of the Shanghu Formation and the Shanghuan age is To1.

The latest known member of the Asian lizard subfamily Polyglyphanodontinae occurs in the upper 80 m of the Nanxiong Formation. It is even more specialized than the previously most specialized form known, *Polyglyphanodon* from the latest Lancian North Horn Formation, just under the Puercan Wagonroad fauna. In view of the conformity between the Nanxiong and Shanghu formations, it seems likely that dinosaurs survived into the Puercan equivalent of South China.

The second zone of the Shanghuan stage contains the anagalid *Linnania,* the mesonychid *Hukoutherium, Prosarcodon* (McKenna and others, 1984), and an undescribed species of *Bemalambda,* best known from Laonan (Xue and Sloan, in preparation). It is approximately To3 to Ti2 in age. The third zone contains *Bemalambda nanshiungensis* and the mesonychid *Yatanglestes.* The fourth and highest zone of the Shanghuan contains *Bemalambda crassa* and B. *shizikouensis,* as well as *Lofochaius,* the smallest and oldest member of the esthonychid tilodonts. Extrapolating from the proposed evolutionary rates of the North American Paleocene and Eocene species of *Esthonyx* (Gingerich, 1979), which appears in North America in Cf1, the best estimate of the top of the Shanghuan is about Ti4.

From the standpoint of crossdating the Nanxiong Paleocene with the North American zones, the most important undescribed specimen is the very primitive small perissodactyl skull collected by Zhang shortly before her death, to be named for the late Leonard Radinsky and described by Zhou, McKenna, and Ting. The skull is complete except for the premaxillaries, both P^3s and the left P^1 and P^2. The length of the skull as preserved is 79 mm, measured from the occipitals to the front of the right maxilla. The molar teeth have a fully developed metaloph, the P^4 is semimolariform, and all cheek teeth resemble *Hyracotherium,* so the specimen is clearly a perissodactyl; but the specimen lacks a diastema, so it just as clearly is not *Hyracotherium*. The length of P^4–M^3 is 19.3 mm; total length P^1 to M^3 is only 37 mm. The skull greatly resembles Matthew's (1937) reconstruction of *Tetraclaenodon* except it is more gracile. The ear region resembles *Homogalax* more than it does *Hyracotherium*. Most likely it is derived from *Desmatoclaenus mearae* (Van Valen, 1978) from the Ti3 of Bison Basin Saddle; the teeth resemble *D. mearae* closely in size and morphology, except for the loss of the saddle between the metaconule and hypocone.

Since the Zuiguikeng Member of the Nongshan Formation from which the skull came lies under the Gashatan Datang Member of the same formation, which has the oldest notoungulate in south China and correlates with Ti5 in the North American sequence, we can tentatively date the Zhuguikeng Member as Ti4. This implies the roughly 850 m of the Shanghu and Zhuguikeng were deposited in roughly 7.7 million years for a rate of deposition of roughly 110 b.

The origin of the order Perissodactyla then was North American from BC to Ti3 and was south Asian from Ti4 to Cf3, when *Hyracotherium* and *Homogalax* expanded northward as the climate warmed and crossed back to North America and when *Hyracotherium* went on to Europe by way of the DeGeer route. A critical 3-million-year interval missing from the North American record of perissodactyl evolution is now filled with the new south China record.

Known times of intermigration between Asia and North America in terms of the North American mammal zones are Lancian, BCa, To1, To2, To3, Ti4, Cf2, and the earliest Wasatchian Eocene. In absolute ages these are approximately 70.0, 67.0, 65.4, 63.9, 63.2, 57.7, 54.9, and 54.0 Ma. At other times the two faunas evolved independently. At times of intermigration there were significant extinctions: of dinosaurs in both continents, of pantodonts in China, and of multituberculates, primates, and ungulates in North America. Known migrations from Asia to North America were gypsonyctopine leptictids and paleoryctids at about 76 Ma, *Stygimys* and *Catopsalis* among multituberculates at about 67 Ma, *Deltatherium* at 63.9 Ma, *Pantolambda* at 63.2 Ma, *Cyriacotherium* at 58.3 Ma, *Esthonyx* at 56.2 Ma, and *Hyracotherium* at about 54 Ma. Known migrations from North America to Asia were mesonychids and mioclaenids at about 65.4 Ma, notoungulates at about 57.7 Ma, and *Prodinoceras* at about 56.2 Ma (in the latter two cases moving through North America from South America). In each of these cases the complete prior evolution of the ancestors of the immigrating species is well documented in the continent from which it emigrated.

Europe

The European Paleocene is still represented primarily by the two now classic faunas best described by Russell (1964)—Cernay-Berru and Walbeck—and the recently discovered Hainin fauna of Belgium (Godfriaux and Thaler, 1972; Thaler, 1977; Vianey-Liaud, 1979). Although these faunas show some generic endemism, they are remarkably closer to standard North American faunas than to those of Asia or South America. On the basis of *Plesiadapis* species, Cernay is clearly of Cfl age—early Clarkforkian—since *Plesiadapis gingerichi* is very similar to the Cernay species *P. tricuspidens*. Walbeck is significantly older and best compares with Ti3, the *P. rex* zone. These close similarities suggest that the North Atlantic bridge was less of a filter than were the Caribbean or Beringia during the Paleocene. The Hainin fauna appears to be correlative with the Torrejonian (Savage and Russell, 1983).

The reason for these differences in penetration rate of Beringia and the North Atlantic bridges may well be correlated with the progressive shift of the Northern Hemisphere continents with respect to the North Pole. The latest Cretaceous pole has an apparent position of about 72°, making the latitude of the southernmost part of Beringia about 12° south of the pole (Sloan, 1970). By the time of To2, the pole had shifted to a present latitude of 76° (Butler and Taylor, 1978), shifting Beringia south by 4°. By the late Wasatchian, the pole had shifted to 84° (Shive and Pruss, 1977), and the south edge of Beringia would be about 24° from the pole. The path of wandering lies between Beringia and the present pole. At the same times, the North Atlantic filter was only about 10° north of the North Dakota localities (Szalay and McKenna, 1971).

The reasons earlier postulated by Sloan (1970) for the migrations of the Clarkforkian and Wasatchian faunas have been borne out by Hickey (1977) in his analysis of the flora of the latest Paleocene-Eocene Golden Valley Formation of North Dakota. Hickey showed that western North Dakota warmed from an average annual temperature of 15°C, with a mean temperature of the coldest month of 6.1°C, during the late Paleocene (about Cf1) to an average temperature of 18.3°C, with a mean temperature during the coldest month of 13°C, during the Graybullian. This significant rise in temperature and equability, coupled with the development of frost-free winters, is more than enough climatic change to account for the increase in intercontinental migration during the Clarkforkian and Graybullian. By any standards, the top of the Clarkforkian and the base of the Graybullian is the appearance of the *Microsyops-Pelycodus-Hyopsodus-Diacodexis-Hyracotherium-Homogalax* fauna in the Big Horn Basin.

Africa

Until recently the oldest Cenozoic mammals known in Africa were middle Eocene forms from the Sahara. This fauna consists of prosimians and hyracoids clearly derived from Europe. Cappetta and others (1978) report a small but significant Montian (Danian) lower Paleocene fauna from Morocco. The locality is near Ouarzazate on the south flank of the High Atlas mountains (7° west, 21° north). At the site late Cretaceous, Senonian (=Campanian + Maastrichtian) red beds are conformably overlain by 12 m of early Montian lacustrine beds containing freshwater gastropods. The mammals occur in a 10-cm-thick marine tidal channel sand 8 m above the Senonian redbeds. The lacustrine beds are overlain by a marine limestone of Montian age and that in turn by another limestone of Thanetian or lower Ypresian age, both well dated by invertebrates. The microvertebrate fish fauna consists of a dozen species of sharks and rays, a lungfish, a pycnodont, and the oldest known characid. The reptilian fauna consists of turtles, lizards, and crocodiles. The mammals in the fauna include two species of paleoryctid insectivores, a possible proviverrine creodont, and a possible miacid carnivore. Only a few of the sharks have been named. Some of the mammals were illustrated in Cappetta and others (1978). One of the paleoryctids is noted as resembling *Batodon* and *Aboletylestes.*

On the basis of stratigraphic position, the best estimate of the age of this locality in terms of the North American land mammal ages is P2 to To1. The fauna is the first partial confirmation of the plausible hypothesis that paleoryctids were ancestral to the Chrysochloridae, Tenrecidae, and Macroscelidae. Although the sample size is very small, the absence of marsupials and multituberculates is significant. The Paleocene mammalian fauna of Africa would appear to be ultimately derived from the late Cretaceous fauna of Asia, perhaps through North America and Europe.

TAXONOMY

Full utilization of the accompanying range charts requires description of seven new species of multituberculates and reillustration of three old species inadequately illustrated in the past. Some of these new species have been previously cited by other authors by initial letter of the species name. The most common mammals at most Paleocene fossil localities are the neoplagiaulacid multituberculates. They were mainly herbivorous, although—like modern rodents—they were perhaps likely to ingest animal tissue as well.

To facilitate the rapid identification of these species in new faunas as they are found, I here present a summary table of dental measurements and cusp counts as well as summary figures (Figs. 10, 11, and 12) to show shape differences and evolution. Although variation is not shown, most cusp numbers vary by ±1, and lengths of teeth vary by up to ±0.2 m.

NEOPLAGIAULACIDAE Ameghino, 1890
MESODMA Jepsen, 1940
Mesodma pygmaea new species
(=*Mesodma* sp. P, Krishtalka, 1973)
(=*Mesodma* sp. P. Krause, 1977)
(=*Mesodma* sp. Holtzman, 1978)
(Fig. 10, Tables 4, 5)

Etymology: Latin *pygmaeus* (dwarf) in allusion to its size.
Holotype: AMNH 35298, left jaw with M_b, M_1, M_2 from Gidley Quarry, To3a.
Referred specimens: PU 16851, left jaw with P_3, M_b, M_1, M_2 from Medicine Rocks I, SW¼Section2,T3N,R58E, Carter County, Montana, To3; MCZ 18781 M_b; MCZ 18789 P^4; and many other specimens, Shotgun Quarry, Wyoming, Ti1; ACM specimens from Bison Basin Saddle locality, Wyoming, Ti3a; UA specimens from Police Point locality, Alberta (Krishtalka, 1973); SPSM specimens from Brisbane locality, North Dakota, Ti3a (Holtzman, 1978); UA specimens from Roche Percée locality, Saskatchewan, Ti4 (Krause, 1977).
Known stratigraphic range: To3a to Ti4.
Diagnosis: The smallest species of Paleocene multituberculate in North America. Probably derived from Lancian *Mesodma hensleighi* from which it differs in uniformly smaller size and relatively larger blade. Modal dimensions in Table 4.

Neoplagiaulax Lemoine 1880
Neoplagiaulax nelsoni new species
(=*Neoplagiaulax* cf. *hunteri,* Rigby, 1980)
(Figs. 11, 13, Tables 4, 5)

Etymology: Named for several unrelated good friends who have been of great help to me in my Montana field work, notably Lyle and Darlene Nelson of Fort Peck, Montana, who have succored our field parties in inclement weather for 25 years; Professor William F. Nelson of the University of Tennessee–Martin, who helped in the discovery of Purgatory Hill and Harbicht Hill and in many other ways; and James Nelson of St. Paul, Minnesota, whose enthusiasm in 1960 and 1961 was infectious.
Holotype: UMVP 6030, left M_b from Purgatory Hill fauna, Purgatory Hill, NW¼Section36,T23N,R43E, McCone County, Montana, P2.
Paratypes: UMVP 6031, right M_b; 6032, 6033, left P^4's from Purgatory Hill.
Referred Specimens: UMVP 6042, many AMNH specimens from Swain Quarry, To3, Wyoming (Rigby, 1980). MCZ 18791 and nine other specimens, M_b's; MCZ 18790 and four other specimens, P^4's from Shotgun fauna, Wyoming, Ti1.
This species differs from its larger descendant *N. hunteri* in smaller sizes, fewer serrations, and lower cusp formulas. The very largest specimens of *N. nelsoni* overlap the very smallest specimens of *N. hunteri* in size and counts, but the mean values are significantly different as are the ages. Several specimens from Rock Bench Quarry (PU 18218 and 18209) and Medicine Rocks I (PU 18208), of zone To3b, are as large as *N. hunteri* but have counts typical of *N. nelsoni.* This suggests that the transition from one species to the other is gradual rather than punctuated. Measurements in Table 5.

Neoplagiaulax hunteri Simpson 1936
(Fig. 11, 14, Table 4)

Neoplagiaulax hunteri is a very common middle Tiffanian species, originally described from Scarritt Quarry. Paratype materials are very well preserved and include all molars and premolars in jaw fragments. Originally the species was very poorly illustrated. Although isolated teeth have been well figured from other localities, there are enough jaws in the type material so that this species

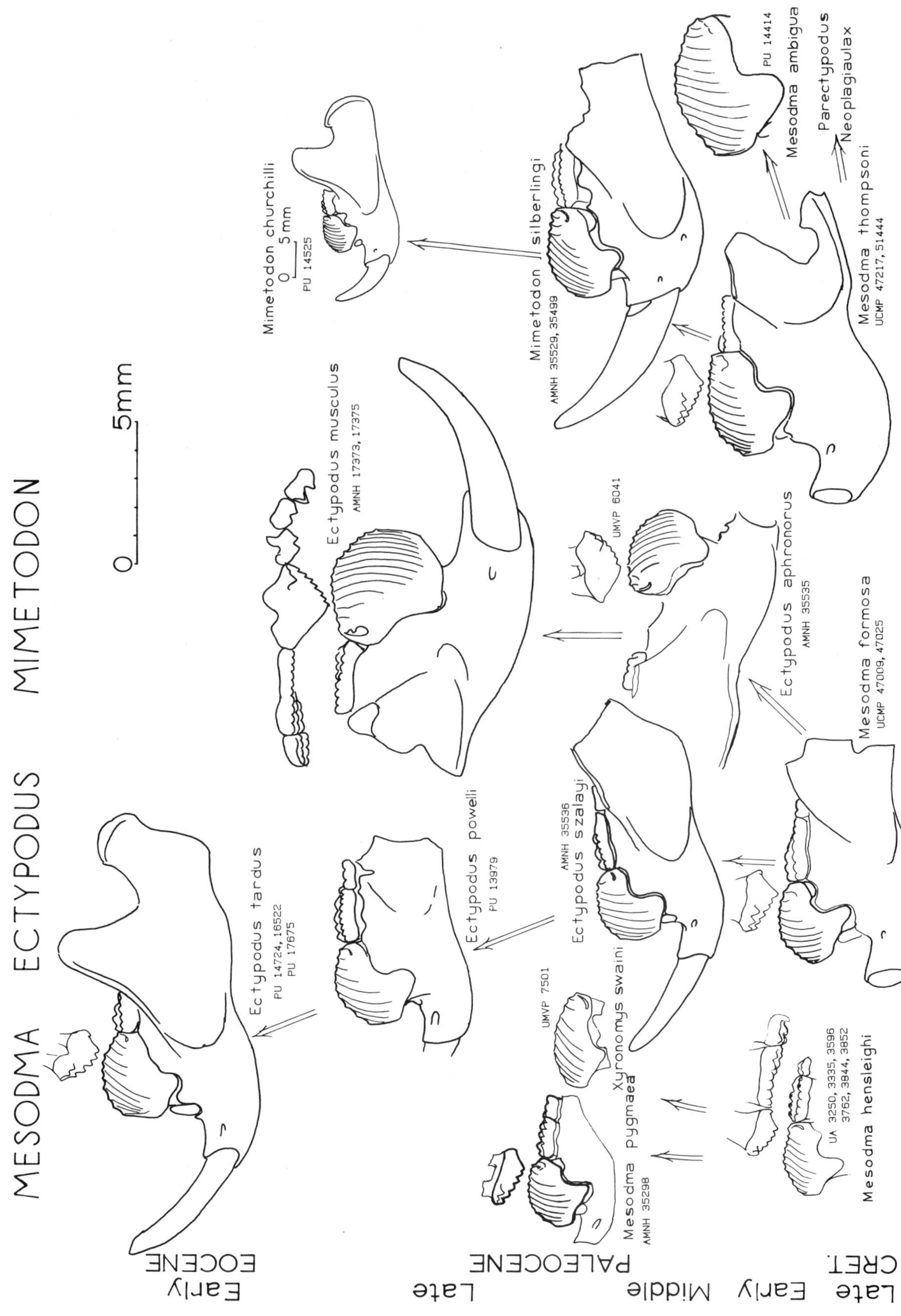

Figure 10. Drawings of latest Cretaceous, Paleocene, and early Eocene species of *Mesodma, Ectypodus,* and *Mimetodon* from western North America.

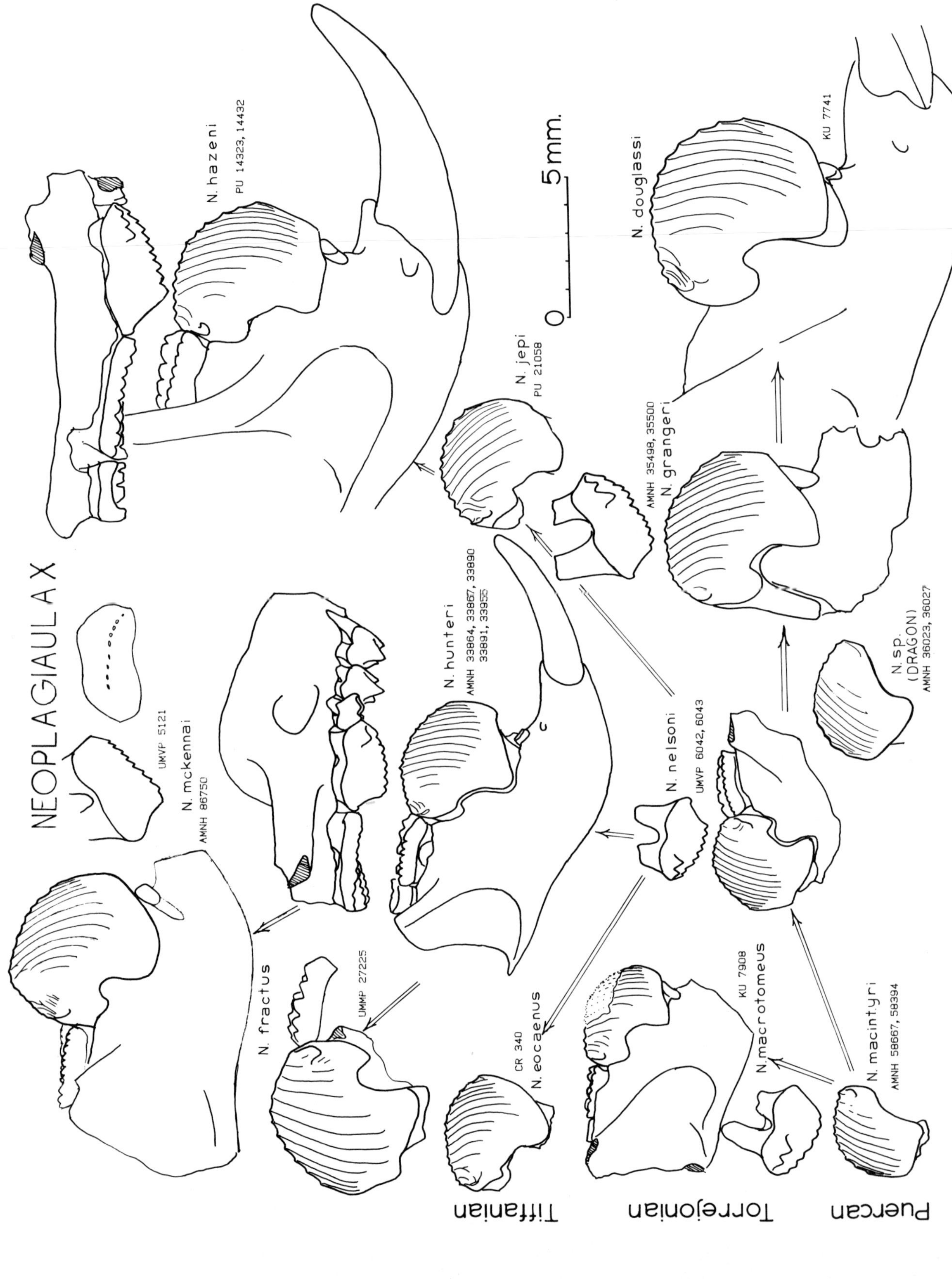

Figure 11. Drawings of Paleocene species of *Neoplagiaulax* from North America and Europe.

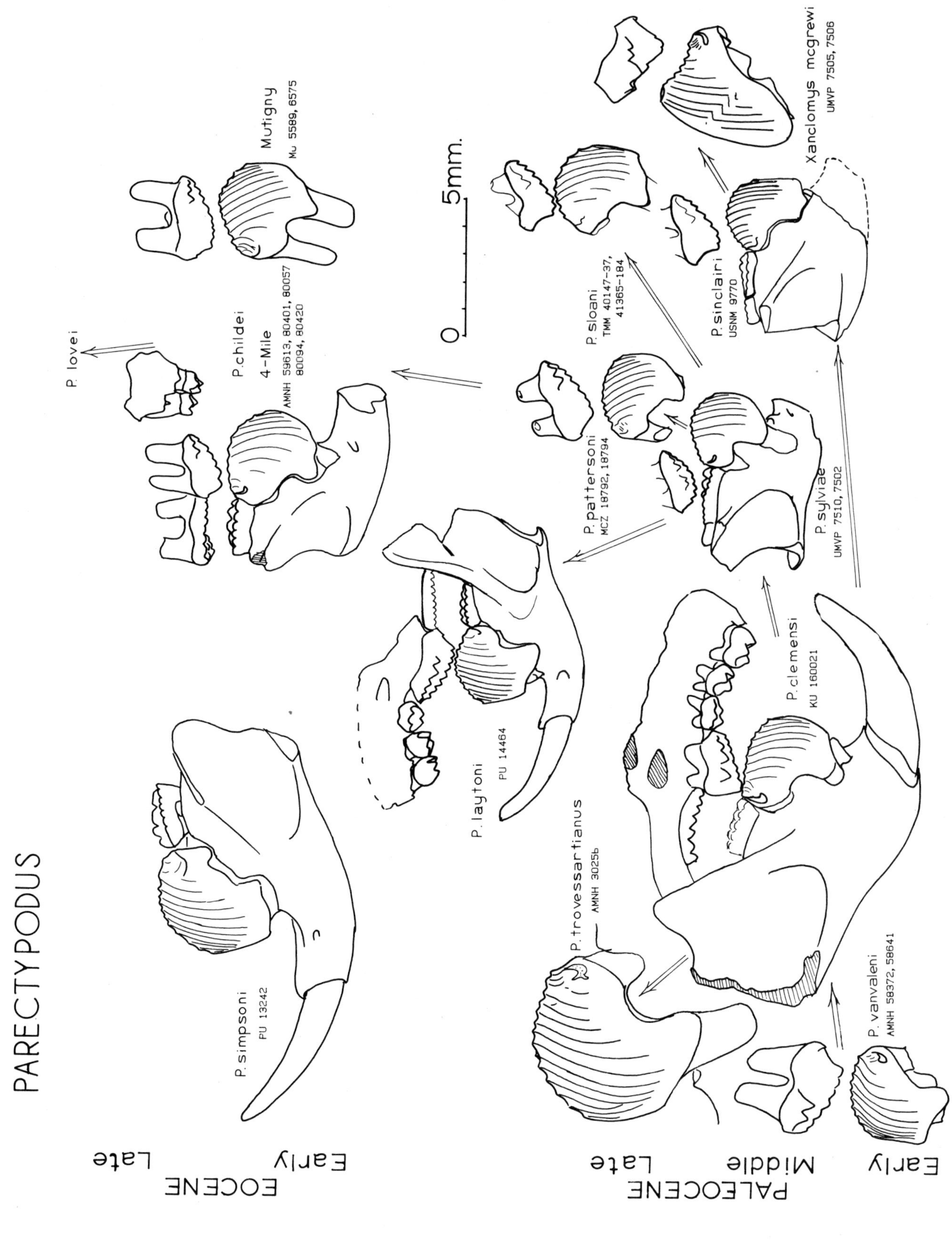

Figure 12. Drawings of Paleocene and early Eocene species of *Parectypodus* from North America and Europe.

TABLE 4. MODAL PRINCIPAL MEASURES OF LATEST CRETACEOUS, PALEOCENE, AND EARLY EOCENE NEOPLAGIAULACID MULTITUBERCULATES FROM NORTH AMERICA

	M_b		M_1			M_2			P^4		M^1			M^2		
	L	Ser.	L	W	Cusp	L	W	Cusp	L	Cusp	L	W	Cusp	L	W	Cusp
Mesodma hensleighi	2.8	11	1.8	0.8	6:4	1.0	0.8	3:2	1.8	1:6	2.2	1.0	7:8:5	1.1	1.1	1:3:3
M. formosa	3.3	11	1.5	0.7	6:4	1.4	1.2	3:2	2.5	3:6						
M. thompsoni	4.0	13	2.5	1.2	6:4	1.6	1.3	3:2	2.9	4:7	3.2	1.3	6:8:5	1.5	1.4	1:3:3
M. ambigua	4.3	14	2.6	1.3	7:5											
M. pygmaea	2.4	10	1.5	0.7	6:4	1.6	1.3	3:2	2.0	2:6						
Xyronomys swaini	3.3	10														
Ectypodus szalayi	2.8	11	2.0	0.9	9:5	1.0	1.1	4:3	2.1	?						
E. powelli	3.2	10	2.0	1.1	8:4	1.3	1.1	4:2								
E. tardus	2.8	10	1.8	0.8	8:4	1.0	0.9	4:2	1.8	3:6	2.5	1.0	8:10:5	1.0	1.0	1:3:3
E. aphronorus	3.1	12	2.3	?	?	1.3	1.0	4:2								
E. musculus	4.0	14	2.5	1.0	8:6	1.4	1.1	5:2	2.7	1:7	3.1	1.4	12:13:10	1.2	1.3	1:3:4
Mimetodon silberlingi	3.3	12	2.3	0.9	9:6	1.2	1.0	5:3	2.4	3:6	2.8	1.2	9:12:6	1.2	1.2	1:3:3
M. churchilli	5.1	12	3.2	1.4	6:5											
Neoplagiaulax macintyrei	3.6	10							3.2	2:7						
N. macrotomeus	3.7	13	1.8	0.9	6:5	1.0	0.9	4:2								
N. nelsoni	4.2	14	2.3	1.0	9:4				3.5	2:9	3.2	1.2	9:13:7	1.3	1.2	1:4:4
N. hunteri	4.6	15	2.4	1.1	9:4	1.5	1.3	5:3	3.0	1:9	3.2	1.4	8:11:8	1.5	1.5	1:4:4
N. mckennai	5.4	15	2.7	1.3	10:4	1.8	1.3	7:2	3.8	0:9	3.3	1.7	8:12:6			
N. fractus	5.2	12	3.0	1.4	6:6											
N. jepi	4.7	12														
N. hazeni	5.1	13	3.2	1.4	8:6	2.2	1.7	5:3	4.2	3:9	4.4	2.0	10:12:6	2.1	1.8	1:3:4
N. grangeri	5.3	14	3.4	1.3	8:7											
N. douglassi	6.6	13	3.7	1.7	6:5	2.0	2.1	4:2								
Parectypodus vanvaleni	3.9	11							3.5	2:7						
P. clemensi	3.4	11	2.2	1.0	6:4	1.4	1.1	3:2								
P. trovessartianus	5.8	14	3.1	1.4	6:4	1.7	1.5	3:2								
P. sinclairi	3.2	12	1.9	0.8	7:4	1.0	0.9	4:2	2.4	2:8						
P. sylviae	2.9	11	2.0	0.8	9:4	1.0	?	4:?	2.2	5:9						
P. pattersoni	3.0	11							2.7	3:7						
P. sloani	3.5	10							2.8	2:8						
P. laytoni	3.1	12	1.7	0.8	7:4	1.0	1.0	3:2	2.6	5:9	2.3	1.1	9:9:5	1.2	1.1	3:3:3
P. childei	3.3	10	1.9	1.0	7:4				2.3	2:6	2.4	1.2	8:9:5			
P. simpsoni	4.2	14	2.3	1.1	8:4											
Xanclomys mcgrewi	4.9	11							3.2	5:5						

Explanation of column headings: L = length, W = width (mm), Ser. = serration count on M_b, Cusp = cusp formula (labial row counted first) for each known tooth type.

is potentially better known than any other species of the genus. The illustration given here (Fig. 11) is a composite of AMNH 33955, right jaw with P_3, M_b, M_1; AMNH 33867 M_2 in jaw fragment; AMNH 33864, maxilla with P^1, P^2, P^3, and alveoli for P^4 and M^1; AMNH 33869, P^4; AMNH 33890 and 33891, M^1 and M^2. AMNH 33933 (Fig. 14) is a left jaw of a very young animal with fully erupted M_b, erupting permanent P_3 and I_1, and crushed crypts for M_1 and M_2. This specimen, with others, demonstrates that the lower blade of multituberculates has no deciduous precursor and is in fact a molar rather than a premolar, hence the designation of M_b rather than P_4. The upper blade appears to have a deciduous precursor and therefore is a premolar on the basis of materials from other species, so I continue to term it P^4.

Neoplagiaulax mckennai new species (Fig. 11, Tables 4, 5)

Etymology: Named for my good friend Malcolm C. McKenna, who collected the type and many other specimens of this and other species and who has been of very great help and support in my studies of multituberculates.
Holotype: AMNH 86750, right jaw with P_3, M_b, alveolus for M_1, and crypt for M_2, Love Quarry, Wyoming, Ti3a.
Paratypes: AMNH 86752, right M_b; AMNH 86760 and 88255, right P^4's; AMNH 88254 and 88204, right and left M^1's; 88210, M^1; AMNH 88222 right M^2 and other specimens from Love Quarry.
Referred specimens: UMVP specimens including 27 P^4's, 9^1's, 7 M^2's, 3 M_bs, 17 M_1's, and 5 M_2s from Olive Quarry, Powder River County, Montana, Ti4. PU specimens from Long Draw and Croc Tooth Quarries, Wyoming, Ti4. SPSM specimens from Wannagan Creek Quarry, North Dakota, Ti4.
Diagnosis: A large *Neoplagiaulax* with a distinct angle in the profile of M_b at about the seventh serration. P^4 distinctly wider than usual and with no external row of cusps, unlike any other species.

Neoplagiaulax jepi new species (Fig. 11, Tables 4, 5)

Etymology: Named for the late Professor Glenn Lowell Jepsen, of Princeton University, who with field parties collected this and many other specimens and who made the most important contributions to solving the maze of Paleocene multituberculates. His familiar name was Jep, from which this name is derived. The only multituberculate species previously named for him is *Stygimys jepseni* (Simpson), named on the basis of a few specimens from a single locality, of uncertain validity and originally

TABLE 5. MEASUREMENTS AND COUNTS OF TEETH OF NEW SPECIES

	M_b		M_1		M_2		p^4	
	L	No. of Serrations	L	Cusp Formula	L	Cusp Formula	L	Cusp Formula
Mesodma pygmaea								
AMNH 35298	2.5	12	1.5	6:4	0.8	3:2		
PU 16851	2.3	10	1.4	6:4	0.8	3:2		
MCZ 18781	2.5	10						
MCZ 18782	2.5	10						
MCZ 18789							2.3	2:7
MCZ means of 6 specimens							2.3	2:7
Neoplagiaulax nelsoni								
UMVP 6030	4.2	14						
UMVP 6031	4.0	14						
UMVP 6032							3.4	0:9
UMVP 6033							3.6	1:8
AMNH 87861	4.6	15						
UMVP 6042	3.7	14	2.3	11:5				
UMVP 6043							3.0	2:3
MCZ 18790	4.4	13						
MCZ means of 10	4.3	13						
MCZ 18790							3.8	2:9
MCZ means of 5							3.8	1-2:9-10
Neoplagiaulax jepi								
PU 21058	4.4	12						
PU 21046	4.6	12						
PU 21053	4.9	12						
PU 21054	4.8	12						
PU 21055	5.0	12						
PU 21056	4.5	12						
PU 21057a	4.7	12						
PU 21057b	4.4	12						
Ectypodus aphronorus								
AMNH 35434	3.4	12	1.9	7:5	1.0	?:2		
AMNH 356435	3.2	11	2.3	-	1.2	4:2		
AMNH 35528	3.1	12						
PU 14707	3.6	12	2.2	6:5	1.4	3:2		
MCZ 18793	3.2	10						
UMVP 6043							2.5	2:6

Explanation of column headings: L = length in mm, Number of Serrations = serration count on lower blade.

named *?Parectypodus jepseni.* This effectively prevents the naming of a valid neoplagiaulacid species with the epithet *jepseni.* The name is also an allusion to the teacher-student relationship between Jepsen and Hazen, whose patronymic species is descended from the present one.

Holotypes: PU 21058, right M_b from Cedar Point Quarry, Wyoming, Ti3a.

Paratypes: PU 21046 and 21051, left and right M_bs from Cedar Point Quarry.

Diagnosis: Derived from the older and smaller *N. nelsoni,* more highly arched in profile than the contemporary *N. hunteri,* and similar in shape to its later, larger descendant, *N. hazeni* of Ti5.

ECTYPODUS Matthew and Granger 1921
Ectypodus aphronorus new species
(Fig. 10, Tables 4, 5)

Etymology: Greek *aphron* (crazy, inane) + *oros* (mountain) in analogy to the type locality in the Crazy Mountains of Montana.

Holotype: AMNH 35435, left jaw with P_3, M_b, M_1, M_2 from Gidley Quarry, Montana, To3a.

Paratypes: AMNH 35535, right jaw with M_b, alveolus, M_1, M_2; AMNH 35528, left M_b; USNM 9797, left jaw with M_b, M_1, M_2; USNM 6575, left jaw with M_b, M_1—all from Gidley Quarry.

Referred specimens: PU 14707, left jaw with P_3, M_b, M_1, and M_2 from Rock Bench Quarry, Wyoming, To3b; MCZ 18780, 18793, left M_bs from Shotgun Quarry, Ti1.

Known stratigraphic range: To3 to Ti1.

Diagnosis: Medium-sized species of *Ectypodus,* distinguished from other species by profile of blade and cusp formulae of teeth. Only known plausible ancestor to *E. musculus* from which it differs in lower profile of M_b and M_b/M_1 ratio. Distinct in M_b/M_1 ratio from the smaller, earlier, and contemporary *E. szalayi.*

PARECTYPODUS Jepsen 1930
Parectypodus sylviae (Rigby) 1980

This species, known originally from Swain Quarry, Washakie

Basin (To3), is the earliest of the species of *Parectypodus* with the high first serration that represent the late Paleocene and Eocene members of the genus. PU 18203a, 18203b, P^4s and 18204, left M_b, are referred specimens known from the Medicine Rocks I fauna of Montana, also of To3 age.

Parectypodus pattersoni new species (Fig. 12, Tables 4, 5)

Etymology: This species is named for my greatly esteemed former professor, the late Bryan Patterson of the Field Museum and the Museum of Comparative Zoology. Besides collecting these and many other specimens, he taught me most of what I needed to know about mammalian paleontology.

Holotype: MCZ 18792, right M_b from Shotgun Quarry, Ti1, Wyoming.

Paratypes: MCZ 18794, left P^4 and six other P^4s and three additional M_bs, all from Shotgun Quarry.

Diagnosis: Larger than both *P. sylviae,* its ancestor, and *P. laytoni* and *P. childei,* its direct descendants. Profile of M_b lower than *P. laytoni* and *P. childei.* Mean length of M_b 2.98 mm; mean length of P^4 2.80 mm.

Parectypodus sinclairi (Simpson) 1935 (Fig. 8, Table 4)

In Simpson's classic work on the Crazy Mountain Field, this species was originally assigned to *Ptilodus,* and most small neoplagiaulacids were assigned to this species. The illustration in Simpson (1937b) was particularly bad so that a completely misleading idea of this species was all that could be had from the literature. The type, USNM 9770 (shown in Fig. 12), a left jaw with P_3, M_b, M_1, M_2, is well preserved. The other specimens from Gidley Quarry referrable to it are USNM 9799 with M_b, M_1; AMNH 35540 M_b, and AMNH 35532, also an M_b. The blade is taller than any other M_b of a Torrejonian species, and the exodaenodont lobe is particularly large, almost rectangular in shape and very diagnostic. No other species can be confused with this rare form. The very peculiar *Xanclomys mcgrewi* from Swain Quarry which is also To3 in age can only be derived from this species. The latest occurrence of this species is at Shotgun, where MCZ 18783 is a typical M_b, and MCZ 18784 and 18788 represent typical P^4s, so the range is from To3 to Ti1.

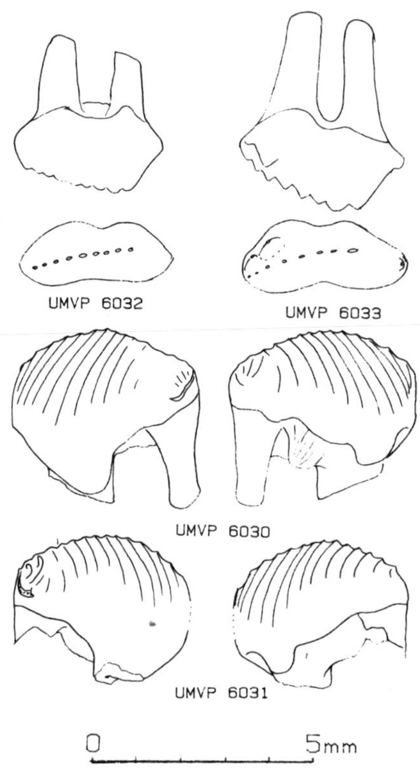

Figure 13. Holotype and paratype specimens of *Neoplagiaulax nelsoni,* new species, from the late Puercan Purgatory Hill fauna, Montana.

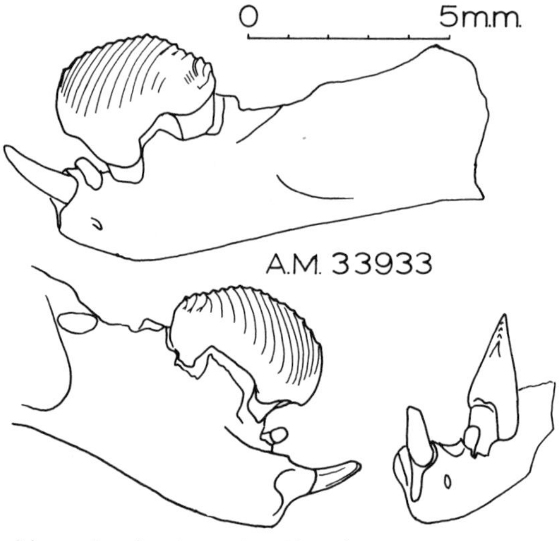

Figure 14. Lower jaw of AMNH 33933, a very immature *Neoplagiaulax hunteri* from the early Tiffanian Scarritt Quarry, Montana. The specimen shows that the lower blade has no deciduous precursor and is therefore a molar. The permanent lower incisor is just erupting; the deciduous incisor has just been shed, but the empty alveolus is readily visible; P_3 is erupting underneath the leading edge of the lower blade; crypts for M_1 and M_2 are visible, showing they have not yet erupted. This specimen is the major (but not the only) evidence for calling the lower blade a molar rather than a premolar.

CIMOLODONTIDAE Marsh 1887
ANCONODON Jepsen 1940
Anconodon lewisi new species (Fig. 15)

Etymology: Named for Arnold "Arnie" Lewis, then chief preparator of paleontology at the Museum of Comparative Zoology, who collected the type and other specimens from Shotgun Quarry.

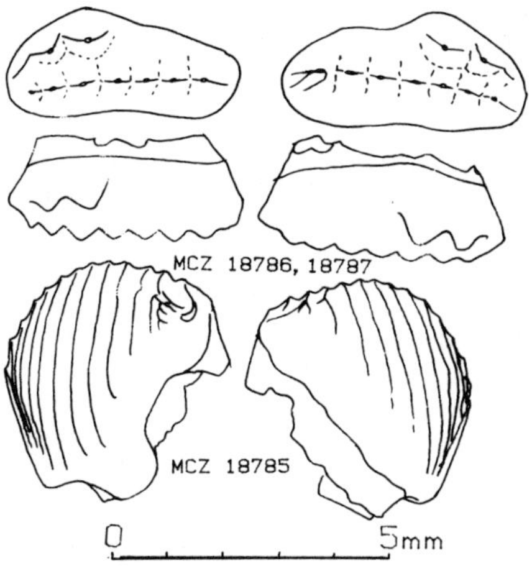

Figure 15. Holotype and paratypes of *Anconodon lewisi,* new species, from the early Tiffanian Shotgun fauna, Wyoming.

Holotype: MCZ 18785, left M_b from Shotgun Quarry, Wyoming, Ti1.
Paratypes: 12 additional M_bs (uncataloged) and MCZ 18786, right P^4 and MCZ 18787, left P^4, all from Shotgun Quarry.
Diagnosis: Smallest species of *Anconodon,* standard length M_b 4.1 to 4.7 mm, mean 4.4 mm, 12 to 15 serrations, 13 modal value. P^4 with straight profile, 4.1 mm long, cusp formula 2:7.
Discussion: Although not the latest *Anconodon* known, since *Anconodon cochranensis* (=*A. russelli*) is known from Cochrane (Ti1) and Bison Basin Saddle locality (Ti3a) as well as from Gidley Quarry and Rock Bench Quarry (To3a and To3b), this small species is a surprise. The largest species *A. gidleyi* is known from To2 to To3b. The medium-sized species *A. cochranensis* is known from To3a to Ti3a, and the smallest species *A. lewisi* is known only from Ti1. This genus shows both diversification and size reduction during the same time interval.

ACKNOWLEDGMENTS AND DEDICATION

I wish to dedicate this paper to the memory of the three pioneers in Paleocene mammal studies who most influenced me.

Bryan Patterson blocked out the evolution of the Pantodonta and the Taeniodonta, studied and collected the late Paleocene of Colorado, and—while the Curator of Fossil Mammals at the Field Museum in Chicago—taught the Fossil Mammal course of the University of Chicago to, among others, me. I knew it was well taught, but I did not realize just how well until much later.

George Gaylord Simpson, while Curator of Fossil Mammals at the American Museum during the 1930s, developed the methods of quantitative paleontology and fleshed out the details of Torrejonian and Tiffanian mammals. Although he had few personal graduate students, he was the theoretical mentor of a whole generation of paleontologists.

Glenn Lowell Jepsen, Scott Professor of Paleontology at Princeton, spent most of his career collecting and describing the Paleocene and Early Eocene mammals of northwestern Wyoming and adjacent Montana. He appreciated the need for detailed stratigraphy and was responsible for developing the techniques of differentiating the species of small multituberculates.

I miss them.

This paper has had a protracted period of growth and criticism. I have had significant input of unpublished data, criticism, and review from J. D. Archibald, D. Baird, W. A. Clemens, J. E. Fassett, P. D. Gingerich, J. H. Hartman, R. C. Holtzman, G. L. Jepsen, P. A. Johnston, D. W. Krause, Li Chuan-Kuei, Li Jinling, M. C. McKenna, M. D. Middleton, D. Parris, J. K. Rigby, Jr., K. D. Rose, J. Schiebout, B. Standhardt, Ting Su-yin, L. H. Taylor, Wang Ban-yue, L. M. Van Valen, Xue Xiang-xu, Zhai Ren-jie, Zhao Zi-kuei, Zhang Yu-ping, and Zhou Ming-zhen. I thank them for their help and hasten to disclaim any responsiblity on their part for my use of their data.

Institutional abbreviations in the systematic portion of this paper are as follows: ACM—Amherst College Museum; AMNH—American Museum of Natural History; CR and Mu—Museum National d'Histoire Naturelle, Paris; KU—University of Kansas; MCZ—Museum of Comparative Zoology, Harvard; PU—Princeton University; TMM—Texas Memorial Museum; UA—University of Alberta; UCMP—University of California Museum of Paleontology, Berkeley; UMMP—University of Michigan Museum of Paleontology; UMVP—University of Minnesota Vertebrate Paleontology. I thank the curators of these organizations for their permission to study their fossils and for their many courtesies.

My wife, Sally Crippa Sloan, drew Figure 3 many years ago for a paper that was never published; many of my colleagues have found copies of this figure useful. I am very pleased to thank her for her hours of cartographic skill put to this good use and for her patience and support for many years.

I thank the Committee on Scholarly Communication with the Peoples Republic of China (National Academy of Science) for the grant for my 1986 trip to China and my hosts of the Institute of Vertebrate Paleontology and Paleoanthropology for their many courtesies.

REFERENCES CITED

Alvarez, W., and Vann, D. W., 1979, Comment *on* 'Biostratigraphy and magnetostratigraphy of Paleocene terrestrial deposits, San Juan Basin, New Mexico': Geology, v. 6, p. 66–67.

Alvarez, W., Arthur, M. A., Fischer, A. G., Lowrie, W., Napoleone, G., Premoli Silva, I., and Roggenthen, W. M., 1977, Upper Cretaceous–Paleocene magnetic stratigraphy at Gubbio, Italy; Type section for the Late Cretaceous–Paleocene geomagnetic reversal time scale: Geological Society of America Bulletin, v. 88, p. 383–389.

Alvarez, L. A., Alvarez, W., Asaro, F., and Michel, H. V., 1980, Extraterrestrial cause for the Cretaceous-Tertiary extinction: Science, v. 208, p. 1095–1108.

Archibald, J. D., 1982, A study of Mammalia and Geology across the Cretaceous-Tertiary boundary in Garfield County, Montana: University of California Publications in the Geological Sciences, v. 122, 286 p.

Archibald, J. D., and Clemens, W. A., 1982, Late Cretaceous extinctions: American Scientist, v. 70, p. 377–385.

Archibald, J. D., Butler, R. F., Lindsay, E. H., Clemens, W. A., and Dingus, L., 1982, Upper Cretaceous–Paleocene biostratigraphy and magnetostratigraphy Hell Creek and Tullock Formations, northeastern Montana: Geology, v. 10, p. 153–159.

Baadsgard, H., Cumming, G. L., Folinsbee, R. E., and Godfrey, J. D., 1964, Limitations of radioactive dating, *in* Osborne, F. F., ed., Geochronology in Canada: Toronto, University of Toronto, p. 20–38.

Baltz, E. H., Ash, S. R., and Anderson, R. Y., 1966, History of nomenclature and stratigraphy of rocks adjacent to the Cretaceous-Tertiary boundary, western San Juan Basin, New Mexico: U.S. Geological Survey Professional Paper 524-D, 23 p.

Barnes, H., 1953, Geology of the Ignacio area, Ignacio and Pagosa Springs quadrangles, La Plata and Archuleta counties, Colorado: U.S. Geological Oil and Gas Investigations OM 138.

Berggren, W. A., 1972, A Cenozoic time-scale; Some implications for regional geology and paleobiography: Lethaia, v. 5, p. 195–215.

Berggren, W. A., McKenna, M. C., Hardenbol, J., and Obradovitch, J. D., 1978, Revised Paleogene polarity time scale: Journal of Geology, v. 86, p. 67–81.

Berggren, W. A., Kent, D. V., and Flynn, J. J., 1985a, Paleogene geochronology and chronostratigraphy, *in* Snelling, N. J., ed., Geochronology and the geologic record: Geological Society of London Special Paper.

Berggren, W. A., Kent, D. V., Flynn, J. J., and Van Couvering, J. A., 1985b, Cenozoic geochronology: Geological Society of America Bulletin, v. 96, p. 1407–1418.

Bonhomme, M., Odin, G. S., and Pomerol, C., 1968, Age de formations glauconieuses de l'Albien de l'Eocene du Bassin de Paris: Paris, France, Memoires du Bureau de Recherches Geologiques et Minieres, v. 58, p. 339–345.

Brandenburg, J., 1983, Late Cretaceous vertebrate faunas from the upper Hell Creek Formation of Montana and North Dakota [M.S. thesis]: Minneapolis, University of Minnesota, 111 p.

Breithaupt, B. H., 1982, Paleontology and paleoecology of the Lance Formation (Maastrichtian) east flank of Rock Springs uplift, Sweetwater County, Wyoming: Laramie, University of Wyoming Contributions to Geology, v. 21, p. 123, 151.

Brown, B., 1910, The Cretaceous Ojo Alamo beds of New Mexico, with description of the new dinosaur genus *Kritosaurus:* American Museum of Natural History Bulletin, v. 28, p. 267–274.

Brown, R. W., 1962, Paleocene flora of the Rocky Mountains and Great Plains: U.S. Geological Survey Professional Paper 375, p. 1–119.

Butler, R. F., 1984, Paleomagnetic and rock-magnetism of continental deposits, San Juan Basin, New Mexico: Geological Society of America Abstracts with Programs, v. 16, no. 4, p. 217.

Butler, R. F., and Lindsay, E. H., 1985, Mineralogy of magnetic minerals and revised magnetic polarity stratigraphy of continental sediments, San Juan Basin, New Mexico: Journal of Geology, v. 93, p. 535–554.

Butler, R. F., and Taylor, L. H., 1978, A middle Paleocene paleomagnetic pole from the Nacimiento Formation, San Juan Basin, New Mexico: Geology, v. 8, p. 495–498.

Butler, R. F., Lindsay, E. H., Jacobs, L. L., and Johnson, N. M., 1977, Magnetostratigraphy of the Cretaceous-Tertiary boundary in the San Juan Basin, New Mexico: Nature, v. 267, p. 318–323.

Butler, R. F., Gingerich, P. D., and Lindsay, E. H., 1981, Magnetic polarity stratigraphy and biostratigraphy of Paleocene and lower Eocene continental sediments, Clark's Fork Basin, Wyoming: Journal of Geology, v. 89, p. 299–316.

Cappetta, H., Jaeger, J. J., Sabatier, M., Sigé, B., Sudre, J., and Vianey-Liaud, M., 1978, Decouverte dans le paleocene du Maroc des plus anciens Mammiferes Eutheriens D'Afrique: Geobios, no. 11, fasc. 2, p. 257–263.

Chow M., 1960, *Prodinoceras* and a summary of mammalian fossils of Sinkiang: Vertebrata Palasiatica, v. 4, p. 99–102.

Chow M., and Li C. K., 1963, A fossil of *Homogalax* from the Eocene of Shantung: Scientia Sinica, v. 12, p. 1411–1412.

Chow M., and Qu T., 1978, Paleocene mammalian fossils from Nomogen Formation of Inner Mongolia: Vertebrata PalAsiatica, v. 16, p. 77–85, 4 plates (Chinese, English summary).

Chow M., and Wang B., 1978, A new pantodont from the Paleocene of south China: Vertebrata PalAsiatica, v. 16, p. 86–90, 2 plates (Chinese, English summary).

Chow (=Zhou) M., Chang (=Zhang) Y., Wang B., and Ting (=Ding) S., 1973, New mammalian genera and species from the Paleocene and Nanhsiung, N. Kwangtung: Vertebrata PalAsiatica, v. 11, p. 31–35 (Chinese, English summary).

Chow M., Qi T., and Li Y., 1976, Paleocene stratigraphy and faunal characters of mammalian fossils of Nomogen Commune, Si-Zi-Wang Qi, Nei Mongol: Vertebrata PalAsiatica, v. 14, p. 228–233 (Chinese, English summary).

Chow (=Zhou) M., Chang (=Zhang) Y., Wang B., and Ting (=Ding) S., 1977, Mammalian fauna from the Paleocene of Nanxiong Basin, Guangdong: Paleontologica Sinica 153, new series C, v. 20, 100 p., +28 plates (Chinese, long English summary).

Cifelli, R. C., 1983a, Eutherian tarsals from the Late Paleocene of Brazil: American Museum of Natural History Novitates 2761, p. 1–31.

——, 1983b, The origin and affinities of the South American Condylarthra and Early Tertiary Litopterna (Mammalia): American Museum of Natural History Novitates, v. 2772, p. 1–49.

Clemens, W. A., 1963, 1965, 1973, Fossil mammals of the type Lance Formation, pts, I, II, and III: University of California Publications in the Geological Sciences, v. 48, 62, and 94.

Curry, D., and Odin, G. S., 1982, Dating the Paleogene, *in* Odin, G. S., ed., Numerical dating of sediments: New York, John Wiley and Sons, p. 593–606.

Dane, C. H., 1936, The La Ventana–Chacra Mesa coal field; pt. 3 of geology and fuel resources of the southern part of the San Juan Basin, New Mexico: U.S. Geological Survey Bulletin 860-C, p. 81–161.

Dashzeveg, D., 1982, A revision of the Prodinoceratinae of Central Asia and North America: Paleontological Journal, v. 16, p. 91–99.

Downs, T., 1961, A study of variation and evolution in Miocene *Merychippus:* Los Angeles County Museum Contributions to Science, v. 45, p. 1–75.

Emry, R. J., 1970, A North American Oligocene pangolin and other addition to the Pholidota: Bulletin of the American Museum of Natural History, v. 142, p. 455–510.

Erickson, B. R., 1972, The Lepidosaurine reptile *Champsosaurus* in North America: Science Museum of Minnesota, Monograph v. 1: Paleontology, 91 p.

——, 1976, Osteology of the early Eusuchian crocodile *Leidyosuchus formidabilis* sp. nov.: Science Museum of Minnesota, Monograph v. 2: Paleontology, 61 p.

Estes, R., 1964, Fossil vertebrates from the Late Cretaceous Lance Formation eastern Wyoming: University of California Publications in the Geological Sciences, v. 49, 187 p.

Everndon, J. F., Savage, D. E., Curtis, G. H., and James, G. T., 1964, Potassium-argon dates and the Cenozoic mammalian chronology of North America: American Journal of Science, v. 262, p. 145–198.

Fassett, J. E., 1979, Comment *on* 'Biostratigraphy and magnetostratigraphy of Paleocene terrestrial deposits, San Juan Basin, New Mexico': Geology, v. 7, p. 69–70.

Flerov, K. K., 1957, Dinocerata of Mongolia (Russian): Trans. Inst, Paleont. Acad. Sci. USSR, English translation, Israel Program for Scientific Translations, Jerusalem, p. 1–84.

Flores, R. M., and Kaplan, S. S., eds., 1985, Cenozoic paleogeography of the West-Central United States: Rocky Mountain Section, Society of Economic Paleontologists and Mineralogists, Denver, Colorado, p. 1–460.

Fox, S. K., and Olsson, R. K., 1969, Danian planktonic foraminifera from the Cannonball Formation in North Dakota: Journal of Paleontology, v. 43, p. 1397–1404.

Gazin, C. L., 1941, The mammalian faunas of the Paleocene of Central Utah with notes on the geology. Smithsonian Institution Proceedings U. S. National Museum, v. 91, p. 1–53.

Gingerich, P. D., 1975a. What is type Paleocene?: American Journal of Science, v. 275, p. 984.

—— 1975b, New North American Plesiadapidae (Mammalia, Primates) and a biostratigraphic zonation of the Middle and Late Paleocene: Contributions from the Museum of Paleontology, University of Michigan, v. 24, p. 135–148.

—— , 1976, Cranial anatomy and evolution of Early Tertiary Plesiadapidae (Mammalia, Primates) University of Michigan, Museum of Paleontology Papers no. 15, p. 144.

—— , 1978, New Condylarthra (Mammalia) from the Paleocene and early Eocene of North America: Contributions from the Museum of Paleontology, University of Michigan, v. 25, p. 1–9.

—— , 1979, *Lambertocyon eximius,* a new arctocyonid (Mammalia, Condylarthra) from the late Paleocene of western North America: Journal of Paleontology, v. 53, p. 524–529.

—— , ed., 1980a, Early Cenozoic paleontology and stratigraphy of the Bighorn Basin, Wyoming: University of Michigan Papers on Paleontology, no. 24, 146 p.

—— , 1980b, Evolutionary patterns in early Cenozoic mammals: Annual Review of Earth and Planetary Science, v. 8, p. 407–424.

—— , 1983, Rates of evolution; Effects of time and temporal scaling: Science, v. 22, p. 159–161.

Gingerich, P. D., and Gunnell, G. F., 1979, Systematics and evolution of the genus *Esthonyx* (Mammalia, Tillodontia) in the early Eocene of North America: Contributions from the Museum of Paleontology, University of Michigan, v. 25, p. 125–153.

Gingerich, P. D., Rose, K. D., and Krause, D. W., 1980. Early Cenozoic mammalian faunas of the Clark's Fork Basin–Polecat Bench area Northwestern Wyoming, *in* Gingerich, P. D., ed., Early Cenozoic paleontology and stratigraphy of the Bighorn Basin, Wyoming: University of Michigan Papers on Paleontology, no. 24, p. 51–68.

Gingerich, P. D., Houde, P., and Krause, D. W., 1983, A new earliest Tiffanian (Late Paleocene) mammalian fauna from Bangtail Plateau, western Crazy Mountain Basin, Montana: Journal of Paleontology, v. 57, p. 957–970.

Godfriaux, I., and Thaler, L., 1972, Note sur la decouverte de dents de mammiferes dans le Montien continental du Hainault (Belgique): Bulletin Academie Royale de Belgique, Classe de Sciences, v. 58, p. 536–541.

Grambast, L., Martinez, M., Mattauer, M., and Thaler, L., 1967, *Perutherium altiplanense,* nov. gen., nov. sp., premier Mammiferé mesozoique d'Ameriqué du Sud: Academie des Sciences, Compte Rendu Hebdomedaires Paris, v. 264 (D), p. 707–710.

Granger, W., 1917, Notes on Paleocene and lower Eocene mammal horizons of northern New Mexico and southern Colorado: American Museum of Natural History Bulletin, v. 34, p. 821–830.

Haldane, J.B.S., 1949, Suggestion as to quantitive measurement of rates of evolution: Evolution, v. 3, p. 51–56.

Hardenbol, J., and Berggren, W. A., 1978, A new Paleocene numerical time scale, *in* Cohee, G. V., Glaessner, M. F., and Hedberg, H. D., eds., Contributions to the geologic time scale: American Association of Petroleum Geologists Studies in Geology, v. 6, p. 213–234.

Harland, W. B., Cox, A. V., Llewellyn, P. G., Pickton, C.A.G., Smith, A. G., and Walters, R., 1982, A geologic time scale: Cambridge University Press, p. 1–131.

Harris, W. B., 1982, Rubidium–strontium glaucony ages, southeastern Atlantic Coastal Plain U.S.A., *in* Odin, G. S., ed., Numerical dating of sediments: New York, John Wiley and Sons, p. 607–688.

Hays, J. D., and Pitman, W. C., 1973, Lithospheric plate motion, sea-level changes, and climatic and ecological consequences: Nature, v. 246, p. 18–22.

Hickey, L. J., 1977, Stratigraphy and paleobotany of the Golden Valley Formation (early Tertiary) of western North Dakota: Geological Society of America Memoir 150, 181 p.

—— , 1980, Paleocene stratigraphy and flora of the Clark's Fork Basin, *in* Gingerich, P. D., ed., Early Cenozoic paleontology and stratigraphy of the Bighorn Basin, Wyoming: University of Michigan Papers on Paleontology, no. 24, p. 33–50.

Holtzman, R. C., 1978, Late Paleocene mammals of the Tongue River Formation, western North Dakota: U.S. Geological Survey Report of Investigations 65, p. 1–88.

Houde, P., and Olson, S. L., 1981, Paleognathous carinate birds from the early Tertiary of North America: Science, v. 214, p. 1236–1237.

Jepsen, G. L., 1930, Stratigraphy and paleontology of the Paleocene of northeastern Park County, Wyoming: Proceedings of the American Philosophical Society, v. 69, p. 463–528.

—— 1940, Paleocene faunas of the Polecat Bench Formation, Park County, Wyoming, Part I: Proceedings of the American Philosophical Society, v. 83, p. 217–314.

Jepsen, G. L., and Woodburne, M. O., 1969, Paleocene hyracotheres from Polecat Bench Formation, Wyoming: Science, v. 164, p. 543–547.

Johnston, P. A., 1980, First record of Mesozoic mammals from Saskatchewan: Canadian Journal of Earth Sciences, v. 17, p. 512–519.

Johnston, P. A., and Fox, R. C., 1984, Paleocene and Late Cretaceous mammals from Saskatchewan, Canada: Palaeontographica A, Bd. 186, p. 163–222.

Krause, D. W., 1973, Paleocene primates from western Canada: Canadian Journal Earth Sciences, v. 15, p. 1350–1371.

—— , 1977, Paleocene multituberculates of the Roche Percée local fauna; Ravenscrag Formation, Saskatchewan, Canada: Palaeontographica, abt. A, bd. 159, p. 1–3.

Krishtalka, L., 1973, Late Paleocene mammals from the Cypress Hills, Alberta: Texas Tech University Special Publication 2, p. 1–77.

Lehman, T., 1983, The Alamo Wash local fauna, a new look at the old Ojo Alamo fauna, *in* Lucas, S. G., Rigby, J. K., Jr., and Kues, B. S., eds., Advances in San Juan Basin paleontology: Albuquerque, Unviersity of New Mexico Press, 393 p.

Lerbekmo, J. F., Evans, M. E., and Baadsgard, H., 1979, Magnetostratigraphy, biostratigraphy, and geochronology of Cretaceous-Tertiary boundary sediments, Red Deer Valley: Nature, v. 279, p. 26–30.

Li C., 1977, Paleocene Eurymyloids (Anagalida, Mammalia) of Qinshan, Anhui: Vertebrata PalAsiatica, v. 15, p. 104–108.

Li C., and Ting S., 1983, The Paleogene mammals of China: Bulletin Carnegie Museum of Natural History, no. 21.

Lillegraven, J. A., 1969, Latest Cretaceous mammals of upper part of Edmonton Formation of Alberta, Canada, and review of marsupial-placental dichomammalian evolution: Lawrence, University of Kansas Paleontological Contributions, art. 50, p. 1–122.

Lindsay, E. H., Jacobs, L. L., and Butler, R. F., 1978, Biostratigraphy and magnetostratigraphy of Paleocene terrestrial deposits, San Juan Basin, New Mexico: Geology, v. 6, p. 425–429.

Lindsay, E. H., Jacobs, L. L., and Butler, R. F., 1979a, Reply *to* Comments *on* 'Biostratigraphy and magnetic polarity stratigraphy of Paleocene terrestrial deposits, San Juan Basin, New Mexico': Geology, v. 7, p. 68–69, 70–71.

—— , 1979b, Reply *to* Comments *on* 'Biostratigraphy and magnetic polarity stratigraphy of Paleocene terrestrial deposits, San Juan Basin, New Mexico:

Geology, v. 7, p. 326.

Lindsay, E. H., Butler, R. F., and Johnson, N. M., 1981, Magnetic polarity zonation and biostratigraphy of late Cretaceous and Paleocene continental deposits, San Juan Basin, New Mexico: American Journal of Science, v. 281, p. 390–435.

Love, J. D., 1973, Harebell Formation (Upper Cretaceous) and Pinyon Conglomerate (uppermost Cretaceous and Paleocene), northwestern Wyoming: U.S. Geological Survey Professional Paper 734-A, 54 p.

Lowrie, W., and Alvarez, W., 1981, One hundred million years of geomagnetic polarity history: Geology, v. 9, p. 392–397.

Lucas, S. G., and O'Neill, F. M., 1981, Occurrence of *Pantolambda* (Mammalia, Pantodonta) in the Torrejonian *Deltatherium* "zone," San Juan Basin, New Mexico: American Journal of Science, v. 28, p. 187–191.

Lucas, S. G., and Rigby, J. K., Jr., 1979, Comment and reply *on* 'Biostratigraphy and magnetic stratigraphy of late Cretaceous and Paleocene terrestrial deposits, San Juan Basin, New Mexico': Geology, v. 7 p. 323–325.

Lucas, S. G., and Schoch, R. M., 1982, Discussion *of* 'Magnetic polarity zonation and biostratigraphy of late Cretaceous and Paleocene continental deposits, San Juan Basin, New Mexico': American Journal of Science, v. 282, p. 920–927.

Lund, S. P., 1976, Continental sediments as paleomagnetic recorders; Evidence from the Upper Cretaceous Hell Creek Formation of the Williston basin [M.S. thesis]: Minneapolis, University of Minnesota.

Marshall, L. G., Butler, R. F., Drake, R. E., and Curtis, G. H., 1981, Calibration of the beginning of the age of mammals in South America: Science, v. 212, p. 43–45.

Marshall, L. G., de Muizon, C., and Sigé, B., 1983, *Perutherium altiplanense,* un notoungulé du Crétacé Supérior du Pérou: Paleovertebrata (Montpellier), v. 13, no. 4, p. 145–155.

Mattauer, M., Matte, Ph., Malavielle, J., Tapponier, P., Maluska, H., Xu Z. Q., Lu Y. L., and Tang Y. Q., 1985, Tectonics of the Qinling Belt; Build up and evolution of eastern Asia: Nature, v. 317, p. 496–500.

Matthew, W. D., 1937, Paleocene faunas of the San Juan Basin, New Mexico: American Philosophical Society, Transactions, n.s., v. 30, p. 1–510.

Matthew, W. D., and Granger, W., 1921, New genera of Paleocene mammals: American Museum of Natural History Novitates 13, p. 2–7.

McDonald, R. E., 1972, Eocene and Paleocene rocks of the southern and central basins, *in* Mallory, W. W., ed., Geologic atlas of the Rocky Mountain region: Denver, Rocky Mountain Association of Geologists, p. 243–256.

McKenna, M. C., 1975, Toward a phylogenetic classification of the Mammalia, *in* Luckett, W. P., and Szalay, F. S., eds., Phylogeny of the primates: New York, Plenum Publishing Co., p. 21–46.

—— , 1980, Late Cretaceous and Early Tertiary vertebrate paleontological reconnaissance, Togwotee Pass area, northwestern Wyoming, *in* Jacobs, L. L., ed., Aspects of vertebrate history; Essays in honor of Edwin Harris Colbert: Flagstaff, Museum of Northern Arizona Press, p. 321–343.

McKenna, M. C., Xue X., and Zhou M., 1984, *Prosarcodon lonanensis,* a new Paleocene Micropternodontine Palaeoryctoid insectivore from Asia: American Museum of Natural History Novitates 2780, p. 1–17.

Middleton, M. D., 1982, A new species and additional materials of *Catopsalis* (Mammalia, Multituberculata) from the western interior of North America: Journal of Paleontology, v. 56, p. 1197–1206.

Napoleone, G., Premoli Silva, I., Heller, F., Cheli, P., Corezzi, S., and Fischer, A. G., 1983, Eocene magnetic stratigraphy at Gubbio, Italy, and its implications for Paleogene geochronology: Geological Society of America Bulletin, v. 94, p. 181–191.

Ness, G., Levi, S., and Couch, R., 1980, Marine magnetic timescale for the Cenozoic and Late Cretaceous; A précis, critique, and synthesis: Reviews in Earth and Space Physics, v. 18, p. 753–770.

Odin, G. S., 1982, Numerical dating of sediments: New York, John Wiley and Sons, 1040 p.

O'Sullivan, R. B., and Beikman, H. M., 1963, Geology, structure and uranium deposits of the Shiprock quadrangle, New Mexico and Colorado: U.S. Geological Survey Miscellaneous Geologic Investigations Map I-345.

Owens, J. P., and Sohl, N. F., 1973, Glauconites from New Jersey–Maryland coastal plain, their K-Ar ages, and application in stratigraphic studies: Geological Society of America Bulletin, v. 84, p. 2811–2838.

Palmer, A. R., 1983, The Decade of North American Geology 1983 time scale: Geology, v. 11, p. 503–504.

Patterson, B., 1939, New Pantodonta and Dinocerata from the upper Paleocene of western Colorado: Field Museum of Natural History Geological series, v. 6,, p. 351–334.

—— , 1949, Rates of evolution in Taeniodonts, *in* Jepsen, G. L., Simpson, G. G., and Mayr, E. E., eds., Genetics, paleontology, and evolution: Princeton, N.J., Princeton Univresity Press, p. 243–278.

Paulo Couto, C., de., 1952, Fossil mammals from the beginning of the Cenozoic in Brazil; Condylarthra, Litopterna, Xenungulata, and Astrapotheria: American Museum of Natural History Bulletin, v. 99, p. 355–394.

—— , 1970, Evolucão de communidades modificacões faunisticas e integracões biocenóticas do vertebrado cenozóicos do Brazil: Actas IV Cong. Latinoamericano Zoologia Caracas, v. 2, p. 907–930.

Percival, S., and Fischer, A. G., 1977, Changes in calcareous nannoplankton in the Cretaceous-Tertiary boundary crisis at Zumaya, Spain: Evolutionary Theory, c. 222, p. 1–36.

Pitman, W. C., 1978, Relations between eustacy and stratigraphic sequences of passive margins: Geological Society of America Bulletin, v. 89, p. 1389–1403.

Plaziat, J. C., 1970a, La' limite crétacé-tertiare en Alava méridionale (Pays basque espagnol); le Rognacien n'y est pas l'equivalent continental du Danien: Societe Geologie de France, Compte Rendu Sommaire des Seances, fasc 3, seance de 2 mars 1970, p. 77.

—— , 1970b, Consequences stratigraphiques de l'interstratification de Rognacien dans le Maestrichtien superieur d'Alava (Espagne): Academie des Sciences, Compte Rendu Hebdomedaires, Paris, T.270, p. 2768–2771, series D.

Pomerol, C., 1969, Rapport sur la limite Paleocene-Eocene, *in* Colloque sur l'Eocene: Paris, Memoires du Bureau de Recherches Geologiques et Minieres, v. 69, p. 447–449.

Qi T., and Huang X. S., 1982, A mesonychid (Mammalia) skull from the Paleocene of Laonan Shaanxi: Vertebrata PalAsiatica, v. 20, p. 18–23.

Radinsky, L., and Ting [=Ding] S. Y., 1984, The skull of *Ernanodon:* Journal of Mammalogy, v. 65, p. 155–158.

Rapp, S. D., Macfadden, B. J., and Schiebout, J. A., 1983, Magnetic polarity stratigraphy of the Early Tertiary Black Peaks Formation, Big Bend National Park, Texas: Journal of Geology, v. 91, p. 555–572.

Reeside, J. B., Jr., 1924, Upper Cretaceous and Tertiary formations of the western part of the San Juan Basin of Colorado and New Mexico: U.S. Geological Survey Professional Paper 134, p. 1–70.

Rigby, J. K., Jr., 1980, Swain Quarry of the Fort Union Formation, Middle Paleocene (Torrejonian), Carbon County, Wyoming; geologic setting and mammalian fauna: University of Chicago Evolution Monograph, v. 3, 179 p.

—— , 1983, A skeleton of *Gillisonchus gillianus* (Mammalia, Condylarthra) from the early Paleocene (Puercan) Ojo Alamo Sandstone, San Juan Basin, New Mexico, with comments on the local stratigraphy of Betonnie Tsosie Wash, *in* Lucas, S. G., Rigby, J. K., Jr., and Kues, B. S., eds., Advances in San Juan Basin paleontology: Albuquerque, University of New Mexico Press, p. 89–126.

Robinson, P., 1972, Tertiary history, *in* Mallory, E. E., ed., Geologic atlas of the Rocky Mountain region: Denver, Rocky Mountain Association of Geologists, p. 233–242.

Romer, A. S., 1966, Vertebrate paleontology: Chicago, Univ. of Chicago Press, 468 p.

Rose, K. D., 1975, The Carpolestidae; Early Tertiary primates from North America: Museum of Comparative Zoology Bulletin, v. 147, p. 1–74.

—— , 1977, Evolution of carpolestid primates and chronology of the North American middle and late Paleocene: Journal of Paleontology, v. 51, p. 536–542.

—— , 1978a, A new Paleocene Epoicotheriid (Mammalia) with comments on the Palaeanodonta: Journal of Paleontology, v. 52, p. 658–674.

——, 1978b, Clarkforkian mammal fauna of the northern Bighorn Basin, Wyoming: Geological Society of America Abstracts with Programs, v. 10, p. 283.

——, 1980, Clarkforkian land mammal age; Revised definition, zonation, and tentative intercontinental correlations: Science, v. 208, p. 744–746.

——, 1981, The Clarkforkian land-mammal age and mammalian faunal composition across the Paleocene-Eocene boundary: University of Michigan Papers on Paleontology, v. 24.

Russell, D. E., 1964, Les Mammiferes paleocenes d'Europe: Paris, Memoires du Museum National d'Histoire Naturelle (Nouvelle Series), Series C, v. 13, p. 1–324.

——, 1967, Le Paleocene continental d'Amerique du Nord: Paris, Memoires du Museum National d'Histoire Naturelle (Nouvelle Series), Series C, v. 16, p. 39–99.

Russell, L. S., 1967, Palaeontology of the Swan Hills area, north–central Alberta: Royal Ontario Museum, Life Sciences Contributions, v. 71, p. 1–31.

Savage, D. E., and Russell, D. E., 1983, Mammalian paleofaunas of the world; Chapter 3, Paleocene: Reading, Mass., Addison-Wesley, p. 19–59.

Schiebout, J. A., 1974, Vertebrate paleontology and paleocology of Paleocene Black Peaks Formation, Big Bend National Park, Texas: Texas Memorial Museum Bulletin, v. 24, p. 1–88.

Schorn, H. E., 1971, What is type Paleocene?: American Journal of Science, v. 271, p. 402–409.

——, 1975, Reply *to* 'What is type Paleocene?': American Journal of Science, v. 275, p. 986–987.

Shapiro, L. H., 1971, Structural geology of the Black Hills region and the implications for the origin of the uplifts of the middle Rocky Mountain province [Ph.D. thesis]: Minneapolis, University of Minnesota, 276 p.

Shaw, A. B., 1964, Time in stratigraphy: New York, McGraw Hill, p. 365.

Shive, P. N., and Pruss, E. F., 1977, A paleomagnetic study of basalt flows from the Absaroka Mountains, Wyoming: Journal of Geophysical Research, v. 82, p. 3039–3048.

Sigé, B., 1972, La faunule de mammifères du Crétacé supérieur de Laguna Umayo (Andes péruviennes): Paris, Bulletin du Muséum National d'Histoire Naturelle, (3) 99 (Sc. Terre 19), p. 375–405.

Simons, E. L., 1960, The Paleocene Pantodonta: Transactions of the American Philosophical Society, n.s., v. 50, pt. 6, p. 1–100.

Simpson, G. G., 1935a, The Tiffany fauna, Upper Paleocene; Parts I, II, III: American Museum of Natural History Novitates, no. 795, p. 1–19; no. 816, p. 1–30; no. 817, p. 1–28.

——, 1935b, Occurrence and relationships of the Rio Chican fauna of Patagonia: American Museum of Natural History Novitates, no. 818, p. 1–21.

——, 1936, A new fauna from the Fort Union of Montana: American Museum of Natural History Novitates, no. 873, p. 1–27.

——, 1937a, Additions to the Upper Paleocene fauna of the Crazy Mountain Field: American Museum of Natural History Novitates, no. 940, p. 1–15.

——, 1937b, The Fort Union of the Crazy Mountain Field, Montana, and its mammalian faunas: Smithsonian Institute, U.S. Natural History Museum Bulletin, v. 169, p. 1–287.

——, 1948, 1967, The beginning of the age of mammals in South America: American Museum of Natural History Bulletin, v. 91, p. 1–232; v. 137, p. 1–259.

——, 1978, Early mammals in South America; Fact, controversy, and mystery: Proceedings of the American Philosophical Society, v. 122, p. 318–328.

Sinclair, W. J., and Granger, W., 1914, Paleocene deposits of the San Juan Basin, New Mexico: American Museum of Natural History Bulletin, v. 33, p. 297–316.

Sloan, R. E., 1970, Cretaceous and Paleocene terrestrial communities of western North America: Proceedings of the North American Paleontological Convention, p. 427–453.

——, 1976, The ecology of dinosaur extinction, *in* Churcher, C. S., ed., Athlon: Royal Ontario Museum of Life Science Contributions, p. 134–154.

——, 1981, Systematics of Paleocene multituberculates from the San Juan Basin, New Mexico, *in* Lucas, S. G., Rigby, J. K., Jr., and Kues, B. S., eds., Advances in San Juan Basin paleontology: Albuquerque, University of New Mexico Press, p. 127–160.

Sloan, R. E., and Van Valen, L., 1965, Cretaceous mammals from Montana: Science, v. 148, p. 220–227.

Sloan, R. E., Rigby, J. K., Jr., Van Valen, L. M., and Gabriel, D., 1986, Gradual dinosuar extinction and simultaneous ungulate radiation in the Hell Creek Formation: Science, v. 232, p. 629–633.

South China "Redbeds" Research Group, 1977, Paleocene vertebrate horizons and mammalian faunas of South China: Scientia Sinica, v. 20, p. 665–678 (English).

Steiger, R. H., and Jäger, E., 1978, Subcommission on geochronology; Convention on the use of decay constants in geo- and cosmochronology: Earth and Planetary Science Newsletter, v. 36, p. 359–362.

Steven, T. A., Smedes, H. W., Prostka, H. J., Lipman, P. W., and Christiansen, R. L., 1972, Upper Cretaceous and Cenozoic igneous rocks, *in* Mallory, W. W., ed., Geologic atlas of the Rocky Mountain region: Denver, Rocky Mountain Association of Geologists, p. 229–232.

Szalay, F. S., and McKenna, M. C., 1971, Beginning of the age of mammals in Asia; The late Paleocene Gashato fauna, Mongolia: American Museum of Natural History Bulletin, v. 144, p. 269–312.

Tang Y., and Yan D., 1976, Notes on some mammalian fossils from the Paleocene of Yunnan and Xuancheng, Anhui: Vertebrata PalAsiatica, v. 14, p. 92–99 (Chinese).

Taylor, L. H., 1981, The Kutz Canyon local fauna, Torrejonian (middle Paleocene) of the San Juan Basin, New Mexico, *in* Lucas, S. G., Rigby, J. K., Jr., and Kues, B. S., eds., Advances in San Juan Basin paleontology, Albuquerque, University of New Mexico Press, p. 242–263.

Taylor, R. S., Mathews, W. H., and Kupsch, W. O., 1964, Tertiary, *in* Geological history of western Canada: Alberta Society of Petroleum Geologists, p. 190–194.

Thaler, L., 1977, Etat des recherches sur la faune de mammiferes du montien de Hainin (Belgique): Geobios, Mem. Spec. v. 1, p. 57–58.

Ting (=Ding) S., 1979, A new Edentate from the Paleocene of Guangdong: Vertebrata PalAsiatica, v. 17, p. 57–64 (Chinese from English summary).

Tomida, Y., 1981, "Dragonian" fossils from the San Juan Basin and the status of the "Dragonian" land mammal "age," *in* Lucas, S. G., Rigby, J. K., Jr., and Kues, B. S., eds., Advances in San Juan Basin paleontology: Albuquerque, University of New Mexico Press, p. 222–241.

——, 1982, A new genus of picrodontid primate from the Paleocene of Utah: Folia Primatologica, v. 37, p. 37–43.

Tomida, Y., and Butler, R. F., 1980, Dragonian mammals and Paleocene magnetic polarity stratigraphy, North Horn Formation, central Utah: American Journal of Science, v. 280, p. 787–811.

Tong, Y., 1973, Late Paleocene mammals of the Turfan Basin, Sinkiang: Institute of Vertebrate Paleontology and Palaeanthropology, Academia Sinica Memoirs no. 13, p. 82–101.

——, 1979, A late Paleocene primate from South China: Vertebrate PalAsiatica, v. 17 p. 65–70 (Chinese with English summary).

Tsentas, C., 1981, Mammalian biostratigraphy of the Middle Paleocene Torrejonian strata of the San Juan Basin; Notes on Torreon Wash and the status of the *Pantolambda* and *Deltatherium* zones: *in* Lucas, S. G., Rigby, J. K., Jr., and Kues, B. S., eds., Advances in San Juan Basin paleontology: Albuquerque, University of New Mexico Press, p. 264–292.

Vail, P. R., Mitchum, R. M., and Thompson, S. III, 1977, Seismic stratigraphy and global stands of sea level, pt. 4; Global cycles of changes of sea levels, *in* Payton, C. E., ed., Seismic stratigraphy; Application to hydrocarbon exploration: American Association of Petroleum Geologists Memoir 26, p. 83–97.

Van Valen, L., 1978, The beginning of the age of mammals: Evolutionary Theory, v. 4, p. 45–80.

Van Valen, L., and Sloan, R. E., 1965, The earliest primates: Science, v. 150, p. 743–745.

——, 1966, The extinction of the multituberculates: Systematic Zoology, v. 15, p. 261–278.

Van Valen, L., and Sloan, R. E., 1977a, Ecology and the extinction of the

dinosaurs: Evolutionary Theory, v. 2, p. 37–64.
——, 1977b, Contemporaneity of late Cretaceous extinctions: Nature, v. 270, p. 37–193.
Vianey-Liaud, M., 1979, Les Mammiferes montiens de Hainin (Paleocene moyen de Belgique), pt. 1; Multitubercules: Paleovertebrata, v. 4, p. 117–139.
Wang B., 1975, Paleocene mammals of Chaling basin, Hunan: Vertebrata Palasiatica, v. 13, p. 154–162 (Chinese).
Weed, W. H., 1896, Ft. Union Formation: American Geologist, v. 18, no. 4, p. 201–211.
Wheeler, W. H., 1961, Revision of the Uintatheres: Yale University Peabody Museum Bulletin, v. 14, p. 1.
Wilder, F. A., 1904, The Laramie and Ft. Union beds in North Dakota: Journal of Geology, v. 12, no. 4, p. 290–293.
Wilson, R. W., 1956, A new multituberculate from the Paleocene Torrejon fauna of New Mexico: Transactions of the Kansas Academy of Science, v. 59, p. 76–84.
Wilson, R. W., and Szalay, F. S., 1972, New paromomyid primate from middle Paleocene beds, Kutz Canyon area, San Juan Basin, New Mexico: American Museum of Natural History Novitates, no. 2499, p. 12.
Wolberg, D. L., 1978, The mammalian paleontology of the late Paleocene (Tiffanian) Circle and Olive localities, McCone and Powder River Counties, Montana [Ph.D. thesis]: Minneapolis, University of Minnesota, 385 p.
Wolfe, J. A., and Hopkins, D. M., 1967, Climate changes recorded by Tertiary land floras in northwestern North America, *in* Hatai, K., ed., Tertiary correlations and climate changes in the Pacific: Symposium Pacific Science Congress Tokyo, v. 25, p. 67–76.
Wood, H. E., and 6 others, 1941, Nomenclature and correlation of the North American continental Tertiary: Geological Society of America Bulletin, v. 52, p. 1–48.
Wood, R. C., 1967, A review of the Clark Fork vertebrate fauna: Breviora, no. 257, p. 1–30.
Xue X. X., 1982, The discovery of Paleocene series and the division of Cenozoic group at Shimen Basin, Shaanxi Province: Journal of Northwestern University (Xian), no. 3, p. 70–80 (Chinese with English summary).
Young C. C., and Chow, M., 1963, Cretaceous and Paleocene vertebrate horizons of North Kwangtung: Scientia Sinica, v. 12, p. 1411.
Zapp, A. D., 1949, Geology and coal resources of the Durango area, La Plata and Montezuma Counties, Colorado: U.S. Geological Survey Oil and Gas Investigations Preliminary Map 109.
Zhai R., 1973, Two new Early Eocene mammals from Sinkiang with remarks on the age of Gashato Formation: Institute of Vertebrate Paleontology and Palaeoanthropology, Academia Sinica Memoirs no. 13, p. 102–104.
Zhang Y. P., 1978, Two new genera of condylarthran phenacolophids from the Paleocene of Nanxiong Basin, Gwangdong: Vertebrata PalAsiatica, v. 16, p. 267–274 (Chinese, with English summary).
Zhou (see Chow).

MANUSCRIPT ACCEPTED BY THE SOCIETY MARCH 24, 1986
PUBLICATION NO. 1025 FROM THE SCHOOL OF EARTH SCIENCES, UNIVERSITY OF MINNESOTA

Typeset by WESType Publishing Services, Inc., Boulder, Colorado
Printed in U.S.A. by Malloy Lithographing, Inc., Ann Arbor, Michigan